Palgrave Studies in the History of Science and Technology

Designed to bridge the gap between the history of science and the history of technology, this series publishes the best new work by promising and accomplished authors in both areas. In particular, it offers historical perspectives on issues of current and ongoing concern, provides international and global perspectives on scientific issues, and encourages productive communication between historians and practicing scientists.

Edward Guimont

The Power of the Flat Earth Idea

History from the Marginalised

palgrave
macmillan

Edward Guimont
Department of History and Cultural Studies
Bristol Community College
Fall River, MA, USA

ISSN 2730-972X ISSN 2730-9738 (electronic)
Palgrave Studies in the History of Science and Technology
ISBN 978-3-032-23980-8 ISBN 978-3-032-23981-5 (eBook)
https://doi.org/10.1007/978-3-032-23981-5

Cover credit: Stephen Wallace/Moment/Getty

This Palgrave Macmillan imprint is published by the registered company Springer Nature Switzerland AG
The registered company address is: Gewerbestrasse 11, 6330 Cham, Switzerland

CLASSROOM IN PROPOSED BRYAN UNIVERSITY OF TENNESSEE

July 18, 1925

Acknowledgements This project originated in a brief aside in my doctoral dissertation, in a chapter I wrote in 2018. Since writing that passage, I finished and defended my dissertation, served as an adjunct at multiple schools, worked through the limitations of the covid pandemic, and then completed the majority after being hired as a tenure-track professor at a community college. Beyond its merit as a work of history, I hope this book demonstrates the work that adjuncts and community college faculty—neither well respected in broader academia—are capable of doing, particularly at a time when the academy is not only coming under heavy political attack, but itself remains ignorant and dismissive of what its own underclasses do. But across that span of time, working in those conditions, it was not easy to turn a few sentences in a dissertation into a book of its own. I could not have done it without the help of the following.

Mónica Belevan gave me the chance to write about Flat Earthers for the first time on her *Lapsus Lima* website in 2019.[1] I further presented on my Flat Earth research at the October 2019 Northeast Conference on British Studies; the January 2021 Colonial Knowledges Research Network: Environment and Logistics in the Creation of Knowledge in British Colonies from 1750–1950 seminar; the July 2021 STEMM and Belief in Diverse Contexts: Publics, Praxis, Policy and Pluralism conference; the Fall 2023 New England Regional World History Association Symposium on Ships and Seafarers in World History (held at the Northeast Maritime Institute in Fairhaven, Massachusetts, several hundred feet from where Joshua Slocum rebuilt and launched the *Spray*); the October 2023 Quarry Farm Symposium on Mark Twain: Invention, Technology, and Science Fiction; and the August 2024 NecronomiCon Providence. All of those conferences allowed me to present and discuss research that was incorporated into this book. I would like to also thank the attendees at those various conferences who made questions and comments, including Max Chapnick, Charlotte Coull, Chafik Graigeur, Neal Palmer, Jennifer Regan-Lefebvre, Lacey Sparks, and Todd Nathan Thompson. Conor Dube's January 2021 Harvard University seminar, "The Globe Before

[1] Edward Guimont, "Squaring the Spheroid," *Lapsus Lima*, June 26, 2019, accessed June 26, 2019, http://www.lapsuslima.com/squaring-the-spheroid/.

the Globe: Medieval Arabic Tales of the Margins," and Emily Steiner's January 2021 lecture, "Medieval People Were Not Flat Earthers," both presented me with useful resources.

I would especially like to thank Tim Binnall, who I met when Loren Coleman invited both of us to participate in his Fourth Annual International Cryptozoology Conference in April 2019. Despite the subject of the conference, we both connected over our interest in Flat Earthers. In the years since, besides inviting me on his *Binnall of America* podcast multiple times, Tim has kept me up to date on developments and stories in the Flat Earth community, and his interviews with prominent Flat Earthers have been invaluable resources for this book. Similarly, Kay Burns of the Museum of the Flat Earth on Fogo Island, Newfoundland, discussed with me both the exhibits and philosophy of the museum, the history of the Flat Earth Society of Canada, and her own history researching the Flat Earther movement.

David Killingray was kind enough to share some of his sources and research on John Edward Quinlan and Emanuel Vaun Mulgrave. Linda Gore and Jill Tartaglino of the Fluvanna County Historical Society forwarded to me copies of their archival records on John Jasper at the height of the pandemic. Kathryn Heffner looked into the records of the International Flat Earth Research Society held by the Science Fiction Foundation. Meredith Mattlin discussed her research on the intersection of Flat Earthers and anti-vaxxers. Olivier Arvisais and Patrice Potvin shared insights on their research into the education policies of ISIS. Ed Papenfuse of the Baltimore City Historical Society provided assistance in navigating relevant areas of the BCHS archives, and Matt Testa of the Arthur Friedheim Library Archives of the Peabody Institute at the Johns Hopkins University provided copies of their Garrett P. Serviss materials. Anna Merlan and Andrew Fishman responded to questions on their coverage of US and Brazilian conspiracy theorists, respectively. Stephanie Halmhofer shared some of her research on pseudoarchaeology and the far-right. Kevin Hooper provided information on Liberia College and the American Colonization Society. Michael J. Bielawa shared his biographical research on George Loring Porter and the potential connection to Hans Powell. Sibusiso Biyela spoke with me about science and its popularization in South Africa. Allison Horrocks shared both thoughts and resources on Spiritualism. The Center for Mark Twain Studies awarded me the 2024 Quarry Farm Fellowship, which allowed me to research astronomical topics in nineteenth-century American literature.

Besides the above, the following offered more general insight, ideas, feedback, or just conversation, all of which contributed toward what will hopefully be seen as the strength of this work: Claudio Daflon, Heydn Ericson, Richard Fallon, Dexter Gabriel, Cian Gill, Adele Guyton, Craig Howe, Brendan Kane, Emily Kern, Bernard Lightman, Kathleen Pearle, Adam Richter, Luke Ritter, Dan Rourke (aka SciManDan), Margaret E. Schotte, Horace A. Smith, Susan Souza-Mort, and David T. Zeppieri. Any errors in this work are obviously the fault of myself and not any of the above heroes of academic labor.

Special thanks are particularly due to Sarah A. Qidwai for both talking to me about Sayyid Ahmad Khan, but also for her role in establishing the Virtual History of Science, Technology, and Medicine Community in March 2020, such an important community of early career scholars especially for these times. I learned much from every participant in the Virtual HistSTM Community, but thanks are especially due to Megan Baumhammer, Jörg Matthias Determann, Kelcey Gibbons, Daniella McCahey, and Mariam Sabri. I likewise need to thank the writing accountability group started in the pandemic by Elaine Ayers, and including Alison Laurence, Adrianna Link, Lydia Pyne, Kathleen Sheppard, and Anna Toledano. They kept me on target for this as well as many other works, and I hope they gained as much inspiration hearing my talk about my research as I gained from listening to them discuss their own projects. Between those two groups, how strange it is to feel nostalgia for those early pandemic days. One post-pandemic group that has helped provide a forum for sounding out ideas is the North Atlantic STS Network, particularly the other participants in its May 2024 meeting: Xan Chacko, Elain Gan, Jennifer Hamilton, Tony Hatch, Lisa Messeri, Kathleen Pierce, Lukas Rieppel, Helen Rozwadowski, and Laura Stark.

For the general research that was carried out, special thanks to the Fairhaven Historical Society (particularly Beth Luey and Allegra Pawlowski); Brown University's John Hay Library (particularly Raymond Butti, Heather Cole, and Laurie Rossi); the New Bedford Whaling Museum (particularly Samantha Santos and Madeline Smith); New-York Historical Society's Patricia D. Klingenstein Library; University of Massachusetts-Amherst W. E. B. Du Bois Library Special Collections and University Archives (particularly Jeremy Smith); University of Connecticut Homer D. Babbidge Library; Wesleyan University Olin Library; Bristol Community College Library (particularly Melanie Johnson, Richard Kugler, Devon Mesler, and Emily Zekan Brown); Fall

River Public Library; and Welles-Turner Memorial Library of Glastonbury, Connecticut.

At Palgrave Macmillan, I would like to thank Phil Getz, Emily Russell, Arunaa Devi, and Eliana Rangel, as well as Palgrave Studies in the History of Science and Technology series editors Roger D. Launius and James Rodger Fleming.

Finally, the best for last. During the span of writing this book, we were joined by Mosi the cat and Cubby the dog, both of whom helped supervise the completion of this manuscript from their favorite perches on top of the keyboard and under my chair, respectively. The writing of this also spanned my engagement, civil wedding, and religious marriage. So I must conclude by sending many thanks for the support and love given by my wife Jessica Rivas. More than once, she voiced her amazement at how I could spend so much time writing about a small group of people who held such strange views, and her skepticism at my response that it was just interesting. I promise that I will make sure the next book is shorter.

Competing Interests The author has no competing interests to declare that are relevant to the content of this manuscript.

Introduction: A Swiftly-Tilting Plane

In 2005, *New York Times* columnist Thomas Friedman (b. 1953) published his fourth book, *The World Is Flat*. One of the author's many odes to globalization, Friedman stated that the idea for the title came from the fact that Christopher Columbus (1451–1506) attempted to sail from Europe to India in his 1492 voyage, but found the Americas in between, deciding to call the natives Indians anyways. Friedman himself went to India from America, but found everyone there imitating American culture due to globalization, which he argued was the reversal of Columbus and his symbolic start of European colonialism. Friedman's title stemmed from the fact that he viewed Columbus's discovery of the Americas as confirmation that the world was spherical.[1]

Friedman's account was littered with errors common among the popular conception of Columbus. He overstates the number of sailors under Columbus's command, as over a hundred instead of the ninety or less. Friedman also mistakes Columbus's destination as the area that is now the country of India, and not "the Indies," the general European term for East Asia at the time. But most substantively, Friedman accepts the stereotypical myth that it was Columbus who proved the Earth was round with his voyage, when the Catholic establishment of his day believed the world to be flat—a legend termed the Flat Error in

[1] Thomas L. Friedman, *The World Is Flat: A Brief History of the Twenty-first Century* (New York: Farrar, Straus and Giroux, 2005), 3–5.

1991. Friedman's broader argument on globalization in *The World is Flat* has likewise been challenged, while others have argued that Friedman's views on Americanization "flattening" the world are, in the twenty-first century, more suitably applied to China—ironic given traditional Confucian cosmology did indeed accept a Flat Earth.[2] In return, pop-psychologist Steven Pinker (b. 1954) dismissed critics of globalization as "not just wrong but flat-earth wrong."[3] Friedman did embody Columbus in at least one respect, however. Just as the explorer brought disease to the Americas in his quest for economic gain, in March 2020, Friedman argued that large numbers of covid deaths should be accepted as the cost of keeping the economy productive during a pandemic.[4]

The World is Flat was published in 2005, the same year that YouTube was launched. A decade later, YouTube would be the vector through which Flat Earth would receive its twenty-first-century renaissance, in line with a general expansion of conspiracy theories after the disputed 2000 US presidential, 11 September 2001 terrorist attacks, and 2003 invasion of Iraq.[5] The jump to the analog world was fast, beginning

[2] Pankaj Ghemawat, "Why the World Isn't Flat," *Foreign Policy*, October 14, 2009, accessed June 3, 2023, https://foreignpolicy.com/2009/10/14/why-the-world-isnt-flat/; Geremie R. Barmé, "China's Flat Earth: History and 8 August 2008," *The China Quarterly*, Vol. 197 (Mar., 2009), 64–86; Christopher Cullen, "A Chinese Eratosthenes of the Flat Earth: A Study of a Fragment of Cosmology in Huai Nan tzu," *Bulletin of the School of Oriental and African Studies, University of London*, Vol. 39, No. 1 (1976), 106–27.

[3] Steven Pinker, "Guess what? More people are living in peace now. Just look at the numbers," *The Guardian*, March 20, 2015, accessed July 10, 2020, https://www.theguardian.com/commentisfree/2015/mar/20/wars-john-gray-conflict-peace.

[4] Thomas L. Friedman, "A Plan to Get America Back to Work," *The New York Times*, March 22, 2020, accessed March 23, 2020, https://www.nytimes.com/2020/03/22/opinion/coronavirus-economy.html.

[5] Mick West, *Escaping the Rabbit Hole: How to Debunk Conspiracy Theories Using Facts,Logic,and Respect* (New York: Skyhorse, 2018), 194–211; Shaheed N. Mohammed, "Conspiracy Theories and Flat Earth Videos on YouTube," *The Journal of Social Media in Society*, Vol. 8, No. 2 (Fall, 2019), 84–102; Simona Stano, "The Internet and the spread of conspiracy content," in Michael Butter and Peter Knight, eds., *Routledge Handbook of Conspiracy Theories* (London: Routledge, 2020), 483–96; Michael D. Gambone, *Modern Conspiracies in America: Separating Fact from Fiction* (Lanham, MD: Rowman & Littlefield, 2022), 83–84; Mike Rothschild, *The Storm Is Upon Us: How QAnon Became a Movement,Cult,and Conspiracy Theory of Everything* (Brooklyn: Melville House Publishing, 2021), 141–58; Will Sommer, *Trust the Plan: The Rise of QAnon and the Conspiracy That Unhinged America* (New York: HarperCollins, 2023), 50–53; Talia

in January 2016 through a particular vector: African-American musicians,in particular B.o.B, Wilson Chandler, Stefon Diggs, Draymond Green, Kyrie Irving,Lord Jamar, Shaquille O'Neal, Geno Smith, Sammy Watkins, and Marcellus Wiley.[6] Even outside the United States with its particular racial politics, prominent Flat Earthers who emerged included German rapper Xavier Naidoo (b. 1971 in South Africa, to parents of Indian ancestry) and the Black Briton Dave Murphy, who goes by the name "Dave Moor." By this point, the Black actor Terrence Howard (b. 1969) had already declared his belief in a new version of mathematics he called Terryology, based on the non-existence of zero. In his 2019 Terryology manifesto, *One Times One Equals Two*, Howard claimed that modern accepted mathematics is a deception, the "Jim Crow Laws of Mathematics," which supposedly requires the maintaining of a Flat Earth cosmology as established by Greek philosopher Pythagoras.[7] Terryology also has some similarities with "vortex math," a fraudulent concept developed by an alternative energy proponent named Marko Rodin in 2012 that claims to show the universe is actually inside a toroidal black hole, and which was subsequently adopted by Flat Earthers several years later to represent a Flat Earth in the middle of a dome overhead and a dome beneath the flat bottom of the world, linked by a central column.[8]

To be clear, Flat Earth in 2016 was neither the first nor last time rappers in particular had become entangled in conspiracism. In the early 1990s, rappers such as Tupac Shakur, Public Enemy, and Jay-Z embraced William Cooper's conspiracy theory magnum opus *Behold a Pale Horse*, and in 2022 Kanye West embraced a number of conspiracy

Lavin, *Culture Warlords: My Journey Into the Dark Web of White Supremacy* (New York: Hachette Books, 2020), 20–21.

[6] Donald R. Prothero, *Weird Earth: Debunking Strange Ideas about Our Planet* (Bloomington: Red Lightning Books, 2020), 22–23; Kelly Weill, *Off the Edge: Flat Earthers,Conspiracy Culture,and Why People Will Believe Anything* (Chapel Hill: Algonquin Books, 2022), 195; West, *Escaping the Rabbit Hole*, 199–200.

[7] Terrence Howard, *One Times One Equals Two* (2019), 11, 21. Accessed 3 June 2024, https://tcotlc.com/wp-content/uploads/2021/10/OTOET_PREVIEW_062_October_03_2021.pdf.

[8] Ben Kissel, Marcus Parks, and Henry Zebrowski, "Episode 334: Flat Earth," *The Last Podcast on the Left*, September 29, 2018, accessed July 23, 2024, https://www.audible.com/podcast/Episode-334-Flat-Earth/B09RKFCLTB; Mark Chu-Carroll, "Zombie Math in the Vortex," *Good Math/Bad Math*, January 2, 2018, accessed July 23, 2024, http://www.goodmath.org/blog/2018/01/02/zombie-math-in-the-vortex/.

views, including Black Israelism and support for Adolf Hitler.[9] It should also be noted that white athletes also endorse conspiracies—especially fueled by the restrictions the covid pandemic placed on public sports from 2020–2022. A prime example is Aaron Rodgers, who attacked Flat Earthers as a technique for legitimizing his own covid conspiracy theories in comparison, but does believe in a Hollow Earth.[10] Outside of athletics, the white actor Millie Bobby Brown, star of the breakout Netflix series *Stranger Things* (heavily inspired by conspiracy notions such as the Montauk Project, remote viewing, and MK-ULTRA) announced on Instagram in 2018 that she believed the Earth was flat, but partially—though not completely—walked that back on TikTok in 2022.[11] In contrast to these beliefs, one of the main opponents of the Black Flat Earthers who emerged in 2016 was African-American astrophysicist Neil DeGrasse Tyson, arguably the most prominent science popularizer in the contemporary United States.[12]

To be clear, the reason so many Black celebrities turned to Flat Earth in 2016 is not due to any intrinsic aspect of their Blackness. A number of studies have been conducted on the particular strains of conspiratorial thinking among African-Americans, stemming from their unique status as oppressed people suspicious of the white-dominated hegemonic power and the institutional science which supports it, and has at times subjected Blacks to genuine scientific conspiracies, such as the Tuskegee Syphilis Study of 1932–1972.[13] It is also worth noting that not all Blacks

[9] Mark Jacobson, *Pale Horse Rider: William Cooper,the Rise of Conspiracy,and the Fall of Trust in America* (New York: Blue Rider Press, 2018), 175–203.

[10] Julian Feeld, Travis View, Jake Rockatansky, and Arif Hasan, "Episode 263: Aaron Rodgers feat Arif Hasan," *QAnon Anonymous*, January 19, 2024, accessed January 21, 2024, https://soundcloud.com/qanonanonymous/episode-263-aaron-rodgers-feat-arif-hasan.

[11] Sophie Thompson, "Millie Bobby Brown says she no longer believes the Earth is flat," *indy100*, November 12, 2022, accessed September 16, 2024, https://www.indy100.com/video/millie-bobby-brown-flat-earth.

[12] Prothero, *Weird Earth*, 23-24; Michael D. Gambone, *Modern Conspiracies in America: Separating Fact from Fiction* (Lanham, MD: Rowman & Littlefield, 2022), 173.

[13] Anna Merlan, *Republic of Lies: American Conspiracy Theorists and Their Surprising Rise to Power* (New York: Metropolitan Books, 2019), 35–56; David Hutchings and James C. Ungureanu, *Of Popes and Unicorns: Science,Christianity, and How the Conflict Thesis Fooled the World* (Oxford: Oxford University Press, 2022), 49–50; Christopher D.

express this reaction to hegemonic conspiracies through conspiracy theories of their own; from among the athletic community in the United States are the examples of Husain Abdullah and Colin Kaepernick, who used their positions to engage in social justice activism. Complicating this narrative as well is someone like Kyrie Irving, who in between his Flat Earth advocacy in 2016 and his covid vaccination conspiracy promotion in 2021, was very active promoting and supporting the 2020 Black Lives Matter protests. There is perhaps an example of this to be made from the fact that in the current US Flat Earth movement, the involvement of other minority ethnic groups—Native American, Latino, Asian—is less prevalent than Blacks. Even the covid pandemic, which brought both a heightening of anti-Asian racism and conspiratorial belief around China, did not seemingly result in an increase of Asian-American Flat Earthers.

For someone whose familiarity with the Flat Earth movement comes from watching the 2018 documentary *Behind the Curve*, with its focus of an overwhelmingly white and conservative evangelical movement, or from witnessing the involvement of Flat Earthers in covid denialism, QAnon, and the 6 January 2021 attack on the US Capitol, the fact that someone like Irving—Black and liberal, supporting social justice movements—might endorse Flat Earth could seem strange. Additionally, works which study the conspiracy views of the Black community in the United States have largely not covered Flat Earth interest.[14] But it is worth noting that the Flat Earthers who emerged out of social media of the mid-2010s are almost entirely unaware of their own movement's past,

Bader, Joseph O. Baker, L. Edward Day, and Ann Gordon, *Fear Itself: The Causes and Consequences of Fear in America* (New York: NYU Press, 2020), 42; Andrew Jewett, *Science under Fire: Challenges to Scientific Authority in Modern America* (Cambridge, MA: Harvard University Press, 2020), 16–21; Nina Gilden Seavey, "True Conspiracies: The Legacy of J. Edgar Hoover, Dr. Martin Luther King, Jr., and James Earl Ray," in Luke Ritter, ed., *American Conspiracism: An Interdisciplinary Exploration* (London: Routledge, 2024), 80–97.

[14] Merlan, *Republic of Lies*, 35–56. Of more anecdotal note is the podcast *My Momma Told Me* (2020-present) which explores conspiracy theories in Black communities, and as of this writing has not covered Flat Earth. However, it is worth noting that the host of *My Momma Told Me*, Langston Kerman (b. 1987), was present recording a show at the NBA game where Kyrie Irving came out as a Flat Earther: Langston Kerman and Troy Walker, "Talk That Ship To Me," *My Momma Told* Me, March 22, 2022, accessed March 22, 2022, https://www.iheart.com/podcast/1119-my-momma-told-me-with-lan-69919303/episode/talk-that-ship-to-me-with-94482894/.

outside of a few select publications and figures. Broadly, they will know that English eccentric Samuel Birley Rowbotham (1816–1884), who went by the name Parallax, published a pamphlet in 1849 titled *Zetetic Astronomy: Earth Not a Globe*, the ur-text that renewed Flat Earth belief. In 1870, spurred on by scientist Alfred Russel Wallace, Rowbotham and his followers performed the "Bedford Level Experiment" along an old canal, attempting to demonstrate that the Earth is indeed flat given that they claimed to be able to see signposts longer than the curvature of the Earth would allow (Fig. 1).

Rowbotham died in 1884, but the year prior he founded the Universal Zetetic Society, an organization of Flat Earthers. The society initially floundered on Rowbotham's death, but was revitalized in 1892 under the leadership of an aristocrat, Lady Blount. Meanwhile, in 1879, Rowbotham's disciple William Carpenter emigrated to Baltimore in the United States, where he began spreading the "Zetetic" doctrine.

That is broadly the limit of what many of the modern Flat Earthers may know of their movement. Many of the Flat Earthers discussed in chapters below—Paul Kruger, John Jasper, John Edward Quinlan—who were leaders of the movement in the nineteenth century are unknown to those who learned about the Flat Earth from YouTube, Twitter, Reddit, and

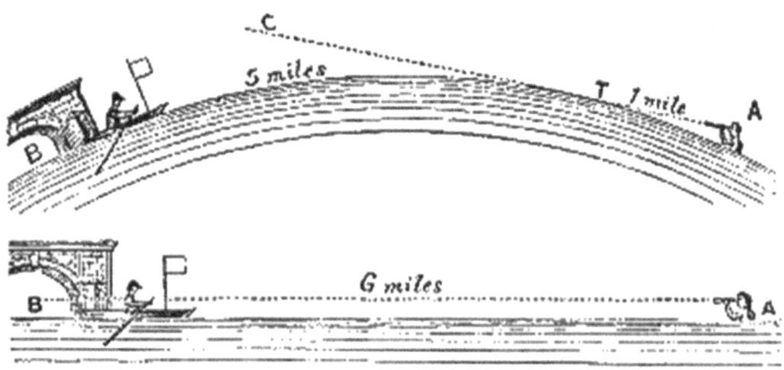

Fig. 1 Combination of figures drawn by Samuel Rowbotham to illustrate his Bedford Level Experiment[15]

[15] Samuel Rowbotham, *Zetetic Astronomy: Earth Not a Globe, Third Edition* (London: Simpkin, Marshall, and Co., 1881), 313.

TikTok.[16] Even relatively recent Flat Earthers are unknown to them, and a cursory exploration of their views will reveal a very different motivation behind the rejection of the Earth's sphericity.

An example is Leo Ferrari (1927–2010), a philosophy professor at St. Thomas University who in 1970 established the Flat Earth Society of Canada (FESC). In a 1975, Ferrari noted "the rise of Globularism on the one hand and the decline of archetypal Femininism on the other... the first step in a vicious and sustained belittling of the Earth, which would transform Her from the Pillar and Ground of our Being to some invisible speck of dust lost in an infinite universe."[17] While Ferrari was being somewhat tongue in cheek here, the general sentiments—FESC's endorsement of the Flat Earth as a liberal critique of male-dominated and hierarchical traditional scientific and educational structures—was genuine. A germ of critique of the established order is even present in modern Flat Earthers, who rail against "elites" responsible for inflation, low wages, and high rents. The conservatism of many modern Flat Earthers will not allow them to blame capitalism for these symptoms of a capitalist economy, and so a conspiratorial explanation has to be found. To them, a spherical Earth is limited, while a Flat Earth, set upon an infinite plane, has endless land, resources, and energy. The same "elites" responsible for the decimation of high-paying jobs due to outsourcing would of course want to hide the truth of infinite space on a Flat Earth from the rubes who would otherwise be able to exploit them and lead to an era of infinite plenty for everyone. After all, infinite land for all would also mean the end of racial and national tensions. It is not a coincidence that Flat Earth tends to be associated with other conspiracy theories involving scientist Nikola Tesla (1856-1943), the Great Reset, and the World Economic Forum.[18]

[16] Tim Binnall and Edward Guimont, "Binnall of America Summer of Strangeness with Dr. Edward Guimont," *Binnall of America*, August 14, 2020 (accessed August 17, 2020), http://www.binnallofamerica.com/boaa081720.html.

[17] Leo Ferrari, "Feminism and education in a Flat Earth perspective," *McGill Journal of Education*, Vol. 10, No. 1 (Spring, 1975), 78–79.

[18] "The "Earth is a Realm" Quote Misattributed to Nikola Tesla," FlatEarth.ws, November 8, 2018, accessed April 9, 2024, https://flatearth.ws/tesla-realm#more-343; Aric Toler, ""No Safe Spaces on the Flat Earth"—Emergent Alt-Right Inspired Flat Earth Online Communities," Bellingcat, June 7, 2017, accessed December 10, 2019, https://www.bellingcat.com/resources/articles/2017/06/07/flat-earth-online-communities/.

It is also not a coincidence that modern Flat Earth is resurrected in the aftermath of the 2008 financial collapse, which showed the limit of neoliberal capitalism and endless expansion, as pioneered by Friedman's conception of Columbus. And in the United States, it was minority, largely Black, populations who were by far hit hardest by the 2008 recession, and had the least rebound in subsequent years.[19] There is some irony that Friedman's book and YouTube also emerged the same year as Hurricane Katrina, when the breaking of the levees holding back the water of Lake Pontchartrain allowed heavy flooding of Black neighborhoods in New Orleans. In addition to leading to the rise of Black conspiracism given the government's seeming lack of interest in protecting or aiding Black citizens, the power lines stretching across the lake would become a major fixture of YouTube videos demonstrating the visual curvature of the Earth's surface[20] (Fig. 2).

There is clearly a failure of Flat Earthers to understand their own past—but also a failure of science educators and popularizers to respond to Flat Earthers in ways that communicate their message effectively rather than propagate difficult to understand concepts. An example is Tyson's 2018 efforts to explain away Flat Earth claims in an easily-digestible YouTube video, much as four years earlier his friend Bill Nye, "the Science Guy," was unable to prove away Young-Earth Creationism in his debate with Answers in Genesis president Ken Ham. Tyson, Nye, and many other public figures who try to combat pseudoscience, pseudohistory, or cryptozoology fundamentally lack understanding not only of the conspiratorial mind, but why simply appealing to science—especially science promoted by government agencies and established academia—fails to engage with the anti-authoritarian conspiracy worldview. This is perhaps not surprising, given Tyson's own focus on STEM education has led him to make statements on academia that completely ignore the role of the humanities, conflating all higher education with technical, scientific study.[21]

[19] Douglas S. Massey and Jacob S. Rugh, "The Great Recession and the Destruction of Minority Wealth," *Current History*, Vol. 117, No. 802 (Nov., 2018), 298–303.

[20] Merlan, *Republic of Lies*, 36-38.

[21] Avi Selk, "Why Neil deGrasse Tyson failed to prove Earth isn't flat," *The Washington Post*, March 12, 2018, accessed March 18, 2018, https://www.washingtonpost.com/news/speaking-of-science/wp/2018/03/12/why-neil-degrasse-tyson-failed-to-prove-earth-isnt-flat/?utm_term=.21e68d784588; Kenneth L. Feder, "Answering Pseudoarchaeology," in Jeb J. Card and David S. Anderson, eds., *Lost City, Found Pyramid: Understanding*

Helping the Tysons and Nyes of the future learn how best to respond to pseudohistorical pseudoscience is not the aim of this book. The goal of this book is also not meant to teach the Flat Earthers of today their own movement's history—although I certainly do hope that many buy copies and use it to gain a new perspective on their world (so to speak). But the goal of this book is to explore that hidden history of the Flat Earthers of yore. Doing so will reveal the history of the marginalized—not only due to their cosmological beliefs, but because many were Black, women, religious minorities, and political fringers, at times when the boundaries for acceptable participation even in democracies like the US and UK were far narrower than they are today. But it also shows the power that came from Flat Earth belief—such a unique idea giving common cause, common identity, utility in providing a lens to critique society. The Flat Earth gave organization to former slaves asserting their power against both former owners and national governments; ethnic and religious groups seeking to preserve their autonomy; critics of empires from both the periphery and metropole. This resistance was not always successful—but it also was not always a failure, and when it did succeed, it is doubtful that it would have done so without some element of Flat Earthism. These are all issues that have particular relevance today in issues of contemporary prominence, such as museum decolonization, statue removal, and indigenous rights.

These issues are also important because of the relevance of Flat Earthers in the time this book is being written. Even ignoring conspiracies around the covid pandemic, public—and government—interest in UFOs has never been more active, and cryptids like Bigfoot and Mothman are likewise more popular than ever.[22] Yet Flat Earth belief is far older than belief in UFOs, Bigfoot, the Loch Ness Monster, chemtrails, or other

Alternative Archaeologies and Pseudoscientific Practices (Tuscaloosa: University of Alabama Press, 2016), 199-210; Sharon A. Hill, *Scientifical Americans: The Culture of Amateur Paranormal Researchers* (Jefferson, NC: McFarland & Company, 2017), 132–55.

[22] Fernanda Pérez-Gay Juárez, Lina Khayyat, Michael Ronca, and Ian Gold, "Viral Belief: The Psychology of COVID Conspiracy Theories," in Luke Ritter, ed., *American Conspiracism: An Interdisciplinary Exploration* (London: Routledge, 2024), 53–79.

ideas blending the conspiratorial and paranormal.[23] Journalist Robert Evans has argued that UFO belief was the first populist anti-government movement, given the US government has always denied their being alien spacecraft despite approximately half the population believing that since the 1950s.[24] However, Flat Earth fits the same bill: people can see with their own eyes the logic of the Earth not being curved or spinning, while scientists and other specialists, often with government funding if not directly employed by the state, deny it in favor of more complicated, less perceptible explanations for the natural world.

Likewise, in his 2021 book on the expansive conspiracy theory QAnon which emerged in 2017, journalist Mike Rothschild asserts that Flat

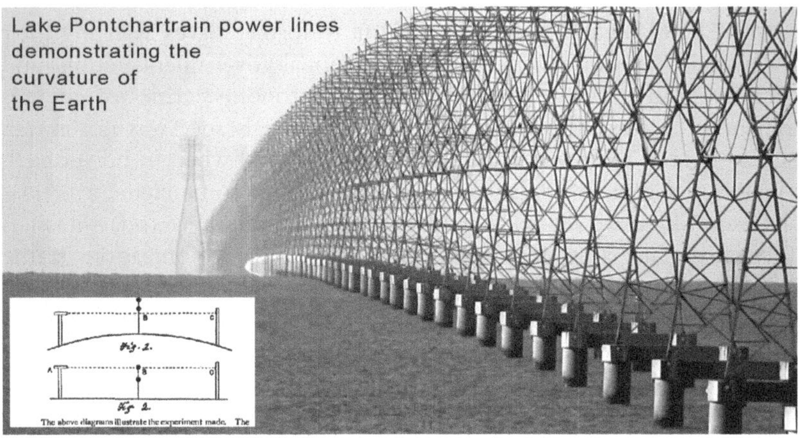

Fig. 2 Photograph of power lines over Lake Pontchartrain, curving with the Earth's surface[25]

[23] Tim Binnall and Mark Sargent, "BoA Holiday Season 2019: Mark Sargent," *Binnall of America*, November 11, 2019 (accessed November 20, 2019), http://www.binnallofamerica.com/boaa111119.html.

[24] Robert Evans and Jamie Loftus, "Part One: Bill Cooper: The Man Who Killed Truth," *Behind the Bastards*, July 14, 2020 (accessed July 20, 2020), https://podcasts.apple.com/us/podcast/part-one-bill-cooper-the-man-who-killed-truth/id1373812661?i=1000484862608.

[25] Mick West, "Soundly Proving the Curvature of the Earth at Lake Pontchartrain," *Metabunk*, July 22, 2017, accessed June 9, 2023, https://www.metabunk.org/threads/soundly-proving-the-curvature-of-the-earth-at-lake-pontchartrain.8939/.

Earthers are less influential than QAnon, lumping them in with 9/11 Truthers, alien abductees, and Apollo Moon landing deniers.[26] This not only severely downplays the influence of Flat Earthers, but also that there is a large overlap between believers in those categories—indeed, Moon landing denialism arguably originated from a British Flat Earther in the mid-1960s, and as will be discussed in chapter seven, QAnon itself spread so quickly by infiltrating the already-extant Flat Earth social media spaces.[27] In his own 2023 study of QAnon, journalist Will Sommer (b. 1988) describes it as "just the start of the all-consuming conspiracy theory movements to come… a glimpse into our future."[28] As will be shown, there is no question that Flat Earth came first, and that it provided a cultural template that was far more expansive than QAnon. To understand our modern culture—not just conspiratorial, but political and hierarchical—Flat Earth offers a versatile, if unlikely, tool.

To be clear, this is not to say that there have not been any studies of Flat Earth before this. Outside of a large number of articles—many of which are cited throughout the following chapters—there are four primary books which cover the topic. In 1991, historian of science Jeffrey Burton Russell (1934–2023) published *Inventing the Flat Earth: Columbus and Modern Historians*. In it, Russell pioneered the term "Flat Error" to describe the false myth that medieval Christians believed the Earth was flat, only to be proved wrong by Columbus. Russell's book charts why that belief emerged and demonstrates how wrong it is using medieval sources. In 2007, another historian of science, Christine Garwood, published *Flat Earth: The History of an Infamous Idea*, a comprehensive overview of ideas surrounding the titular notion from the ancient world until the death of Charles K. Johnson (1924–2001), President—and sole remaining member—of the International Flat Earth Research Society (IFERS). Johnson's death followed by one year that of science journalist and skeptic Robert Schadewald (1943–2000), who extensively interviewed and covered advocates of fringe science. On his death, he had been working on his own comprehensive history of the Flat Earth concept, begun in 1984. In 2015, his sister Lois, widow Wendy,

[26] Rothschild, *The Storm Is Upon Us*, 199.

[27] Christine Garwood, *Flat Earth: The History of an Infamous Idea* (New York: Thomas Dunne Books, 2007), 219–79; Weill, *Off the Edge*, 64–67.

[28] Sommer, *Trust the Plan*, 5.

friend Bob Forrest, and publisher Michael Behrend edited and revised the manuscript, publishing it as the open access *The Plane Truth: A History of the Flat Earth Movement*. The fourth book is the 2022 *Off the Edge: Flat Earthers, Conspiracy Culture,and Why People Will Believe Anything*, written by journalist Kelly Weill (b. 1994) and focusing much more on the post-2015 US Flat Earth movement.

These are all good works which have their uses, but all are limited to some extent. Russell's is not focused on Flat Earth believers, but those who invented imaginary Flat Earthers. Garwood and Schadewald's works were published prior to the current Flat Earth revival; conversely, Weill, being a journalist, focuses almost entirely on the modern movement. But beyond their scopes, there are also omissions and errors which limit the political analysis of the movement. A key example is that in the fin de siècle United States, undoubtedly the most important Flat Earther was the former slave preacher John Jasper, the central figure in chapter two below. Jasper is absent from Russell and Weill, and given a single paragraph each in Garwood and Schadewald's books; further, those both position Jasper merely as the friend and supporter of Rowbotham's disciple William Carpenter, a relation that evidence does not support even existed.[29] These limitations are mirrored when Flat Earth figures are covered in mainstream works as well. In his groundbreaking work *Reconstruction*, historian Eric Foner only gives one mention of Jasper, misspelling his name ("Jaspar") and only referring to him as a geocentrist, removing an important element of Jasper's liberation theology and wider appeal at the time.[30]

Beyond the existing Flat Earth scholarship, this book also takes inspiration from the work and methodologies of a number of scholars—historians, anthropologists, literary critics, and archaeologists—including David S. Anderson, Bill Black, Jeb J. Card, Jason Colavito, Michael D. Gordin, Kenneth L. Feder, Aaron Gulyas, Sharon A. Hill, Brian Regal, Blake Smith, and Karen Stollznow. It also takes inspiration from the work of a number of journalists who have covered how conspiracy theories, broadly-categorized "pseudoscience," paranormal beliefs, and political culture have increasingly become entwined in the twenty-first century:

[29] Robert J. Schadewald, *The Plane Truth: A History of the Flat Earth Movement* (Wendy S. Schadewald, 2015), 95; Garwood, *Flat Earth*, 163–64.

[30] Eric Foner, *Reconstruction: America's Unfinished Revolution, 1863–1877* (New York: Harper & Row, 1988), 101.

Brad Abrahams, Liv Agar, Vic Berger IV, Tim Binnall, Jamelle Bouie, Andrew Callaghan, Susannah Crockford, Julian Feeld, Joseph L. Flatley, John Ganz, Cian Gill, Jared Holt, Annie Kelly, Anna Merlan, Chuck Mertz, Jake Rockatansky, Mike Rothschild, Luke Savage, Robert Skvarla, Will Sloan, Will Sommer, Asawin Suebsaeng, Travis View, and Kelly Weill.

In line with that somewhat heterodox list of influences, this book might be described as, to use a term from futurist Buckminster Fuller (1895–1983), a "comprehensivist" work, one which studies the interrelationships between various fields of study. Modern Flat Earth belief incorporates elements of the Flat Error with its laudatory framing of Columbus and the European conquest of the Americas. To fully explore Flat Earth's origins and impact, one has to look at it through a decolonial lens—a lens which looks at European empires not only as political structures, but knowledge structures, which frame themselves as radiating out from the metropole but like all imperial constructs inevitably take and mix as well. From the Flat Error on, the stereotypical view of both Flat Earthers and those they deem "Globularists" has been that the basis of modern science in establishing the sphericity of the Earth originates with the imperial projects of the so-called Age of Discovery, in contrast with both earlier European and non-European cosmologies. In her 2010 article "When Science Became Western: Historiographical Reflections," historian of Middle Eastern science Marwa Elshakry rejects the idea that a "Western science" became the basis for "global" science.[31] If Flat Earthers reject the existence of an implicitly Western-centric view of modern science, then it makes sense to explore the global perception of cosmological politics on their own turf, so to speak.

The primeval spark for this work was the strange confluence of the emergence of Black celebrity Flat Earthers in 2016, at the exact same time I encountered references to Flat Earthers in colonial Southern Africa as part of my research for my dissertation on pseudohistorical interpretations of the Great Zimbabwe ruins. The academic atmosphere of this project was further shaped by movements like Rhodes Must Fall and Confederate monument removal campaigns, the 2020 George Floyd protests, and the debates around decolonizing archives and museum holdings.[32] While this

[31] Marwa Elshakry, "When Science Became Western: Historiographical Reflections," *Isis*, Vol. 101, No. 1 (Mar., 2010), 98–109.

[32] For the latter, as expressed by Ariella Aïsha Azoulay, *Potential History: Unlearning Imperialism* (New York: Verso Books, 2019).

book was pitched, somewhat tongue in cheek and somewhat earnestly, as a global history of the Flat Earth movement, it has evolved somewhat. While "global" Flat Earthers—in China, Canada, India, Australia, and elsewhere—will be discussed somewhat, the focus is on those in the African continent, as well as in the African Diaspora abroad—the US, UK, Brazil, and Caribbean—as well as the non-Africans who interacted, debated, collaborated with, or fought with them.

This work could then perhaps also be classified among the recent works on the field of the Black Enlightenment by scholars such as Robin D. G. Kelley, Fred Moten, Cedric Robinson, Britt Rusert, and Sylvia Wynter. It is also a history of science. The combination of those two fields brings to mind the 2015 article "Three Questions for Historians of Science in the Modern Middle East and North Africa" by international historian Cyrus Schayegh (b. 1972), whose titular three queries were: "What was science, who was involved in it, and where did it unfold in the modern Middle East and North Africa?"[33] Those are useful guidelines for this work, and for thinking of how to approach such a potentially problematic query. This also benefits from the new study of place attachment.[34] This is most obvious in the broadest sense of Flat Earthers' attachment to a Flat Earth, but also in particular case studies, such as Boers' attachment to (apartheid) South Africa or Salafist scholars' negative projection of Western intrusion.

Attempting to look at the development of knowledge from a non-Western perspective, centered in Africa and with an emphasis on place study, creates a perfect method to analyze the development of the earliest geographical, astronomical, and cosmological ideas, which then propagated out of Africa, both to Asia in the east and Europe to the north. This brings us directly to the concepts explored at the start of chapter one.

[33] Cyrus Schayegh, "Three Questions for Historians of Science in the Modern Middle East and North Africa," *International Journal of Middle East Studies*, Vol. 47, No. 3 (Aug., 2015), 588.

[34] Alexander C. Diener and Joshua Hagen, "The power of place in place attachment," in Alexander C. Diener and Joshua Hagen, eds., *The Power of Place in Place Attachment* (Abingdon: Routledge, 2023), 1–2.

Praise for *The Power of the Flat Earth Idea*

"Flat earthism is an extreme test case for historians and communicators of science who seek to make sense of heterodox scientific perspectives. Guimont's eye-opening book succeeds brilliantly in revealing flat earth science not as mere anachronism or conspiracy, but as a multifaceted phenomenon that has a truly global [sic] history. Taking in slavery, PanAfricanism, UFOs and more, this is also a story that sheds light on many features of mainstream science, power and culture, both in the past and in our confusing present. A triumph!"

—Charlotte Sleigh, *Lecturer in Social Studies of Science, University College London, Dept of Science & Technology Studies; former editor of the British Journal for the History of Science; and former president of the British Society for the History of Science.*

Contents

Abbreviations

AIN	"Alternative Influence Network"
AMA	"Ask Me Anything"
ARIG	"Amateur Research Investigation Group"
ATAC	"ancient, technologically advanced civilization"
BJP	Bharatiya Janata Party (Indian People's Party)
BSAC	British South Africa Company
CPAC	Conservative Political Action Conference (US)
FESC	Flat Earth Society of Canada
FWP	Federal Writers' Project (US)
GAR	Grand Army of the Republic (US)
IFERS	International Flat Earth Research Society
ISIS	Islamic State of Iraq and Syria
IUM	Islamic University of Medina
MMS	"Miracle Mineral Solution" (bleach)
NASA	National Aeronautics and Space Administration (US)
NYZS	New York Zetetic Society
PT	Partido dos Trabalhadores (Workers' Party; Brazil)
PUFT	"Paranormal Unified Field Theory"
UZS	Universal Zetetic Society (UK)
VWP	Virginia Writers' Project
YEC	Young-Earth Creationist

List of Figures

CHAPTER 1

The Dark Ages

In the stereotypical perception, the idea that the Earth is flat was common in the so-called "Dark Ages." Take for example the 2020 comedy show *Miracle Workers: Dark Ages*, a *Monty Python*-esque parody of the medieval era in which schooling consists entirely of being told that the Earth is flat and the Devil is real.[1] Monty Python's own 1975 fantasy epic parody *Monty Python and the Holy Grail* fares slightly better; a scene depicts Sir Bedevere explaining to King Arthur how the "new learning" is "how we know the earth to be banana-shaped."[2]

In academic circles, the term Dark Ages is as anachronistic as the Flat Error, as it overly-emphasizes the loss of knowledge—at least among elites—from the Roman era while under-emphasizing the intellectual circles which existed at the time.[3] Because of this, what might have

[1] *Miracle Workers: Dark Ages*, ep. 1, "Graduation" (dirs. Jarrad Paul & Andrew Mogel, writer Simon Rich), TBS, 28 January, 2020.

[2] *Monty Python and the Holy Grail*, directed by Terry Gilliam and Terry Jones. London: Python (Monty) Pictures, 1975.

[3] James Hannam, *God's Philosophers: How the Medieval World Laid the Foundations of Modern Science* (London: Icon Books, 2009); Violet Moller, *The Map of Knowledge: How Classical Ideas Were Lost and Found: A History in Seven Cities* (New York: Doubleday, 2019); Seb Falk, *The Light Ages: The Surprising Story of Medieval Science* (New York: W. W. Norton & Company, 2020); Charles Freeman, *The Reopening of the Western Mind: The Resurgence of Intellectual Life from the End of Antiquity to the Dawn of the Enlightenment* (New York: Alfred A. Knopf, 2023).

E. Guimont, *The Power of the Flat Earth Idea*,
Palgrave Studies in the History of Science and Technology,
https://doi.org/10.1007/978-3-032-23981-5_1

previously been called the Dark Ages tends now to be described as the medieval period or Middle Ages. But even that terminology is imprecise, not specifically affixed to a particular starting or ending event. If that dating of the European Middle Ages is imprecise, then trying to determine an African Middle Ages is even more so, especially if attempting to date the era based on events in the continent rather than by the European definition of the Middle Ages. The French archaeologist François-Xavier Fauvelle (b. 1968) argued for a definition of an African Middle Ages beginning with the Muslim conquest of North Africa in the late seventh century CE and ending with Vasco da Gama's circumnavigation of the continent in 1498 at the dawn of the "European Age of Discovery."[4] Notably, writing as late as 1893, a British Flat Earther of the Universal Zetetic Society argued that da Gama's circumnavigation actually helped prove the Earth was *not* a sphere:

> The author of a book entitled "The Land of the Midnight Sun" says:— "At the pole the observer seems to be in the centre of a grand spiral movement of the Sun." Why does this NOT occur in the southern regions? These few facts alone are sufficient to prove that the world is not a globe, but a vast irregular plane stretching out from a north centre to a south circumference in every direction; surrounded by water and ultimately in the extreme south by impassable barriers of ice. Vasco-de-Gama in his "Voyage to the South" says:— "The waves rise like mountains in height. The winds are piercing cold and so boisterous that the pilot's voice can seldom be heard, whilst a dismal and almost continual darkness adds greatly to the danger." How far this gloom and darkness and storms extends is not known. All we know is that the most daring have been stopped at the entrance to this gloomy, and what seems to be, forbidden region of the world.[5]

Regardless of the interpretation of da Gama's voyage by Flat Earthers four hundred years after the fact, Fauvelle's African Middle Ages spans from approximately 700–1500 CE. In this eight-hundred-year date range, there is one event, closely tied with one individual, which had more of an impact on European and Arab observers than any other in the African Middle Ages.

[4] François-Xavier Fauvelle, trans. Troy Tice, *The Golden Rhinoceros: Histories of the African Middle Ages* (Princeton: Princeton University Press, 2018), 1–15, 232–47.

[5] A. E. Skellam, *The Shape of the World* (London: Harrison & Sons, 1893), 4.

Around the year 1312, Musa I (d. c. 1337) became the mansa—often translated as sultan—of the Mali Empire, a vast and expansionist state in West Africa whose rise was fueled by the immense gold trade its ruling dynasty controlled. Mali was an Islamic state, established in a way typical for such states in the Sahel in the African Middle Ages: initiating the conversion process by converting the king. In the Sahel, the strategy failed from one point of view because the societies of the region were not centralized enough to convert their followers until the mass popular Islamic movements of the nineteenth century (which will be discussed in chapter six). However, conversion to Islam succeeded from another point of view, as it managed to attract the elites. Islam was popular for Malian elites because it not only offered a connection to the vast intellectual and religious networks that spanned the Islamic world, but it also welcomed external Muslim merchants, demonstrating that Mali and the other Sahel states were safe for trade.[6]

As part of his Muslim faith, in 1324 Mansa Musa undertook the hajj pilgrimage to Mecca, stopping in Cairo along the way. After his departure, the Arab historian Shihab al-Omari (1301–1349) compiled accounts of Musa and his reign drawn from those the mansa had talked with during his stay. Notably, al-Omari briefly recounted Musa's tale of how he came to power from the vanishing of his predecessor mansa:

> The ruler who preceded me did not believe that it was impossible to reach the extremity of the ocean that encircles the earth [Atlantic], and wanted to reach that (end) and obstinately persisted in the design. So he equipped two hundred boats full of men, like many others full of gold, water and victuals sufficient enough for several years. He ordered the chief [admiral] not to return until they had reached the extremity of the ocean, or if they had exhausted the provisions and the water. They set out. Their absence extended over a long period, and, at last, only one boat returned. On our questioning, the captain said: 'Prince, we have navigated for a long time, until we saw in the midst of the ocean as if a big river was flowing violently. My boat was the last one; others were ahead of me. As soon as any of them reached this place, it drowned in the whirlpool and never came out. I sailed backwards to escape this current.' But the Sultan would not believe him. He ordered two thousand boats to be equipped for him and for his men, and one thousand more for water and victuals. Then he conferred on me

[6] Fauvelle, *The Golden Rhinoceros*, 69–74.

> the regency during his absence, and departed with his men on the ocean trip, never to return nor to give a sign of life.[7]

The identity of just who the mansa was who undertook this voyage is contested. Most popularly it is attributed to the figure of Abu Bakr II—but no such figure ever reigned in Mali. Musa's predecessor was named Muhammad ibn Gao, and it may have been him who undertook it—if it happened at all. There is no specific reason to doubt it, even if the large round numbers given are suspect. Writing in 1856, John Leighton Wilson (1809–1885), a South Carolina abolitionist and Presbyterian minister who was one of the first American missionaries to West Africa, noted that the Mpongwe people of modern Gabon built canoes that could carry up to ten tons of cargo for almost five hundred kilometers along the coast, and might conceivably reach South America.[8] British explorer Richard Francis Burton (1821–1890) concurred with Wilson's opinion in 1876.[9] Even in the twenty-first century, canoes of over thirty-five meters in length can be found along the Niger River which was at the heart of the Mali Empire; at least some could have conceivably crossed the Atlantic, though were unlikely to return to due to the currents.[10] But regardless of the fact that Abu Bakr never existed, the story of the voyage attributed to him and possibly undertaken by Muhammed ibn Gao leads to three major points of extrapolation.

The first, and most common, has to do with Musa's activities during his pilgrimage. According to popular legend, Musa handed out so much gold during his hajj that he caused inflation in the gold market for the entire Mediterranean, leading to a decade-long depression. This is almost

[7] Ibn Fadl Allah al-Omari, trans. Gaudefroy Demombynes, *Masalik al Absar fi Mamalik el-Amsar* (Paris: Lib. Orientaliste Paul Geuthner, 1927), 74–75.

[8] J. Leighton Wilson, *Western Africa: Its History, Condition, and Prospects* (New York: Harper & Brothers, 1856), 288.

[9] Richard F. Burton, *Two Trips to Gorilla Land and the Cataracts of the Congo, Vol. I* (London: Sampson Low, Son, & Co., 1876), 83.

[10] David Northrup, "West Africans' Transatlantic Voyages and Trade, 1000–1800" (lecture, Ships and Seafarers in World History Symposium, Fairhaven, MA, 28 October, 2023). For more on West African canoe culture, see Kevin Dawson, *Undercurrents of Power: Aquatic Culture in the African Diaspora* (Philadelphia: University of Pennsylvania Press, 2018), 136–337; and Mary E. Hicks, *Captive Cosmopolitans: Black Mariners and the World of South Atlantic Slavery* (Chapel Hill: University of North Carolina Press, 2024), 65–68.

certainly hyperbole, although it contains a kernel of truth. Al-Omari notes how "the people of Cairo earned incalculable sums from [Musa] whether by buying and selling or by gifts. So much gold was current in Cairo that it ruined the value of money." It is further true that Musa's spending habits during the pilgrimage caused political trouble for him when he returned to Mali. But claims that he collapsed the Mediterranean economy, or that he was the richest human in history, are exaggeration.[11]

Further, Mansa Musa's hajj was not the first known one by a West African. That honor belonged to the ruler of the Kanem Empire, Mai Dunama bin Umme (1092–1150), who made his first at the end of the eleventh century, his second in middle of the twelfth century, and then died while attempting to complete a third. However, Musa's hajj was the most famous, and perhaps the most important for the region. While he took large amounts of gold with him, it was what Musa brought back that had a major impact in West Africa. Musa returned with large numbers of both books and scholars, dramatically escalating the existing intellectual sphere in West Africa.[12] The wealth of West Africa had allowed for a heavy degree of urbanization, dominated by the three largest cities of Gao, Djenné, and Timbuktu, whose populations peaked in the upper tens of thousands by the fifteenth century. The wealth of these cities allowed them to support elite activities, including scholarship, which particularly became associated with Timbuktu.[13] Under Musa's rule, these cities were integrated into the Malian state. When combined with his direct patronage, as well as the scholars and materials be brought with him on his hajj, a flowering of knowledge in Muslim West Africa resulted.[14]

[11] A. G. Hopkins, *An Economic History of West Africa, Second Edition* (Abingdon: Routledge, 2019), 92; Nawal Morcos Bell, "The Age of Mansa Musa of Mali: Problems in Succession and Chronology," *The International Journal of African Historical Studies*, Vol. 5, No. 5 (1972), 221–34; Sebastian Major, "Episode #114 – Who Was Mali's Richest King? (Part I)," *Our Fake History*, July 8, 2020, accessed July 8, 2020, https://ourfakehistory.com/index.php/season-5/episode-114-who-was-malis-richest-king-part-i/; Fauvelle, 191–99.

[12] Ousmane Oumar Kane, *Beyond Timbuktu: An Intellectual History of Muslim West Africa* (Cambridge, MA: Harvard University Press, 2016), 51.

[13] Hopkins, *Economic History of West Africa*, 62–63.

[14] Felipe Fernández-Armesto, *Before Columbus: Exploration and Colonization from the Mediterranean to the Atlantic, 1229–1492* (Philadelphia: University of Pennsylvania Press, 1987), 144–47; Isaac Samuel, "a brief note on Africa's Scientific Manuscripts," *African*

This renaissance became exemplified by Musa's construction of the pyramidal Sankoré Mosque in Timbuktu. With its massive library, Sankoré became the center of what would become termed the University of Sankoré in 1895 by French journalist Félix Dubois (1862–1945), who thought the Sankoré district of Timbuktu was a parallel to the Latin Quarter in Paris. Indeed, when British soldier Alexander Gordon Laing (1794–1826) became the first confirmed European to reach Timbuktu in 1826, the Sankoré Mosque and its library were one of the few things that impressed him about the city. It is true that by the nineteenth century, shifting political and economic conditions had left Timbuktu far from its prime—but it had certainly enjoyed quite a prime. One hundred and eighty schools were built in Timbuktu by Musa and his successors, turning it into the largest precolonial—or more accurately, pre-*European* colonial—center of learning in sub-Saharan Africa. By 1450, one quarter of the city's 100,000 residents were Muslim scholars from as far away as Cairo. In contrast, by Laing's arrival, Timbuktu's population was only 12,000.[15] The benefit to the city—and the world-class scholarship it produced—from its conquest by Mali under Musa is quite clear.

Which brings us to the second point of extrapolation that emerges from al-Omari's account of Musa's visit to Cairo: what exactly the mansa's expedition found on the "extremity of the ocean." In a three-volume monograph series published from 1920 to 1922, the Russo-American linguist Leo Wiener (1862–1939) proposed that the mansa had not only been the first Old World figure to cross the Atlantic, but that the Olmec civilization of modern Mexico derived from his sailors.[16] Wiener's hypothesis, termed Afrocentric hyperdiffusionism, became popularized in 1976 with the publication of *They Came Before Columbus* by the

History Extra, September 30, 2023, accessed September 30, 2023, https://www.africanhistoryextra.com/p/a-brief-note-on-africas-scientific.

[15] Timothy F. Garrard, *Akan Weights and the Gold Trade* (London: Longman Group, 1980), 10–12; Frank T. Kryza, *The Race for Timbuktu: In Search of Africa's City of Gold* (New York: HarperCollins, 2006), 231; Charlie English, *The Storied City: The Quest for Timbuktu and the Fantastic Mission to Save Its Past* (New York: Riverhead Books, 2017), 267–70; Elias N. Saad, *Social History of Timbuktu: The Role of Muslim Scholars and Notables 1400–1900* (Cambridge: Cambridge University Press, 1983).

[16] Leo Wiener, *Africa and the Discovery of America, Vols. I-III* (Philadelphia: Innes & Sons, 1920–1922).

Guyanese Africana historian Ivan Van Sertima (1935–2009).[17] Van Sertima's proposal of West African origins of Mesoamerican civilization, with Abu Bakr successfully crossing the Atlantic almost two centuries before Columbus, became popular as a sign of African achievement during a period of intellectual and political radicalism throughout the African Diaspora. However, by attributing Mesoamerican achievements to West Africans, the theory erases the actual development of indigenous Americans no less than the similarly-ahistorical claims that Norse, Welsh, or Irish explorers were responsible for the North American Native mounds and pictographs.[18] Despite this, Van Sertima's vision of Abu Bakr as an African Columbus has remained popular; for example, both the African-American historian Ibram X. Kendi (b. 1982) and the United Church of Christ cited *They Came Before Columbus* in support of the 1619 Project on African-American history.[19]

If the Malian sailors did not in fact settle precolonial Mexico, therefore, we can still use their expedition to illustrate the third point to be taken from al-Omari's account. In it, we can see that a cornerstone of geography was already well known, for al-Omari describes how the Atlantic Ocean "encircles the earth"—in other words, that West Africans of the fourteenth century knew that the Earth was a sphere. Indeed, the belief in the world being a sphere with the ocean curving around was part of the mansa's worldview, but also his vanishing in the attempt to circumnavigate it was part of Musa's claim to the right to rule. Such knowledge was not exclusive to Mali or the Muslim world in general. However, the Malian expedition serves as continuity of a long history of knowledge production in northern Africa. The world's oldest known

[17] Ivan Van Sertima, *They Came Before Columbus: The African Presence in Ancient America* (New York: Random House, 1976); Brian Regal, *The Battle over America's Origin Story: Legends, Amateurs, and Professional Historiographers* (London: Palgrave Macmillan, 2022), 53–76, 167–96; Kenneth L. Feder, *Encyclopedia of Dubious Archaeology: From Atlantis to the Walam Olum* (Santa Barbara, CA: ABC-CLIO, 2010), 1–3, 263–64.

[18] Edward Guimont, "From King Solomon to Ian Smith: Rhodesian Alternate Histories of Zimbabwe" (PhD diss., University of Connecticut, 2019), 206–13; Fauvelle 160–68.

[19] Ibram X. Kendi, "The Hopefulness and Hopelessness of 1619," *The Atlantic*, August 20, 2019, accessed August 21, 2019, https://www.theatlantic.com/ideas/archive/2019/08/historical-significance-1619/596365/; Velda Love, "400 Years Commemoration 1619 – 2019 Africans in America," United Church of Christ, 2019, accessed August 13, 2020, https://www.ucc.org/400_years_commemoration_1619_2019_africans_in_america.

astronomical site, Nabta Playa, dates from approximately 5000 BCE in modern Egypt, just one of many pioneering works of African scientific and technological advancement that were eclipsed even before the colonial era.[20] Among these developments are also what have been interpreted as the oldest mathematical tools, the Lebombo bone (c. 40,000 BCE, Eswatini) and the Ishango bone (c. 20,000 BCE, Democratic Republic of the Congo).[21] Potentially the oldest known observatory—meaning a building dedicated entirely to study of the night sky—is located in the city of Meroe, capital of the Nubian kingdom, dated to approximately the first century BCE.[22] The oldest university in the world is also in north Africa; the University of al-Qarawiyyin was founded in Fez in modern Morocco in the year 859 CE, by an educated Arab woman named Fatima al-Fihri (c. 800-c. 880).[23] But the ancient north African site with perhaps the greatest connection to the notion of the Earth's sphericity is the Library of Alexandria.

The Library of Alexandria was the third great center of learning in the ancient Greek world, after the Academy (established by Plato in 387 BCE) and the Lyceum (established by Plato's student Aristotle in 334 BCE). But whereas the Academy and the Lyceum were both in Athens, the Library was (obviously) in Alexandria, the city that became the capitol of Hellenistic and Roman Egypt. Alexandria is located on the delta of the Nile River, where it flows into the Mediterranean. The Nile held a special place for the future Flat Earthers of the fin de siècle, a combination of its Biblical association and the fact that it flows from south to

[20] Clare Oxby, "A Review of African Ethno-Astronomy: With Particular Reference to Saharan Livestock-Keepers," *La Ricerca Folklorica*, No. 40 (Oct., 1999), 55–64; J. McK Malville, R. Schild, F. Wendorf, and R. Brenmer, "Astronomy of Nabta Playa," *African Skies/Cieux Africains*, No. 11 (Jul., 2007), 2–7; Bertram Mapunda, "Debate: Why Study Precolonial African Technology and Material Culture?," *Technology and Culture*, Vol. 64, No. 3 (Jul., 2023), 665–76.

[21] Sindiso M. Nleya and Siqabukile Ndlovu, "Audit of Mathematical Concepts in Pre-colonial Africa," in Artwell Nhemachena, Nokuthula Hlabangane, and Joseph Z. Z. Matowanyika, eds., *Decolonising Science, Technology, Engineering and Mathematics (STEM) in an Age of Technocolonialism: Recentring African Indigenous Knowledge and Belief Systems* (Bamenda: Langaa RPCIG, 2020), 128–29.

[22] Thomas J. Logan and Bruce Williams, "On the Meroe Observatory," *Beitraege zur Sudanforschung*, No. 7 (2000), 62–67; Samuel, "a brief note on Africa's Scientific Manuscripts."

[23] Kane, *Beyond Timbuktu*, 45–46.

north. To many Flat Earthers—from the nineteenth into the twenty-first centuries—the association of north with "up" and south with "down" is engrained enough that the idea a river can flow north "up" a sphere seems to break the rules of physics and prove the Earth has to be flat.[24] This, of course, would not be an issue with Islamic cartography (and many of the African mapping systems derived from the Muslim system) where maps are oriented with north at the bottom and south at the top, and it was medieval Muslims who first comprehensively mapped much of the Nile, though without reaching the fabled source.[25] Of note also is that until the fourteenth century nearly all European maps had the east on top.[26]

While the Library of Alexandria was not the first library of the ancient world, it and its associated Museum was the largest, and the first state-run research institute whose staff salaries were paid by first the Egyptian, and later Roman, governments. Named after Alexander the Great, Alexandria was founded by its namesake in 331 BCE, but after the first few months of construction, the conqueror left and never returned during his life (although it would become the site of his tomb after his 323 BCE death). The Library was the brainchild of Demetrius of Phalerum (c. 350–280 BCE), former governor of Athens and student of Theophrastus (c. 371–c. 287 BCE), Aristotle's successor as the head of the Lyceum. Demetrius was one of the philosophers invited to Egypt by Ptolemy I (367–282 BCE), Alexander's general who became pharaoh after the warlord's death, and convinced the new king to establish the Library. Demetrius ultimately died in prison under Ptolemy's son Ptolemy II (309–246 BCE) as he had argued for the succession of a different son, but even so it was likely only under the reign of Ptolemy II that the Library grew to an appreciable size. The Byzantine scholar John Tzetzes (c. 1110–1180) estimated the

[24] William Carpenter, *One Hundred Proofs that the Earth is Not a Globe, Fifth Edition* (Baltimore: William Carpenter, 1886), 6–7; A. Hottentot, "'Parallax' Triumphant," *The Earth not a Globe Review*, No. 3 (Jul., 1893), 19; Charles K. Johnson, "Rivers," *Flat Earth News* (Sep., 1979), 2; Charles K. Johnson, "Scientists claim their word is "Truth"," *Flat Earth News*, No. 84 (Dec., 1992), 1.

[25] Mick Ashworth, *Why North is Up: Map Conventions and Where They Came From* (Oxford: Bodleian Library Publishing, 2019), 7–12; Jim al-Khalili, *The House of Wisdom: How Arabic Science Saved Ancient Knowledge and Gave Us the Renaissance* (New York: Penguin Books, 2011), 88–89; Karen C. Pinto, *Medieval Islamic Maps: An Exploration* (Chicago: University of Chicago Press, 2016), 52–53.

[26] Evelyn Edson, *The World Map, 1300–1492: The Persistence of Tradition and Transformation* (Baltimore: Johns Hopkins University Press, 2007), 20.

ultimate size of the Library to be 400,000 mixed books and 90,000 single books.[27]

Astronomical knowledge had obviously existed among the Greeks before the establishment of the Library of Alexandria, with varying degrees of accuracy. At the very start of the Classical period, Homer (fl. eighth century BCE) was one of several Greek poets who believed the Earth to be flat, describing it as disc in the *Iliad* and the *Odyssey*. Anaximander of Miletus (c. 610–546 BCE), credited with creating potentially the first world map, is sometimes stated as being a Flat Earther, but that is not quite accurate. He believed the Earth to be cylindrical, with a breadth three times that of its depth and what the Greeks saw as their world being one of the two flat ends.[28] Anaximander was not only involved with early theorizing of the Earth's (non-flat) shape, but made an early proposal of evolution, including potentially the first idea that life originally emerged in the sea before moving onto land, and that early humans were of a very different form from modern humans.[29] As will be seen in future chapters, this connection is apt, given that in the nineteenth century Flat Earth advocacy would become connected with Young-Earth Creationists opposing Darwinian evolution.

Anaximander's home of Miletus was located in Ionia, the region of Greek settlement in what is now the west coast of Turkey. Because of their presence at the end of the Persian Royal Road and its Assyrian

[27] M. R. Wright, *Cosmology in Antiquity* (New York: Routledge, 1995), 28; David Deming, *Science and Technology in World History, Volume 1: The Ancient World and Classical Civilization* (Jefferson, NC: McFarland & Company, 2010), 122–24; John Freely, *Aladdin's Lamp: How Greek Science Came to Europe Through the Islamic World* (New York: Alfred A. Knopf, 2009), 38–40; John Freely, *Light from the East: How the Science of Medieval Islam Helped to Shape the Western World* (London: I. B. Tauris & Co., 2011), 15–16; Olaf Pedersen, trans. Richard North, *The First Universities:* Studium Generale *and the Origins of University Education in Europe* (Cambridge: Cambridge University Press, 1997), 12–19; Moller, *The Map of Knowledge*, 19–22.

[28] James S. Romm, *The Edges of the Earth in Ancient Thought: Geography, Exploration, and Fiction* (Princeton: Princeton University Press, 1992), 179–83; Avan Judd Stallard, "Origins of the Idea of Antipodes: Errors, Assumptions, and a Bare Few Facts," *Terrae Incognitae: The Journal of the Society for the History of Discoveries*, Vol. 42, No. 1 (2010), 40; Philippa Lang, *Science: Antiquity and its Legacy* (London: I. B. Tauris, 2015), 140–42, 144; Wright, *Cosmology in Antiquity*, 39.

[29] Adrienne Mayor, *The First Fossil Hunters: Paleontology in Greek and Roman Times* (Princeton: Princeton University Press, 2000), 214–15; Lang, *Science: Antiquity and its Legacy*, 28–31.

antecedents, the Ionian cities were exposed to a wide range of cultural influences from the non-Greek peoples of the Near East. Miletus itself was the center of the influential Ionian School of Greek philosophy, which largely sought to determine the physical foundations and nature of matter. Additional Ionian School philosophers included Anaxagoras, Heraclitus, and Thales; other Ionian Greeks not associated with the school included the philosopher Pythagoras and historian Herodotus. Pythagoras and his teachings were extremely influential in Magna Graecia, the Greek colonies of southern Italy. Another major philosopher there was Parmenides of Elea, who in the early fifth century BCE seems to have been the first Greek philosopher to argue that the Earth was a sphere. Pythagoras also came to that conclusion, likely shortly after Parmenides.[30]

Much later, another Ionian city was the site of the Library of Pergamon, the greatest rival to the Library of Alexandria. Established approximately in the period of 220–160 BCE by the Attalid dynasty (another of the successors to Alexander), its maximum size was approximately 200,000 books. Much of its collection may have been given to the Library of Alexandria by Roman general Mark Antony (83–30 BCE) after the Alexandrian library was accidentally damaged by Julius Caesar in 48 BCE. However, the Library of Pergamon itself still existed for centuries more, albeit in greatly reduced influence and no longer a rival to Alexandria.[31]

Even outside of the schools of philosophy, Greeks of the Classical period may have speculated on the size and shape of the world, including the nature of the southern hemisphere and how the Sun would be perceived there, by incorporating information from Phoenician efforts to circumnavigate Africa and explore the Atlantic. In particular, there was the story repeated by Herodotus (though he doubted its veracity) that Egyptian pharaoh Necho II (r. 610–595 BCE) commissioned a Phoenician crew to circumnavigate Africa, which they supposedly accomplished over the course of three years. Several Greek sources also related a fifth century BCE voyage of the Phoenician Hanno the Navigator around the west coast of Africa.[32] However, European and Near Eastern navigation down

[30] Daniel W. Graham, *Explaining the Cosmos: The Ionian Tradition of Scientific Philosophy* (Princeton: Princeton University Press, 2006), 1–27, 294–308.

[31] Deming, *Science and Technology in World History*, 124; Moller, *The Map of Knowledge*, 12–15.

[32] Duane W. Roller, *Through the Pillars of Herakles: Greco-Roman Exploration of the Atlantic* (New York: Routledge, 2006), 22–43; J. R. Masson, "Geographical Knowledge

the West African coast beyond the Canary Islands was extremely limited until the fifteenth century due to the difficulty of navigating the wind and ocean currents. Likewise, expeditions deep into the Atlantic were also rare. At least some Greeks did voyage north, most famously Pytheas of Massalia in c. 325 BCE, who described a frigid northern region he named Thule which has been variously identified with Ireland, Norway, Iceland, or the Orkneys (whose name comes from Pytheas' name for Scotland, Orcas). Regardless of just where Thule was, Pytheas also circumnavigated Britain, reached the Baltic coast, and encountered ice floating upon the northern sea.[33]

In addition to the naval voyages of Necho and Pytheas, the Greeks, particularly of Ionia, were also familiar with the accounts of merchants and traders who agreed that the landmass of Africa became warmer as one went further south, and the land of Europe became colder the further north one went. This led to the notion that the oikoumene, the connected landmass of Europe, Africa, and Asia—all of whose true extents were not known by the Greeks, Romans, or later Muslims—was placed on a small slice of the spherical Earth proposed by Parmenides. Plato's student, the philosopher Aristotle (384–322 BCE), would eventually propose that this region in which humans could comfortably live formed a temperate band encircling the Earth, even if the landmass of the oikoumene did not. North of this was an Arctic region, which was too cold to support life; it only made sense that below the temperate zone, the heat would gradually become too high to live, and perhaps even high enough that the oceans themselves might boil. This was termed the torrid zone. Because Aristotle valued symmetry and believed it symbolized divine perfection in nature, he rationalized that beneath the torrid zone might be another inhabitable band called the Antipodes mirroring the oikoumene, and beneath it an Antarctic region that mirrored the Arctic of the far north. The concept of the Antipodes as a southern landmass below the equator of the spherical Earth—an idea also applicable to

and Maps of Southern Africa Before 1500 A.D.," *Terrae Incognitae: The Journal of the Society for the History of Discoveries*, Vol. 18, No. 1 (1986), 1–4; Nicholas Nicastro, *Circumference: Eratosthenes and the Ancient Quest to Measure the Globe* (New York: St. Martin's Press, 2008), 23–24; Romm, *Edges of the Earth*, 19–20.

[33] Roller, *Through the Pillars of Herakles*, 57–91; Romm, *Edges of the Earth*, 22, 156–59.

the cylindrical world model of Anaximander—would remain a notion in Western thought into the early modern period.[34]

These were the notions of Greek geography, astronomy, cartography, and cosmology which were brought to the Library of Alexandria upon it and the Museum's establishment. There, Greek science was blended with the thinking of other cultures, particularly Egyptian and Babylonian. The ancient Egyptians and Babylonians (and nearby Israelites) did conceive of the Earth as a flat disc covered by a dome and surrounded by a river, but their systems of astronomy had theoretical frameworks much less systematized than those of the Greeks. Although Babylonians were extremely proficient in astronomy, to the point that "Babylonian" and "Chaldean" (a specific Babylonian people) were used as shorthand for "astronomer" and "astrologer" by the Greeks and Romans, their astronomy was mostly interested in using visible observations to time religious events, and was never developed into a theoretical framework.[35] In Egypt, there was no systematized school of cosmological thought and very little interest in astronomy overall.[36] The Ionians were well suited to synthesize the knowledge of these Near Eastern cultures with Greek scholastic systems. Anaximander, for example, had drawn influence from both Babylonian and Persian Zoroastrian astronomy influencing his design of the universe as a thick disc and the Sun actually being a hole letting the light shine through from beyond.[37]

Babylonian cartography had two longstanding influences in the Western world. The first was their depiction of the world, in a map dated to approximately the sixth century BCE. The challenger to Anaximander for being the earliest known attempt to depict a map of the

[34] Rudolph Simek, trans. Angela Hall, *Heaven and Earth in the Middle Ages: The Physical World before Columbus* (Woodbridge: The Boydell Press, 1996), 39–55; C. Philipp E. Nothaft, "Zaccaria Lilio and the shape of the earth: A brief response to Allegro's "Flat earth science"," *History of Science*, Vol. 55, No. 4 (2017), 490–98; Stallard, "Origins of the Idea of Antipodes," 34–51.

[35] Nicastro, *Circumference*, 29–30; Pinto, *Medieval Islamic Maps*, 95–98; L. Sprague de Camp and Willy Ley, *Lands Beyond: A Fascinating Expedition Into Unknown Lands* (New York: Rinehart & Company, 1952), 281; Karine Gadré, "Le Lever du Soleil sur la Terre d'Egypte," *African Skies/Cieux Africains*, No. 6 (Dec., 2001), 6–9.

[36] Pinto, *Medieval Islamic Maps*, 82–83; Nicastro, *Circumference*, 30–31; De Camp and Ley, *Lands Beyond*, 281–82; Freely, *Light from the East*, 1–3.

[37] Nicastro, *Circumference*, 31; Freely, *Light from the East*, 3–8; Pinto, *Medieval Islamic Maps*, 86–95.

world, it is circular with the outer edge representing the encircling ocean; a horizontal line near the bottom seems to represent the Mesopotamian marshes; and a roughly straight line bisecting the map from the top represent the Zagros Mountains. The Babylonian world map was discovered in 1882 and translated in 1889. In 2016, scholar of Islamic religion Karen Pinto argued that predecessors of the Babylonian world map could be found in Near Eastern sites dating back to the Neolithic period, such as Göbekli Tepe.[38] Likewise, in 2012, historian Carlo Zaccagnini proposed that the Babylonian cartographic system represented in the recovered map inspired—via an unknown form of continuity or recovery of knowledge after a millennium and a half—the T-O maps of medieval Europe.[39] The T-O map was a cartographic system introduced by the post-Roman archbishop of Seville Isidorus Hispalensis, better known as Isidore of Seville (c. 560–636). The Earth depicted is a disc, generally agreed by scholars not to indicate that Isidore was a Flat Earther but that it depicted the northern hemisphere of the Earth, the Antipodes being unknowable. The disc is bisected by a "T" representing the Mediterranean (the base of the T), the Nile (the top right spoke), and the Don River (the top left spoke). The hemisphere at the top is Asia, the lower left quarter Europe, and the lower right quarter Africa. The intersection of all at the center is Jerusalem. Although nearly useless for practical usage, the T-O map had an important theological role, especially when integrated into the interpretation of the Book of Genesis that Noah's three sons Shem, Japheth and Ham became the ancestors for, respectively, Asians, Europeans, and Africans (Fig. 1.1).[40]

[38] Pinto, *Medieval Islamic Maps*, 80–82.

[39] Carlo Zaccagnini, "Maps of the World," in *Leggo! Studies Presented to Frederick Mario Fales on the occasion of his 65th birthday*, eds. Daniele Morandi Bonacossi et al. (Wiesbaden: Harrassowitz Verlag, 2012), 865–74; Pinto, *Medieval Islamic Maps*, 83–86.

[40] Francis S. Betten, "The Knowledge of the Sphericity of the Earth during the Earlier Middle Ages," *The Catholic Historical Review*, Vol. 9, No. 1 (Apr., 1923), 83–86; John Williams, "Isidore, Orosius and the Beatus Map," *Imago Mundi*, Vol. 49 (1997), 7–32; Toby Lester, *The Fourth Part of the World: The Race to the Ends of the Earth, and the Epic Story of the Map That Gave America Its Name* (New York: Free Press, 2009), 27–29; Mick Ashworth, *Why North is Up: Map Conventions and Where They Came From* (Oxford: Bodleian Library Publishing, 2019), 145; Jonathan Lyons, *The House of Wisdom: How the Arabs Transformed Western Civilization* (New York: Bloomsbury Press, 2009), 49; Jeffrey Burton Russell, *Inventing the Flat Earth: Columbus and Modern Historians* (New York: Praeger, 1991), 17–19, 27; Christine Garwood, *Flat Earth: The History of an Infamous Idea* (New York: Thomas Dunne Books, 2007), 25; Nicolás Wey Gómez, *The Tropics*

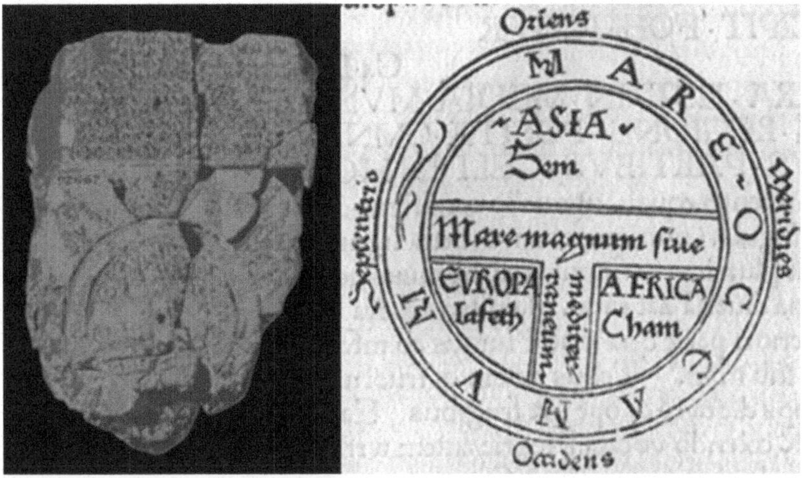

Fig. 1.1 The Babylonian world map (left) and the T-O map of Isidore of Seville (right) (Wikimedia Commons)

Although the general consensus is that Isidore believed in a spherical Earth, there have been some dissenting views that the T-O map was meant to depict a Flat Earth, or that Isidore was ambiguous on the issue because it was neither important to him or wider Christian theology at the time.[41] Perhaps because of this, Isidore is a marginal figure amongst the Flat Earthers of the nineteenth and twentieth centuries, although at least one did mention Isidore's belief, not in a Flat Earth, but that the remnants of Noah's Ark could still be seen in the mountains of Armenia.[42] However, the geography of Isiodore's potential Babylonian inspiration proved to be much more lasting in the modern Flat Earth imaginary.

In the 1881 edition of the ur-text of modern Flat Earthism, *Zetetic Astronomy*, Samuel Rowbotham (to be discussed more next chapter) noted that "Tables of the places of the sun and moon, of eclipses, and of

of Empire: Why Columbus Sailed South to the Indies (Cambridge, MA: The MIT Press, 2008), 114–15; Pinto, *Medieval Islamic Maps*, 122–42; Edson, *The World Map*, 17–18.

[41] Russell, *Inventing the Flat Earth*, 19; Lyons, *The House of Wisdom*, 46, 128–29.

[42] Egerton Sykes, "Noah's Ark: The Secret of Mount Ararat," *Atlantean Research: Official Organ of the Research Centre Group*, Vol. 1, No. 6 (Mar., 1949), 87.

kindred phenomena, have existed for thousands of years, and were formed independently of each other, by the Chaldean, Babylonian, Egyptian, Hindoo, Chinese, and other ancient astronomers," and that the Babylonian notion of the Earth as a flat disc floating on an ocean corresponded with early Hebrew—and therefore Biblical—notions, a claim repeated in an article in the 1896 British Flat Earther journal.[43] In Flat Earther Ebenezer Breach's 1896 pamphlet *Fifty Scientific Facts For Surrender to Nature's Fixed Truths*, the twenty-eighth of the titular facts is that "Though the prediction of eclipses are correct by the metonic cycle, as used by the Chaldeans, the arrangement for them by modem astronomy is out of all proportions, and rendered impossible," and he ended it signed "In the name of the Chaldeans, the real founders of Astronomy; the Egyptians, Chinese, Greeks, and Romans, the true promoters; in the name of Hipparchus, the father and prince of Astronomers; in the name of Ptolemy, whose system continued 1400 years unopposed... and in the name of Lord Bacon, and all that have not been misled, this day and for ever."[44]

Nor were the Babylonian references limited to the British Flat Earthers of the fin de siècle. Charles K. Johnson (1924–2001), the American who led the International Flat Earth Research Society (IFERS) in the late twentieth century, discussed the Babylonian world map in a December 1992 article in his newsletter, interpreting it (correctly) as evidence that the Babylonians saw the Earth as a flat disc, and that they in turn influenced the Old Testament.[45] A year later, he argued that the towering NASA space shuttle launch complex in Florida was a new Tower of Babel, and the (according to him, alleged) exploration of space was actually a means of promoting pagan sun worship—a strange comparison given his earlier connection of Babylonian paganism to the Old Testament.[46]

[43] Samuel Rowbotham, *Zetetic Astronomy: Earth Not a Globe, Third Edition* (London: Simpkin, Marshall, and Co., 1881), 154, 364–65; Balaam's Ass, "Faith and Science," *The Earth not a Globe Review*, Vol. III, No. 2 (May, 1896), 18–19.

[44] Ebenezer Breach, *Fifty Scientific Facts For Surrender to Nature's Fixed Truths* (London: John Williams, 1896), 8, 12.

[45] Charles K. Johnson, "The Wings of the Earth," *Flat Earth News*, No. 82 (Dec., 1992), 2.

[46] Charles K. Johnson, "Babylon-Babel Revisited," *Flat Earth News*, No. 87 (Dec., 1993), 1.

The durability of the knowledge of Babylonian astronomy in the West—and the likely vector that allowed for whatever reception to it that Isidore received—was largely due to its aforementioned integration with Greek, Egyptian, Persian, and other knowledge at the Library of Alexandria. The Library, while established by Hellenistic Greeks under a Hellenistic Greek government, did not derive all of its organization nor all of its staff solely from the Greek world.[47] Outside of a variety of individual archives and libraries in pre-Hellenistic Egypt, there was the tradition of the House of Life institution of temple libraries and archives, which were incorporated into the Library of Alexandria.[48] Like Alexandria itself, the Library was an imperial construct, not merely because it and its staff received funding from pharaonic and Roman governments, but because it relied on government agents searching boats which docked at Alexandria's ports, seizing any books onboard so they could be copied—the copies being returned to the original owners and the original kept in the Library.[49]

Likely the most famous chief librarian of Alexandria was Eratosthenes of Cyrene (c. 276–195 BCE), who held the position from 245 BCE until his death. In 240 BCE, Eratosthenes calculated the circumference of the Earth through measuring the shadow of a pole at noon in both Alexandria and the city the Greeks called Syene—but which native Egyptians called Swenett at the time, and modern Egyptians call Aswan—located in the south of Egypt, along the border with the Kingdom of Kush. Using the government survey records in the Library, Eratosthenes knew the precise distance from Alexandria to Syene, and was able to calculate the curvature of the Earth based on the differences in the rod's shadow. Eratosthenes' circumference was 252,000 stadia, a Greek unit of measurement which is not precisely known but was between 150 and 200 meters. With an estimated likely length of approximately 157 meters, Eratosthenes was extremely close, coming to 39,564 kilometers in contrast to the actual circumference of 40,075 kilometers. In a sign of the knowledge lost from

[47] John Pappademos, "An Outline of Africa's Role in the History of Physics," in Ivan Van Sertima, ed., *Blacks in Science: Ancient and Modern* (New Brunswick, NJ: Transaction Books, 1983), 188–91.

[48] Fayza M. Haikal, "Private Collections and Temple Libraries in Ancient Egypt," in Mostafa El-Abbadi and Omnia Fathallah, eds., *What Happened to the Ancient Library of Alexandria?* (Leiden: Brill, 2008), 39–54; Pedersen, *The First Universities*, 2–4.

[49] Deming, *Science and Technology in World History*, 122–24.

the classical world, Eratosthenes wrote his findings in the book *On the Measure of the Earth*, which also divided the Earth into five zones of latitude and longitude, as well as roughly calculating the distances of the Sun and Moon from Earth. *On the Measure of the Earth* is now lost and only known through its description in the book *On the Circular Motions of the Celestial Bodies* by the later Greek astronomer Cleomedes, whose biographical and chronological information is entirely unknown.[50]

But the greatest astronomer and cartographer associated with the Library was Claudius Ptolemaeus (c. 100–170), often Anglicized to Ptolemy. He was of no relation to the Hellenistic pharaohs who established the Library half a millennium earlier, although some later Muslim scholars believed he was their descendent. In reality, little is known of Ptolemy's biography. Physicist and Afrocentric scholar John Pappademos (1925–2007) argued that Ptolemy had Black African ancestry and incorporated Black African knowledge into his worldview, claims which have come under attack; however, the guess as to his ancestry has as much validity as any other ethnic claim given the lack of actual knowledge.[51] Ptolemy's most important legacies are two of his publications. One, the *Geography*, dates from c. 150 and was the first to develop a system of latitude and longitude, as well as the first determination of terrestrial locations through astronomical observations. It was the basis for an actual map of the world by a largely unknown figure, Agathodaemon of Alexandria, likely within a few decades of the *Geography*'s publication (Fig. 1.2).

The other, more important publication of Ptolemy was his *Syntaxis Mathematica*, better known by its Arabic title *Almagest*. This created a geocentric model of the universe, with the Earth at the center and the Sun, Moon, and all planets orbiting the Earth. This would become the standard model in both Christian and Muslim worlds, only fully being refuted nearly 1500 years later by Nicolaus Copernicus (although

[50] Russell, *Inventing the Flat Earth*, 24–26; Garwood, *Flat Earth*, 20–22; Gómez, *Tropics of Empire*, 180–88; Nicastro, *Circumference*, 24–28; Deming, *Science and Technology in World History*, 133–36; Wright, *Cosmology in Antiquity*, 29–30; De Camp and Ley, *Lands Beyond*, 287–89; Lang, *Science: Antiquity and its Legacy*, 146–47; Freely, *Aladdin's Lamp*, 40–42.

[51] Pappademos, "Outline of Africa's Role," 190; Robert Palter, "Black Athena, Afrocentrism, and the History of Science," in Mary R. Lefkowitz and Guy MacLean Rogers, eds., *Black Athena Revisited* (Chapel Hill: University of North Carolina Press, 1996), 225.

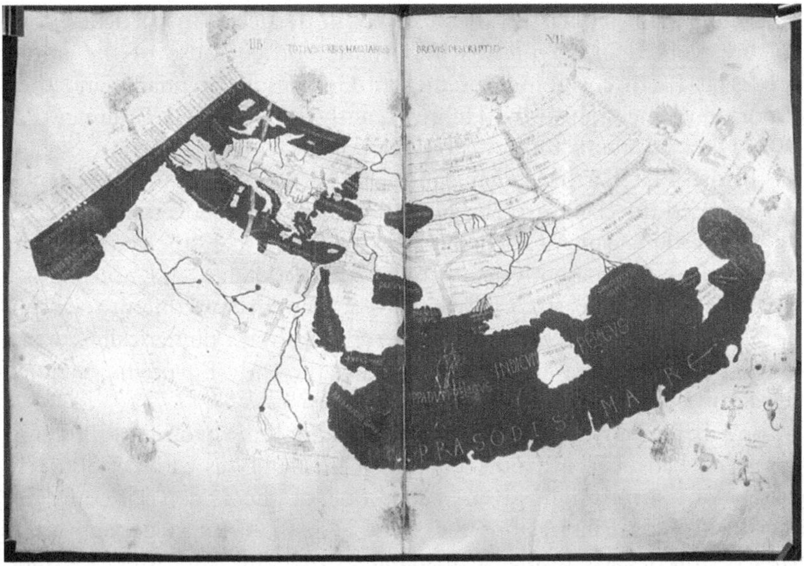

Fig. 1.2 Fifteenth century recreation of Ptolemy's world map from the *Geography* (Wikimedia Commons)

minority views endorsing heliocentrism went at least as far back as Aristarchus of Samos, four centuries before Ptolemy). Both the *Geography* and the *Almagest* were based on a spherical world, with the 'world' map of the former only depicting a slice of the oikoumene. Ironically, this led to some questioning of Ptolemy's maps by Europeans in the "Age of Discovery," over how to reconcile their flatness with the world's sphericity. As with other philosophers like Pythagoras, Ptolemy's geography and astronomy were vital component parts of the broader philosophy through which he saw the world.[52]

As mentioned above via the 1896 statement of Breach, even though Ptolemy's worldview was based on a spherical Earth, his geocentrism

[52] Patrick Gautier Dalché, "The Reception of Ptolemy's *Geography* (End of the Fourteenth to Beginning of the Sixteenth Century)," in David Woodward, ed., *The History of Cartography, Volume 3: Cartography in the European Renaissance* (Chicago: The University of Chicago Press, 2018), 327–33; Jacqueline Feke, *Ptolemy's Philosophy: Mathematics as a Way of Life* (Princeton: Princeton University Press, 2018); Freely, *Aladdin's Lamp*, 53–54; Moller, *The Map of Knowledge*, 31–34.

made him useful to Flat Earthers of the fin de siècle. In particular, that Ptolemy worked in Egypt was important, both because of the undisputed Flat Earth cosmology of the pre-Hellenistic Egyptians, and their purported connection to the Hebrews, authors of the Old Testament. In 1903, Lady Elizabeth Blount, President of the Universal Zetetic Society (the British Flat Earther organization, to be described more in chapter five) disputed the claims of a Bible scholar, J. B. Dimbleby, that the Egyptians believed in a spherical Earth partly by noting that "the system of the great Ptolemy," apparently using Egyptian knowledge, did not endorse "the motions of the earth." However, Lady Blount was content to accept Dimbleby's claim that Egyptian "records are always numerically correct from the date of Creation is evidence that the early Egyptians obtained their knowledge from the Adamic line of patriarchs."[53]

Long before British Christians would adopt the works of Ptolemy, Alexandria had been a major center of Christianity as early as 200 CE, perhaps not surprisingly given it was one of the largest cities in not only the Roman Empire, but the world. North Africa in general was a major region of early Roman Christianity, with Saint Augustine of Hippo (354–430)—likely a Berber from what is now Algeria—being only the best-known example in the modern day. As early as the First Council of Nicaea in 325, Alexandria was recognized as one of the greatest centers of Christianity in the Empire on par with Rome, and whose bishop had authority over much of the most densely populated regions of North Africa.

In book XVI, Chapter 9 of *The City of God* (c. 426), Augustine explicitly states that he accepts the sphericity of the Earth, but rejects the concept of the Antipodes, believing that the only inhabited regions were the oikoumene already known to Roman geography. Augustine's reasoning was if there were inhabited Antipodes, their inhabitants could not be descended from Adam and Eve via Noah and his descendants, completely separated from the oikoumene as they would have to be.[54] In this, Augustine was fairly standard for not only Christians of the Roman Empire, but post-Roman Western Europe as well. For example, in 748, the Irish astronomer and clergyman Vergilius of Salzburg (c. 700–784)

[53] Elizabeth Blount, "Does the Earth Rotate?," *The Earth*, Vol. III, Nos. 37 & 38 (Aug./Sep., 1903), 262–63.

[54] St. Augustine, trans. Henry Bettenson, *Concerning the City of God against the Pagans* (London: Penguin Books, 1984), 664–65; De Camp and Ley, *Lands Beyond*, 293–94.

argued that the Earth was a sphere, with the opposite hemisphere inhabited. Vergilius received official condemnation from Pope Saint Zachary (679–752; r. 741–752) not for describing a spherical Earth, but for his notion of inhabited Antipodes.[55] Medieval opposition to the concept of Antipodes was religious, not scientific or geographical; later scientific notions like the Bering or Thule land bridges would help solve the conundrum.[56]

However, two of the most notable exceptions to this acceptance of the globe in not only early Christianity, but Christianity overall until the nineteenth century, were also North Africans. The first, and more prominent, was Lucius Caecilius Firmianus Lactantius (c. 250–325). Like Augustine, Lactantius was a Berber and likely from the Roman providence of Numidia, corresponding to modern Tunisia. Many specific details on his early life and career are unknown, but he was a student of Arnobius, a Christian Berber rhetor at Sicca Veneria (modern El Kef, Tunisia). It is likely that the Emperor Diocletian (c. 242–312; r. 284–305) hired Lactantius to serve as chair of Latin rhetoric (roughly equivalent to a modern university professor) at Nicomedia, then the capitol of the Eastern Roman Empire, in the mid-290s. Lactantius likely converted to Christianity at Nicomedia rather than in Africa; by then large numbers of Christian converts came from educated classes of Roman society and so were well represented among Roman intellectuals. During his time as rhetor, Lactantius likely met and spoke many times on a casual basis with the future emperor Constantine (c. 272–337) and his mother, the future Saint Helena (c. 246-c. 330). At this time, Diocletian had established a system whereby the Roman Empire was governed by four emperors, a senior and junior emperor for both East and West, with the senior Eastern emperor the unofficial dominant figure. On 24 February 303, Diocletian issued an anti-Christian edict inaugurating what became known as the Great Persecution, which likely led to Lactantius resigning his position of rhetor. On 1 May 305, Diocletian resigned as emperor, at which point Lactantius likely left the East to join the court of Constantine's father, the Western Emperor Constantius Chlorus (c. 250–306), at Augusta Treverorum (modern Trier, Germany), possibly returning to North Africa

[55] John Carey, "Ireland and the Antipodes: The Heterodoxy of Virgil of Salzburg," *Speculum*, Vol. 64, No. 1 (Jan., 1989), 1–10.

[56] Russell, *Inventing the Flat Earth*, 19–20; Garwood, *Flat Earth*, 23–25.

in between. The next year, Constantius Chlorus unexpectedly died and Constantine was acclaimed his successor by his father's army.[57]

In his time at the court of the new emperor, Lactantius held a prominent position, including serving as tutor of Constantine's son Crispus (300–326). It is possible that Helena had already converted to Christianity, and that Lactantius secured his prominence through networking with her.[58] In 310, civil war broke out as the co-Western emperor, Maxentius (c. 283–312), attempted to overthrow Constantine. Their conflict culminated in the Battle of the Milvian Bridge outside of Rome in 312. Prior to the battle, Constantine received a divine vision interpreted as being from the Christian God. Two versions of the vision exist, the first from Lactantius and a later one from Eusebius Pamphilus (c. 260–339), future bishop of Caesarea. In Lactantius' account, in a dream Constantine was told to put the sign of the cross on his soldiers' shields, and they would be victorious in the coming battle; to Eusebius, the vision came in the sign of a Chi-Rho against the sun (which was later adopted by UFO believers as evidence of alien intervention in Earthly religion). Eusebius' account is the far better-known one. Lactantius wrote his version in his c. 316 *On the Deaths of the Persecutors*, which was only rediscovered by scholars in 1678 and not proven to be by Lactantius until 1902. Nevertheless, Lactantius was the only firsthand account of the event which started Constantine on the path to conversion, as not only was Constantine's army victorious, but Maxentius died in the battle.[59]

Constantine would ultimately become sole emperor from 324 until his death thirteen years later, but the victory at the Milvian Bridge secured

[57] H. A. Drake, *Constantine and the Bishops: The Politics of Intolerance* (Baltimore: The Johns Hopkins University Press, 2000), 207, 223; Timothy D. Barnes, *Constantine: Dynasty, Religion and Power in the Later Roman Empire* (Chichester: Wiley-Blackwell, 2011), 8–9, 176–78; Peter Heather, *Christendom: The Triumph of a Religion, AD 300–1300* (New York: Alfred A. Knopf, 2023), 5; David Hutchings and James C. Ungureanu, *Of Popes and Unicorns: Science, Christianity, and How the Conflict Thesis Fooled the World* (Oxford: Oxford University Press, 2022), 65–66.

[58] Barnes, *Constantine*, 44–45.

[59] Drake, *Constantine and the Bishops*, 178–79, 202–03; Barnes, *Constantine*, 79–80; Heather, *Christendom*, 8–9; Jacques Vallee and Chris Aubeck, *Wonders in the Sky: Unexplained Aerial Objects from Antiquity to Modern Times and Their Impact on Human Culture, History, and Beliefs* (New York: Jeremy P. Tarcher/Penguin, 2009), 388–90; Timothy D. Barnes, "Lactantius and Constantine," *The Journal of Roman Studies*, Vol. 63 (1973), 29–32, 39–40.

his authority and in 313 he made peace with his co-emperor Licinius. Together, they issued what has been termed the Edict of Milan, a likely informal agreement to cease persecutions and begin toleration, and in some cases patronage, of Christians. Again, information on the edict comes from both Lactantius and Eusebius, with marked differences on specifics. Lactantius helped guide Constantine in the development of his views of Christianity and policies toward Christians, but details of Lactantius' later life are scarcer than his early life. In 326, Constantine executed his son Crispus, Lactantius' former student, on charges of incest with his stepmother, Constantine's second wife Fausta (who was also executed). Later historians speculated that Fausta and Crispus were executed for attempting to organize a coup against Constantine. If that is true, it is possible, but entirely speculative, that Lactantius was also executed due to participation, although it must be emphasized that there is no evidence for this. Certainly, as late as 324, Lactantius was still close to Constantine; his conversations on Christianity with Constantine likely influenced the emperor's speech to the Assembly of the Saints in Nicomedia at Easter 325.[60]

Besides his duties advising Constantine on matters of Christian policy and theology, Lactantius also wrote apologetics, turning his use of rhetoric toward a defense of the faith just as his teacher Arnobius had. The most significant of Lactantius' theological works was *The Divine Institutes*, written approximately between 303 and 311. It is possible that Lactantius had begun conceptualizing *The Divine Institutes* during a return to Africa between leaving Nicomedia and arriving at Constantine's court. *The Divine Institutes* was aimed at educated pagans in a conciliatory way as a means of urging them to both convert and see Diocletian's Great Persecution as unjust. The book urges pagans to debate Christians, but positions the debate as based around Christian Scripture and not logical reason, tilting the advantage heavily toward the Christian.[61] In this, *The Divine Institutes* is a predecessor to later Flat Earth debate methodology, based as it tends to be around religious convictions and not evidence—something that is not surprising, as *The Divine Institutes* is itself a Flat Earth work.

[60] Barnes, *Constantine*, 118.

[61] Drake, *Constantine and the Bishops*, 207–11; Barnes, *Constantine*, 178.

The Flat Earth elements of *The Divine Institutes* come from Book III, appropriately titled *Of the False Wisdom of Philosophers*. In chapter three, "Of What Subjects Philosophy Consists, and Who Was the Chief Founder of the Academic Sect," Lactantius argues that.

> For to investigate or wish to know the causes of natural things — whether the sun is as great as it appears to be, or is many times greater than the whole of this earth; also whether the moon be spherical or concave; and whether the stars are fixed to the heaven, or are borne with free course through the air; of what magnitude the heaven itself is, of what material it is composed; whether it is at rest and immoveable, or is turned round with incredible swiftness; how great is the thickness of the earth, or on what foundations it is poised and suspended — to wish to comprehend these things, I say, by disputation and conjectures, is as though we should wish to discuss what we may suppose to be the character of a city in some very remote country, which we have never seen, and of which we have heard nothing more than the name. If we should claim to ourselves knowledge in a matter of this kind, which cannot be known, should we not appear to be mad, in venturing to affirm that in which we may be refuted?[62]

This is made more explicit in chapter 24, "Of the Antipodes, the Heaven, and the Stars," which is entirely devoted to the topic:

> How is it with those who imagine that there are antipodes opposite to our footsteps? Do they say anything to the purpose? Or is there any one so senseless as to believe that there are men whose footsteps are higher than their heads? Or that the things which with us are in a recumbent position, with them hang in an inverted direction? That the crops and trees grow downwards? That the rains, and snow, and hail fall upwards to the earth? And does any one wonder that hanging gardens are mentioned among the seven wonders of the world, when philosophers make hanging fields, and seas, and cities, and mountains? The origin of this error must also be set forth by us. For they are always deceived in the same manner. For when they have assumed anything false in the commencement of their investigations, led by the resemblance of the truth, they necessarily fall into those things which are its consequences. Thus they fall into many ridiculous things; because those things which are in agreement with false

[62] Lucius Lactantius, "Divine Institutes, Book III (Of the False Wisdom of Philosophers)," *New Advent*, 2021, accessed June 16, 2023, https://www.newadvent.org/fathers/07013.htm.

things, must themselves be false. But since they placed confidence in the first, they do not consider the character of those things which follow, but defend them in every way; whereas they ought to judge from those which follow, whether the first are true or false.

What course of argument, therefore, led them to the idea of the antipodes? They saw the courses of the stars travelling towards the west; they saw that the sun and the moon always set towards the same quarter, and rise from the same. But since they did not perceive what contrivance regulated their courses, nor how they returned from the west to the east, but supposed that the heaven itself sloped downwards in every direction, which appearance it must present on account of its immense breadth, they thought that the world is round like a ball, and they fancied that the heaven revolves in accordance with the motion of the heavenly bodies; and thus that the stars and sun, when they have set, by the very rapidity of the motion of the world are borne back to the east. Therefore they both constructed brazen orbs, as though after the figure of the world, and engraved upon them certain monstrous images, which they said were constellations. It followed, therefore, from this rotundity of the heaven, that the earth was enclosed in the midst of its curved surface. But if this were so, the earth also itself must be like a globe; for that could not possibly be anything but round, which was held enclosed by that which was round. But if the earth also were round, it must necessarily happen that it should present the same appearance to all parts of the heaven; that is, that it should raise aloft mountains, extend plains, and have level seas. And if this were so, that last consequence also followed, that there would be no part of the earth uninhabited by men and the other animals. Thus the rotundity of the earth leads, in addition, to the invention of those suspended antipodes.

But if you inquire from those who defend these marvellous fictions, why all things do not fall into that lower part of the heaven, they reply that such is the nature of things, that heavy bodies are borne to the middle, and that they are all joined together towards the middle, as we see spokes in a wheel; but that the bodies which are light, as mist, smoke, and fire, are borne away from the middle, so as to seek the heaven. I am at a loss what to say respecting those who, when they have once erred, consistently persevere in their folly, and defend one vain thing by another; but that I sometimes imagine that they either discuss philosophy for the sake of a jest, or purposely and knowingly undertake to defend falsehoods, as if to exercise or display their talents on false subjects. But I should be able to prove by many arguments that it is impossible for the heaven to be lower than the earth, were it not that this book must now be concluded, and that some things still remain, which are more necessary for the present work. And since it is not the work of a single book to run over the errors of each

> individually, let it be sufficient to have enumerated a few, from which the nature of the others may be understood.[63]

What follows from that final pronouncement are several more selections downplaying the alleged wisdom of the pagan philosophers and claiming that true knowledge can only come from the revelations of the Christian God.

From these passages come several arguments that would become standard go-tos of Flat Earthers in the nineteenth century onward. Unlike the later Augustine, Lactantius does not merely doubt the existence of *inhabited* Antipodes, but the Antipodes themselves, based on a misunderstanding of gravity and how it would be perceived by those dwelling on the opposite hemisphere (and inadvertently correctly criticizing the inaccurate notion of gravity established by Aristotelian physics). In line with discussions earlier in the chapter, Lactantius cites the Hanging Gardens of Babylon, long associated with the Tower of Babel. He notes that even if observation indicates the heavens are round, it does not follow that the Earth is; and even the observation of the pagans is in doubt given their association of constellations with their gods, meaning their philosophy must be questioned by all true Christians. Finally, he ends with an appeal to (specifically Christian, spiritual) authority before declaring he does not have the time to continue talking about scientific topics any further, and pivots back to purely religious discussion.

As such, there is an argument to be made that *The Divine Institutes* is the first Flat Earth work, and Lactantius the first Flat Earther. Obviously, the scholars of pre-Hellenistic Egypt and Confucian China, among others, believed the Earth was flat and wrote about it. But Lactantius lived in a society that firmly believed the Earth was spherical, had proven it scientifically, and even had the mainstream of his own religion accept it as such. By using religious arguments to try and debunk those longstanding views, Lactantius pioneered not only the methodology, but some of the specific claims, used by his successors down to the present. Indeed, aiding that claim of primacy is that unlike Lactantius' later *On the Deaths of the Persecutors*, *The Divine Institutes* remained relatively well known, to the

[63] Lactantius, "Divine Institutes, Book III."

degree that Lactantius was identified as a Flat Earther by both proponents and skeptics of the flatness of the Earth in the modern era.[64]

Even at the height of the post-Roman period stereotypically referenced as the Dark Ages, Lactantius was read by influential Christians, but his Flat Earth message was not incorporated. The Venerable Bede (c. 672–735), the greatest early English Christian theologian, likely had copies of Lactantius' *The Works of God* (c. 304) and *The Divine Institutes* in his monastic library, but he also had works by Augustine and Isidore, and it was the latter who inspired Bede's endorsement of the spherical Earth in his *The Reckoning of Time* (c. 725).[65] However little credence his Flat Earth views were given, therefore, Lactantius was still widely read by both early and modern Christians as an important figure of the early Church.

Much less well known than Lactantius—his writings were never translated into Latin, and were not translated into English until 1897, when an exaggerated view of his reception among his contemporary Christians was added—was the sixth-century Byzantine monk Cosmas Indicopleustes.[66] Even less is certain about Cosmas than Lactantius, but it is known that he was from Alexandria, likely born in the late fifth or early sixth century CE, in the period between the collapse of the Western Roman Empire and the rise of Islam. Indicopleustes means "Sailed to India," and before turning to a life of religion, Cosmas was a merchant on the Red Sea, visiting the East African kingdom of Axum (predecessor of modern Ethiopia), India, and Sri Lanka. Around the year 550, while in a monastery in Egypt's Sinai Peninsula, Cosmas wrote the *Christian Topography*, which—going by the standards of *The Divine Institutes*—can be seen as the second (surviving) Flat Earth publication (Fig. 1.3).

[64] B. Charles Brough, "History of Cosmography," *The Zetetic: A Monthly Journal of Cosmographical Science*, No. 1 (Jul. 1872), 7; Betten, "Knowledge of the Sphericity of the Earth," 79–80; Garwood, *Flat Earth*, 8–10; Nothaft, "Zaccaria Lilio," 492; James J. Allegro, "The bottom of the universe: Flat earth science in the Age of Encounter," *History of Science*, Vol. 55, No. 1 (2017), 67, 76; Robert J. Schadewald, *The Plane Truth: A History of the Flat Earth Movement* (Wendy S. Schadewald, 2015), 6.

[65] Pedersen, *The First Universities*, 58; Russell, *Inventing the Flat Earth*, 20; Garwood, *Flat Earth*, 25; M. L. W. Laistner, "The Library of the Venerable Bede," in Alexander Hamilton Thompson, ed., *Bede, His Life, Times, and Writings: Essays in Commemoration of the Twelfth Centenary of His Death* (Oxford: The Clarendon Press, 1935), 256–65.

[66] Russell, *Inventing the Flat Earth*, 33–35; Hutchings and Ungureanu, *Of Popes and Unicorns*, 66–68.

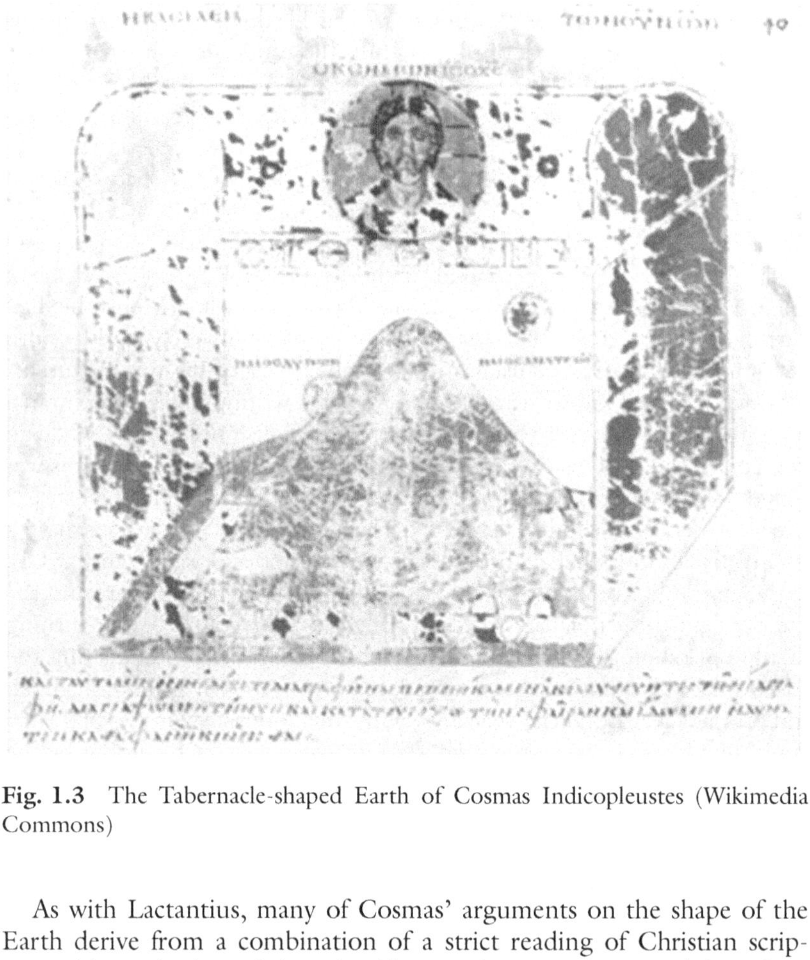

Fig. 1.3 The Tabernacle-shaped Earth of Cosmas Indicopleustes (Wikimedia Commons)

As with Lactantius, many of Cosmas' arguments on the shape of the Earth derive from a combination of a strict reading of Christian scripture with a rejection of the scientific experiments of pagan philosophers proving the Earth's sphericity. Unlike Lactantius, Cosmas incorporated his personal knowledge of the East into his world map, including many unique firsthand accounts of India that make his work useful outside of religious or cosmological interests. His combination of religious interpretations and personal observations also sets a standard that modern Flat Earthers would incorporate, starting with the Bedford Level Experiment. To Cosmas, the Earth was in the shape of a box with a curved lid, modeled on the Tabernacle as described by God in the Book of

Exodus—certainly one of the most unique takes on a non-spherical Earth. The world is located at the bottom of the box, surrounded by ocean, and dominated by a large mountain to the north around which the Sun orbited.[67] Notably, Cosmas dismisses Babylonian astronomy, arguing that the "erroneous" observations of the heavens which led to the idea of the Earth's sphericity had to have been carried out from the top of the Tower of Babel as part of "their warfare with God," and therefore safely able to be dismissed.[68] The tenth book of the *Christian Topography* contains a list of "Passages from the Christian Fathers confirming the Author's views," but interestingly Lactantius is not included there.[69] This is perhaps a reflection that Lactantius' works had come to be seen as apocryphal for a time, considering their inclusion as such in the so-called "Gelasian Decree," a Christian text (originally believed to have been a Papal decree) written around the same time as the *Christian Topography*.[70]

While not placed on any actual or claimed Papal prohibition, the *Christian Topography* underwent its share of criticism as well. Almost immediately Cosmas' Flat Earth beliefs were challenged by John Philoponus (c. 490-c. 570), his contemporary (and much more prominent) Christian theologian of Alexandria. Several centuries later, Cosmas was likewise condemned by none other than Saint Photios I, Patriarch of the Orthodox Church (c. 810–893; r. 858–867, 877–886).[71] Perhaps because of his late translation to English, Cosmas was not as influential among the Flat Earthers of the fin de siècle. Among surviving Zetetic Society publications, he is mentioned once, in an article by Lady Blount, who simply states that "Cosmos" [*sic*] was a more righteous Christian

[67] Cosmas, trans. and ed. J. W. McCrindle, *The Christian Topography* (Edinburgh: The Hakluyt Society, 1897), 129–38.

[68] Cosmas, *Christian Topography*, 91.

[69] Cosmas, *Christian Topography*, 331–58.

[70] Ernst von Dobschütz, "The 'Decretum Gelasianum de Libris Recipiendis et non Recipiendis'," *The Tertullian Project*, December 2, 2000, accessed June 22, 2023, https://www.tertullian.org/decretum_eng.htm.

[71] Birgitta Elweskiöld, "John Philoponus against Cosmas Indicopleustes: A Christian Controversy on the Structure of the World in Sixth-Century Alexandria" (PhD diss., Lund University, 2005); Roger Pearse, "Cosmas Indicopleustes, Christian Topography. Preface to the online edition.," *The Tertullian Project*, July 5, 2003, accessed June 22, 2023, https://www.tertullian.org/fathers/cosmas_00_0_eintro.htm.

than modern Christians. Written in 1906, it is late in her own Flat Earth career.[72]

Among critics of the Flat Earth, Cosmas is slightly better represented, although again in illustratively inaccurate ways. Writing in 2000, the Young-Earth Creationist, geocentric—but explicitly anti-Flat Earther—astronomer Gerardus D. Bouw (1945-2023) mentioned Cosmas, arguing he was more influenced by the Greek pagan philosophers Anaximander and his student Anaximenes of Miletus (c. 585-c. 525 BCE) than Scripture. More intriguingly, Bouw argued that Cosmas' cosmology is similar to that of the Hindu Puranas, implying that Cosmas' Flat Earthism was a heretical Hindu import to Christianity from his merchant travels to India.[73] This seems to be a misinterpretation of the cosmology of the Puranas, based entirely around associating the mythical Mount Meru with Cosmas' sun-orbiting mountain. Others, however, have made more solid arguments for the Christian Cosmas drawing from both Zoroastrian cosmology and the geography described by Plato in *Timaeus*.[74] To the degree the Puranas have been interpreted as supporting a Flat Earth, the shape is a bowl or more traditional disc, not Cosmas' rectangle.[75] But perhaps the height of Cosmas' misuse can be seen in this 1984 passage by chronicler of the outré John Michell (1933–2009), who in a chapter on Flat Earth belief claimed that the.

> *Christian Topography* was well received by the Church, whose policy at the time was to eradicate all previous knowledge and establish itself as the sole authority in religion, philosophy and science. The flat-earth theory, hitched on to the geocentric cosmology of Ptolemy, prevailed among clergymen (if not among navigators) until the sixteenth century, when Copernicus called

[72] Elizabeth Blount, "Ancient and Modern Astronomy," *The Earth*, Vol. VI, Nos. 72 & 73 (Jul./Aug., 1906), 100.

[73] Gerardus D. Bouw, "Flat Earth Mythology and Fact," *The Biblical Astronomer*, Vol. 10, No. 94 (Fall, 2000), 24.

[74] Benjamin B. Olshin, *Lost Knowledge: The Concept of Vanished Technologies and Other Human Histories* (Leiden: Brill, 2019), 205–07.

[75] Toke Lindegaard Knudsen, "Three Purāṇic Statements on the Shape of the Earth," *History of Science in South Asia*, Vol. 9 (2021), 129–31; Pinto, *Medieval Islamic Maps*, 98–109.

it in question by venturing the idea that the earth is a planet orbiting the sun.[76]

As shown above, the opinions of Cosmas—and Lactantius, for that matter—were anything but well-received by the leaders of the Church, and even modern defenders of geocentrism from a Christian perspective have no problem denouncing Cosmas.

While Cosmas' late translation into English no doubt hampered his appeal to fin de siècle Anglophone Flat Earthers given their generally non-specialist education, he was better known among the educated and progressive advocates of science in late nineteenth-century Britain. In the decade of the 1870s, British textbooks began promoting the Flat Error notion that the medieval Church believed in the Flat Earth, due to the development of the debate over Darwinian evolution in that decade. The Flat Error emerged as it did because it could be used by middle-class liberal progressives not only to depict their heroes of the past as sharing their ideals but as a weapon against Christian opponents of Darwinian evolution—a debate which in Britain came to be seen as not just science against religion, but a stand-in for Protestantism against Catholicism (and implicitly, Protestant British unionism against Irish Catholic nationalism).[77]

One such advocate of this notion was Andrew Dickson White (1832–1918), an American scholar who was the first president of both Cornell University (1866–1885) and the American Historical Association (1884–1885), as well as an early advocate of the conflict thesis of historical hostility between science and religion, along with British scientist John William Draper (1811–1882). White published *A History of the Warfare of Science with Theology in Christendom* in 1896, which identified Cosmas as the architect of a widely-held Christian rejection of the sphericity of the Earth. Predicting Bouw, White viewed pre-Hellenistic Egyptian cosmology as the origin of Cosmas' worldview, with the *Christian Topography* his (in White's erroneous argument, successful) effort to bring that pagan cosmology into the Christian mainstream. Draper, while devoting less writing to the matter, had characterized early Christians as Flat

[76] John Michell, *Eccentric Lives and Peculiar Notions* (London: Thames and Hudson, 1984), 21.

[77] Russell, *Inventing the Flat Earth*, 29–32, 35–49.

Earth zealots the same way.[78] But perhaps the strongest proponent of this view was William Whewell (1794–1866), Master of Trinity College, Cambridge and priest of the Church of England from 1841 until his death, and a liberal progressive defender of science. To Whewell, the African priests, Lactantius and Cosmas, were the two main Christian villains of the Flat Error. Highlighting their Flat Earth views, while overexaggerating their prominence in the early Church, aided Whewell's attacks on his contemporary conservative Christian opponents.[79]

An outgrowth of this dominance of the conflict thesis on the warfare between classical Greco-Roman science and Christianity was the belief that Christians, whether institutionally or through populist mob action, were responsible for the destruction of the Library of Alexandria. This was often conflated with a series of events in the late fourth and early fifth centuries: the anti-pagan policies of the Christian Emperor Theodosius I (r. 379–395), particularly a decree against pagan temples in 391; the subsequent destruction of the Serapeum of Alexandria (a philosophical school and former "daughter library" of the Great Library) that year by Christians; and the 415 murder of Hypatia (b. c. 370), a female Alexandrian pagan philosopher, by a Christian mob. Even among Christians, Hypatia's murder was shocking due to the longstanding Roman perception of philosophers as inviolable during periods of political strife. Among her work, Hypatia edited a section of Ptolemy's *Almagest*. The connection of Hypatia's murder to the destruction of the Library of Alexandria was widely popularized by Carl Sagan's (1934–1996) 1980 *Cosmos*; inspired by *Cosmos*, the 2009 biopic *Agora* also depicted Hypatia (played by Rachel Weisz) as endorsing heliocentrism and developing the theory of elliptical orbits, well ahead of the 1609 proposal by Johannes Kepler (1571–1630). However, the destruction of the Library at the hands of Christians, let alone the same Christians who murdered Hypatia—while convenient to nineteenth century liberals for the same reasons as the Flat Error—is as inaccurate as *Agora*'s expansion of Hypatia's astronomical hypothesizing. In reality, the Library had likely ceased to exist, at least as a functioning organization of education and research, even prior to the

[78] Andrew Dickson White, *A History of the Warfare of Science with Theology in Christendom* (New York: D. Appleton and Company, 1896), 91–97; Hutchings and Ungureanu, *Of Popes and Unicorns*, 53–58.

[79] Russell, *Inventing the Flat Earth*, 31–33.

ascent of Theodosius, likely due not to religious persecution but bureaucratic neglect and funding cuts—all too familiar to modern institutions of learning.[80]

However, those seeking a religious alternative to the Christian destruction of the Library had another scapegoat: early Muslims. In 639, Arab Muslim forces entered Roman Egypt and, after a six-month siege, captured the capitol of Alexandria in 641. Muhammed himself had only died in 632, and the institution of the successor Caliphate was still new. The reigning Caliph, Umar ibn al-Khattab (r. 634–644), was only the second to hold that position in what is called the Rashidun Caliphate, a period when the Muslim empire's political and religious policies were still in a state of flux. In 1203, the Arab historian Abd al-Latif al-Baghdadi (1162–1231) invented the idea that in 642, Caliph Umar had ordered the destruction of the Library of Alexandria with the apocryphal justification, "If those books are in agreement with the Qu'ran, we have no need of them; and if these are opposed to the Qu'ran, destroy them." Al-Baghdadi's claim entered into Western scholarship only in 1663 with the translation of an Arabic text by Edward Pococke (1604–1691), Laudian Professor of Arabic at Oxford; already by 1713, the truthfulness of the account was being questioned by the French orientalist Abbé Eusèbe Renaudot (1646–1720).[81] Even so, the claim had a certain resonance; as late as 1972, the British astronomer and science popularizer Patrick Moore (1923–2012), while rejecting the story of Umar ordering the Library destroyed, still believed the Arab occupation of Egypt was responsible for its demise.[82] It is not clear exactly how or why al-Baghdadi created the myth of the Muslim destruction of the Library, as it was

[80] Carl Sagan, *Cosmos* (New York: Random House, 1980), 333–37; *Agora*, dir. Alejandro Amenábar (Madrid: Telecinco Cinema, 2009); Peter Harrison and David C. Lindberg, "Early Christianity," in John Hedley Brooke and Ronald L. Numbers, eds., *Science and Religion Around the World* (Oxford: Oxford University Press, 2011), 69; Freely, *Light from the East*, 21–22; Moller, *The Map of Knowledge*, 41–43; Hutchings and Ungureanu, *Of Popes and Unicorns*, 95–97, 108–12. For a general overview of her life and work, see Michael A. B. Deakin, *Hypatia of Alexandria: Mathematician and Martyr* (Amherst, NY: Prometheus Books, 2007).

[81] Bernard Lewis, "The Arab Destruction of the Library of Alexandria: Anatomy of a Myth," in Mostafa El-Abbadi and Omnia Fathallah, eds., *What Happened to the Ancient Library of Alexandria?* (Leiden: Brill, 2008), 213–15.

[82] Patrick Moore, *Can You Speak Venusian? A Guide to the Independent Thinkers* (New York: W. W. Norton & Company, 1972), 13.

originally created and spread not as an anti-Muslim Christian screed but among Muslims by Muslims. It was possibly the result of a folkloric interpretation of the breakup of a Fatimid Isma'ili library in Cairo by the military commander Saladin (c. 1137–1193) upon his 1169 conquest of Egypt.[83]

In 661, the last of the Rashidun caliphs was assassinated, leading to the establishment of the Umayyad Caliphate. Sometime between 723 and his ascent, the future Umayyad caliph al-Walid II (r. 743–744) built a castle, Qusayr 'Amra, in what is modern Jordan. The only complete medieval Islamic celestial map comes from the dome of Qusayr 'Amra's surviving bathhouse, likely copied by an itinerant Byzantine artist from Byzantine star maps. Karen Pinto has argued that what is generally interpreted as a figure of Nike offering a crown in the famous palace fresco termed the "Painting of the Six Kings" is not actually a crown, but a map of the Earth, a two-dimensional representation of a sphere.[84] Along with the use of a dome to represent the night sky, this demonstrates how the sphericity of the Earth was well understood by mainstream Islamic culture from an early period, not least of which was due to their acceptance of the writings of Ptolemy among other Greek philosophers. It was, after all, the Arabs who gave his greatest work the title of *Almagest*. The Ptolemaic sphericity of both the Earth and heavens was important for early Muslims in North Africa due to the pairing of locations on Earth to the sky to provide precise navigation of the Sahara Desert, which was viewed as a sandy ocean with oases as islands, further equaling oceanic navigation with land navigation.[85] In the realm of metaphorical maps, the medieval Muslims also adopted the more contemporary European T-O map developed by Isidore.[86]

[83] Lewis, "Arab Destruction of the Library of Alexandria," 216–17.

[84] Karen Pinto, "Fit for an Umayyad Prince: An Eighth- Century Map or the Earliest Mimetic Painting of the Moon?," *The Medieval Globe*, Vol. 4.2 (2018), 32–42; Pinto, *Medieval Islamic Maps*, 47–51.

[85] Zayde Antrim, *Routes and Realms: The Power of Place in the Early Islamic World* (Oxford: Oxford University Press, 2012), 88–89; Abbas Hamdani, "An Islamic Background to the Voyages of Discovery," in Salma Khadra Jayyusi, ed., *The Legacy of Muslim Spain* (Leidein: E. J. Brill, 1992), 278; Al-Khalili, *The House of Wisdom*, 206–10.

[86] Pinto, *Medieval Islamic Maps*, 143–46; Karen C. Pinto, "Ecce! A Ninth-Century Isidorean T-O Map Labeled in Arabic," in Albrecht Classen, ed., *Globalism in the Middle Ages and the Early Modern Age: Innovative Approaches and Perspectives* (Berlin: Walter de Gruyter GmbH, 2023), 189–220.

The adoption of Western, particularly Greek, philosophy by Islamic scholars is well documented, particularly after the overthrow of the Umayyads across much of the empire by the Abbasid Caliphate in 750. The Umayyads remained in power in the Iberian Peninsula, called al-Andalus by Muslims, until 1031, ruling from the city of Córdoba. Even after the fall of the Umayyads of al-Andalus, Córdoba remained a major center of Islamic learning, including astronomy, until the final Christian Reconquista of al-Andalus in 1492.[87] But as great a center of learning as Córdoba was, it was initially eclipsed by the new Abbasid capitol Baghdad, founded in 762 by the Caliph al-Mansur (r. 754–775) near the ancient ruins of Babylon.

As with many cities founded early in the Islamic era, Baghdad was built with the consultation of astrologers, a sign of the prominent role of astronomy in Muslim scholarship.[88] It was also a more cosmopolitan city, built with Persian and Central Asian expertise. Early in Baghdad's history, it became the home to the Bayt al-Ḥikmah, or House of Wisdom. This began as a large private library and academic center funded by al-Mansur which was then expanded into a larger academy by the Caliph Harun al-Rashid (r. 786–809). Because the geographic extent of the Abbasid Caliphate was much further east than the Roman Empire had been, the Abbasid philosophers incorporated a much greater extent of Babylonian, Zoroastrian, and Indian philosophy into their work, along with Greek and Roman sources.[89] One of the bridges of this transition was the Academy of Gondishapur, which rose to prominence under the Persian emperor Khosrau I (r. 531–759) and offered refuge for Neoplatonist and Nestorian Christian scholars persecuted in the Roman Empire alongside Persian Zoroastrian teachings. When the House of Wisdom was established, Gondishapur served as a model.[90]

[87] Stephen P. Blake, *Astronomy and Astrology in the Islamic World* (Edinburgh: Edinburgh University Press, 2016), 51–64; Freely, *Light from the East*, 122–32; Moller, *The Map of Knowledge*, 86–118.

[88] Ali A. Olomi, "Baghdad: The Once and Future City of Stars," Public Books, February 8, 2022, accessed February 8, 2022, https://www.publicbooks.org/baghdad-medieval-muslim-astrology/.

[89] Al-Khalili, *The House of Wisdom*, 67–78; Pedersen, *The First Universities*, 117; Freely, *Light from the East*, 36–47; Moller, *The Map of Knowledge*, 46–84.

[90] J. D. Bernal, *Science in History, Volume 1: The Emergence of Science* (Cambridge, MA: The MIT Press, 1971 [1954]), 270–71; S. Frederick Starr, *Lost Enlightenment: Central*

The House of Wisdom was part of a broader Abbasid cultural development, termed the Translation Movement, whereby various philosophical, religious, and cultural works were translated into Arabic from Greek, Syriac, Persian, and Sanskrit, leading to a flowering of intellectual development. Many of these works had commentary added to them by Muslim scholars; these commentaries were retained when the works were translated back into European languages later, meaning many surviving works of Greek philosophy incorporate later Muslim scholarship.[91] The House of Wisdom was destroyed with Baghdad and the Abbasid Caliphate itself in 1258 when the city was sacked by the Mongols. In the popular understanding, this is seen as ending the so-called Islamic Golden Age, including the era of Muslim innovation and development of science. However, much like the notion of the fall of the Roman Empire leading to a European Dark Age, this is inaccurate. Islamic scientific development continued after the sacking of Baghdad, and had also occurred throughout the Muslim world beyond Abbasid territory both before and after the destruction of the House of Wisdom.[92] Cairo had emerged as a major intellectual center in the early eleventh century. Córdoba had already reached the height of its scholarship at the time of the Sack of Baghdad, with its royal library housing approximately half a million volumes. Timbuktu would reach its height in the following century.

It is true that until the Renaissance, astronomy, particularly the observation of stars, was far less dynamic in Europe than it was in the Muslim world.[93] But while the popular narrative is that Muslim knowledge (and the Greek knowledge preserved by them) was only reincorporated into Europe with the start of the Renaissance and the capture of Córdoba in 1492, this is yet another oversimplified and inaccurate notion. Even in the

Asia's Golden Age from the Arab Conquest to Tamerlane (Princeton: Princeton University Press, 2013), 89–91; Freely, *Light from the East*, 24–25.

[91] Al-Khalili, *The House of Wisdom*, 35–48; Freely, *Light from the East*, 28–35; Hayrettin Yücesoy, "Translation as Self-Consciousness: Ancient Sciences, Antediluvian Wisdom, and the 'Abbāsid Translation Movement," *Journal of World History*, Vol. 20, No. 4 (Dec., 2009), 523–57.

[92] Ahmad S. Dallal, "Early Islam," in John Hedley Brooke and Ronald L. Numbers, eds., *Science and Religion Around the World* (Oxford: Oxford University Press, 2011), 120–47; Pedersen, *The First Universities*, 117.

[93] Freeman, *Reopening of the Western Mind*, 175.

Iberian context, the court of Alfonso X, Christian king of Castile (r. 1252–1284) sponsored a school for translating Arabic works into Latin.[94] Not only did Alfonso's school focus heavily on astronomical texts, but the last Islamic work of astronomy in al-Andalus was completed in his Christian court at Toledo.[95] In 1295, the Byzantine monk Maximus Planudes (c. 1260–1305) discovered a copy of Ptolemy's *Geography* in the library of the Chora monastery in Constantinople. The *Geography* was believed to be lost for centuries, and the rediscovery of Ptolemy's work was major for shaping European Christians' worldview. Although this was centuries after the *Geography* was translated into Arabic and spread across the Muslim world in the ninth century, it was also centuries before Europeans gained access to the libraries of Córdoba.[96] But the direct translation of mathematical and astronomical works from Arabic to Latin had already begun in the tenth century, rapidly accelerating in the twelfth. Geographical works were among the first Arabic texts translated into modern European languages, due to their utility in imperial projects.[97]

The influence on European learning from outside Christendom went beyond mere translations from Arabic. Robert Grosseteste (c. 1168–1253) was Bishop of Lincoln in England (r. 1235–1253) and an important early philosopher at Oxford University. Among his works, published c. 1217–1219, was an introductory astronomy textbook titled *De sphera* (*On the Sphere*). Grosseteste wrote it using sources that had been translated from Greek and Arabic into Latin, particularly Aristotle and Ptolemy but also the Muslim Cairo astronomer Ibn al-Haytham (c. 965–1040). In *On the Sphere*, Grosseteste viewed astronomy as a liberal art, which he defined as something which helped people correct errors. *On the Sphere* also used evidence of astronomical observation of the Pole Star in the Indian city of Ujjain to demonstrate the Earth's sphericity. Ujjain, Latinized as Arin or Arym, had since the mid-second century been, in the words of historian of mathematics David Pingree, "the Greenwich of Indian astronomers" and a source for astronomical knowledge for

[94] Moller, *The Map of Knowledge*, 120–50.

[95] Owen Gingerich, *The Eye of Heaven: Ptolemy, Copernicus, Kepler* (New York: American Institute of Physics, 1993), 115–28; Blake, *Astronomy and Astrology in the Islamic World*, 128; Lindberg, "Early Christianity," 77.

[96] Freeman, *Reopening of the Western Mind*, 2–7.

[97] Pedersen, *The First Universities*, 119–20.

both medieval European and Arab scholars. This exchange of knowledge was the completion of a circuit, as the first observers at Ujjain had utilized the Sanskrit translation of a now-lost Greek text on astronomy from Alexandria.[98] Curiously, several European authors, including Roger Bacon (c. 1219-c. 1292) and Pierre d'Ailly (1351–1420), believed Ujjain to be the same city as Syene, where Eratosthenes had measured the Earth's shape.[99] Grosseteste's knowledge of Ujjain and the observations there came via the book *Dialogi contra Iudaeos* (*Dialogue Against the Jews*) by Petrus Alphonsi (fl. twelfth century), a Jewish convert to Christianity from al-Andalus; the first book of Alphonsi's screed was about the cosmos.[100]

Medieval Muslims and Christians did not merely take knowledge out of India for their use. Muslims in particular were very active in northern India, particularly during the reigns of the Delhi Sultanate (1206–1526) and Mughal Empire (1526–1857), Muslim states which incorporated Hindu and other non-Muslim religions and cultures. The Mughal Empire would be one of the last major Islamic states that defined the post-Mongol revitalization of the Muslim world, with power now centered outside of Arabia in non-Arab dynasties. However, the greatest Muslim power of this era was the Ottoman Empire, which had its origins in the late thirteenth century with Turkish tribes along the Byzantine frontier in Anatolia. Within a century, the Ottomans had expanded into Christian Europe with the conquest of the Balkans, followed by the dramatic end of the Roman Empire in 1453 with the capture of Constantinople—which the Ottomans took as their own capitol, and would eventually be renamed again to Istanbul. This was followed by the 1514 victory at Chaldiran over the Shia Muslim Safavid dynasty of Persia, and the 1517 conquest of Mamluk Egypt, which also gave the Ottomans control over the holy cities of Mecca, Medina, and Jerusalem. These actions cemented the Ottoman Sultan's claim to being the new, post-Abbasid Caliph.

By the time of the Ottoman Empire's ascent, the greatest achievements of Islamic astronomy had already peaked. By the early 1420 s

[98] David Pingree, "Astronomy and Astrology in India and Iran," *Isis*, Vol. 54, No. 2 (Jun., 1963), 234–35; Jayant V. Narlikar, *The Scientific Edge: The Indian Scientist from Vedic to Modern Times* (London: Penguin Books, 2003), 34–35.

[99] Edson, *The World Map*, 130.

[100] Giles Gasper, "A Grand Tour of the Medieval Cosmos" (lecture, British Society for the History of Science Global Digital History of Science Festival, 10 July, 2020).

the future Timurid sultan Ulugh Beg (r. 1447–1449) began building an observatory in Samarkand, in modern Uzbekistan. Besides its extensive observation, the astronomers and mathematicians at Ulugh Beg's observatory wrote a star catalogue, *Zij-i Sultani*. Published in 1439, it was the first such catalogue since the *Almagest*. The 1449 assassination of Ulugh Beg and subsequent destruction of the observatory ended this great peak of Islamic astronomy. Afterwards, seven more observatories were built in the Muslim world (an Ottoman observatory in Istanbul in 1577, and six in the Mughal Empire in the span of the years 1728–1734), but none substantially surpassed Ulugh Beg's achievements.[101]

Even before the deep expansion of Muslim states into India, Muslim scholars had journeyed into the subcontinent to study Hindu science and philosophy. The most prominent of these was the Persian scholar Abu Rayhan Muhammad ibn Ahmad al-Biruni (973-c. 1050), who lived in the early part of his life in the city of Kath in modern Uzbekistan. Al-Biruni compared his observations from Kath of the lunar eclipse of 27 May 997 with those conducted by the Persian astronomer Abu al-Wafa' al-Buzjani (940–998) in Baghdad, allowing them to calculate the longitude of both cities upon the spherical Earth.[102] In 1017, Kath was conquered by the Ghaznavid Empire and al-Biruni was made court astrologer to the sultan Mahmud of Ghazni (r. 998–1030). When Mahmud subsequently invaded northern India, al-Biruni accompanied him, extensively studying Hindu customs, including views on religion and science, over the next few years. While Mahmud's army was camped at the hilltop fort of Nandana in modern Pakistan, al-Biruni used observations echoing those of Eratosthenes to accurately calculate the circumference of the Earth. Along with his calculations on longitude, al-Biruni developed a cartographic technique now known as the azimuthal equidistant projection, a circular map where the north pole is at the center and the south pole is stretched to form the outer edge.[103] The flag of the United Nations is the most prominent example of its usage today, but ironically—both for al-Biruni's

[101] Blake, *Astronomy and Astrology in the Islamic World*, 80–95; Freely, *Light from the East*, 162–63; Starr, *Lost Enlightenment*, 496–98.

[102] Nicastro, *Circumference*, 170; Lyons, *The House of Wisdom*, 78–99; Al-Khalili, *The House of Wisdom*, 181–84; Amelia Carolina Sparavigna, "The Science of Al-Biruni," *International Journal of Sciences*, Vol. 2, No. 12 (Dec., 2013), 52–55.

[103] Sparavigna, "The Science of Al-Biruni," 54–56; Al-Khalili, *The House of Wisdom*, 184–88; David Pingree, "Brahmagupta, Balabhadra, Pṛthūdaka and Al-Bīrūnī," *Journal*

acceptance of the Earth's globe, but also the prevalence of Islamophobia among many modern Western Flat Earthers—it has also become very popular amongst modern Flat Earthers, who view it as confirmation that Antarctica is actually an ice wall which encircles the outer edge of the disk-shaped Earth, and prevents the ocean from flowing over the edge into the unknown beyond.[104]

In his *Kitab al-Qanon al-Mas'udi* (1037), al-Biruni hypothesized the existence of an inhabited continent in the vast stretch of water west of Europe and east of Asia, not only independent of the simultaneous Norse settlement of Vinland but more accurately than the Norse who conceptualized it as a mere series of islands. In the same work, al-Biruni emphasized his commitment to geocentrism, although with some criticisms of Ptolemy's model incorporating movement of the Sun in contrast to a fixed path. However, he collaborated with Abu Sa'id al-Sijzi (c. 950–1020), a Persian astronomer in Baghdad, who proposed a heliocentric model with a rotating Earth, and in light of al-Sijzi's work, al-Biruni adopted a public agnostic stance on attacking heliocentric models.[105] However, as with his fellow Muslims, al-Biruni was firmly committed to the evidence of the spherical Earth. This was in concert with contemporary Christians, but at odds with a number of the Hindus of India whom al-Biruni studied. To be sure, this was not a uniform sentiment; the influential Indian astronomer and mathematician Aryabhata (476–550) endorsed the idea of a spherical Earth and universe, calculated the Earth's circumference within a percent of its accurate measure, and developed a heliocentric theory of the cosmos. Aryabhata's spherical world and universe, but not his heliocentrism, was accepted by Brahmagupta (c. 598-c. 668), a major astronomer and mathematician at Ujjain, and were in turn accepted by Chaturveda Prithudaka Swami (c. 830-c. 890),

of the American Oriental Society, Vol. 103, No. 2 (Apr.-Jun., 1983), 353–54; Pinto, *Medieval Islamic Maps*, 11–18.

[104] Garwood, *Flat Earth*, 204–05; Edward Hendrie, *The Greatest Lie on Earth: Proof That Our World Is Not a Moving Globe* (Great Mountain Publishing, 2016), 273–85; Stuart J. Robbins, "Episode 149 - Modern Flat Earth Thought, Part 2 (U.N. Flag and Airplane Flights)," *Exposing PseudoAstronomy*, October 1, 2016, accessed November 21, 2019, https://podcast.sjrdesign.net/shownotes_149.php.

[105] Sparavigna, "The Science of Al-Biruni," 52; Al-Khalili, *The House of Wisdom*, 182–8; Starr, *Lost Enlightenment*, 375–68.

a mathematician who wrote commentaries on Brahmagupta's work. Al-Biruni read both the works of Brahmagupta and Prithudaka, the latter of whom used the work of Aryabhata to reject the flat shape of the Earth mentioned in the Hindu Puranas. Al-Biruni, who was in general critical of the state of Hindu science compared to that of both Muslims and ancient Greeks, used these commentaries to mock the Hindu belief in a Flat Earth. However, as he was not extremely proficient in Sanskrit, and also had a tendency to confuse the authorship of his sources, al-Biruni's analysis contained several errors that seemingly resulted in him believing Flat Earth views were more common in India than they actually were. These included confusing discussions of the spiritual realm with the physical one, conflating the shape of Mount Meru for the shape of the Earth, and believing a philosopher quoted by Brahmagupta to be Buddha.[106] Flat Earth existed in India, therefore, but was likely not as prevalent as al-Biruni—and modern Flat Earthers seeking ancient justification—thought.

Astronomy was as important to the Mughals as to earlier Arab and Persian Muslim dynasties. The Mughals used astrolabes for navigation and timekeeping alike, a practice begun earlier in the Muslim world. This origin of precision timekeeping from astronomy and astronomical instruments, as opposed to the needs of industrial capitalism and imperial coordination driving European timekeeping, has important ramifications on literal worldviews that will be expressed by the tensions between British and Boers, and Westerners and Salafists, that will be explored in subsequent chapters.[107] The importance of the Mughals as a culmination of Islamic astronomy can be seen from the fact that the last observatory built and astronomical treatise written in the Islamic world was by a Hindu Rajput ruler, Sawai Jai Singh (1688–1743; r. 1699–1743), for the Mughal Emperor Muhammed Shah (1702–1748; r. 1719–1748). Jai Singh constructed five observatories in northern India from 1721 to 1738 and compiled the *Zij-i Jadid-i Muhammed Shah* in 1731–1732. In contrast to the achievements of Ulugh Beg, both the observatories and

[106] Pingree, "Brahmagupta, Balabhadra, Pṛthūdaka and Al-Bīrūnī," 357; Knudsen, "Three Purāṇic Statements," 135; Freely, *Light from the East*, 27; Narlikar, *The Scientific Edge*, 73–75.

[107] See for example the seminal article by E. P. Thompson, "Time, Work-Discipline, and Industrial Capitalism," *Past & Present*, No. 38 (Dec., 1967), 56–97.

the treatise were old-fashioned, the latter retaining Ptolemy's geocentrism.[108] Even with the proliferation of observatories in Mughal India, geocentrism remained prominent in the subcontinent among both Hindu and Muslims even after the dissolution of the Mughal Empire by the British following the Indian Revolt in 1857. Of note is that geocentrism among post-Mughal Indian Muslims in particular was based on lingering adherence to Aristotelianism rather than adopted from the views of their Hindu countrymen, and was clung to as a form of intellectual resistance against the British support for Copernicanism in ways we will see also occur among modern Saudi and Nigerian Salafists in chapter six.

Representative of this change was Indian Muslim scholar Sayyid Ahmad Khan (1817–1898). In an 1848 text, he endorsed geocentrism; by 1865, he rejected it, while also noting that Earth's spherical shape was not doubted even by Indian Muslims who critiqued heliocentrism. As noted by historian of science Sarah Qidwai, Khan's conversion was not a case of a colonial subject learning scientific truth from an "Enlightened West" but a combination of personal experiments (in the vein of Flat Earthers contemporary to him in metropolitan Britain) and a search for the supposed scriptural basis for Islamic geocentrism. Khan's conclusion was that geocentrism was not found in the Qur'an, but was a relic of Muslims' devotion to ancient Greek science; Aristotle was venerated to the extent that it became accepted to view his philosophy as part of the Qur'an. To Khan, adopting heliocentrism was therefore necessary to move past the heretical elevation of Aristotle and turn to an Islamic interpretation that was not only in line with modern (European) science, but also purer in its sole dedication to the Qur'an.[109]

Jai Singh did not consciously exclude the works of Copernicus, or of Galileo or Kepler—Singh did not know of their work to begin with. The Copernican revolution was not only a revolution of astronomical science in the West; it heralded the end of the superiority of Muslim

[108] Blake, *Astronomy and Astrology in the Islamic World*, 96, 118.

[109] Sarah A. Qidwai, "Reexamining Complexity: Sayyid Ahmad Khan's Interpretation of "Science" in Islam," in Bernard Lightman, ed., *Rethinking History, Science, and Religion: An Exploration of Conflict and the Complexity Principle* (Pittsburgh: University of Pittsburgh Press, 2019), 50–57; Christian W. Troll, *Sayyid Ahmad Khan: A Reinterpretation of Muslim Theology* (New Delhi: Vikas Publishing House, 1978), 147–49, 154–60; Pankaj Mishra, *From the Ruins of Empire: The Intellectuals Who Remade Asia* (New York: Farrar, Straus and Giroux, 2012), 92–95; Narlikar, *The Scientific Edge*, 193–99.

over European astronomy.[110] Copernicus was of course not working in a vacuum, but was the culmination of astronomical speculation dating from the medieval period, if not earlier. The irony is that Islamic Arab and Persian astronomical knowledge was a major part of that tradition that Copernicus drew from to develop the heliocentric astronomical models discussed in his magnum opus, *On the Revolutions of the Heavenly Spheres* (1543).[111] Copernicus was only interested in interrogating the theory of geocentrism; the need to defend the sphericity of the Earth was unnecessary because of its universal acceptance. But even so, Flat Earth is not absent from *On the Revolutions of the Heavenly Spheres.* In the preface, dedicated to Pope Paul III (1468–1549, r. 1534–1549), Copernicus predicts that when it comes to the reception of his book,

> There may be triflers who though wholly ignorant of mathematics nevertheless abrogate the right to make judgements about it because of some passage in Scripture wrongly twisted to their purpose, and will dare to criticise and censure this undertaking of mine. I waste no time on them, and indeed I despise their judgement as thoughtless. For it is well known that Lactantius, a distinguished writer in other ways but no mathematician, speaks very childishly about the shape of the Earth when he makes fun of those who reported that it has the shape of a globe. Mathematics is written for mathematicians, to whom this work of mine, if my judgement does not deceive me, will seem to be of value to the ecclesiastical Commonwealth over which your Holiness now holds dominion.[112]

Copernicus takes it as a given that the geographical assumptions of Lactantius are so clearly wrong, that instead of attacking him, he uses the mockery of him to his own ends, stating that geocentrists are no

[110] Blake, *Astronomy and Astrology in the Islamic World*, 124, 130.

[111] Nicolaus Copernicus, trans. A. M Duncan, *Copernicus: On the Revolutions of The Heavenly Spheres: A New Translation from the Latin by A. M. Duncan* (Newton Abbot: David & Charles, 1976), 47–48; Syed Nomanul Haq, "That Medieval Islamic Culture Was Inhospitable to Science," in Ronald L. Numbers, ed., *Galileo Goes to Jail and Other Myths about Science and Religion* (Cambridge, MA: Harvard University Press, 2010), 38; Abdelhamid Sabra, "An Eleventh-century Refutation of Ptolemy's Planetary Theory," in Erna and Frank D. Grande, eds., *Science and History: Studies in Honor of Edward Rosen* (Wrocław: The Polish Academy, 1978), 117–31; Freely, *Light from the East*, 172–80; Simek, *Heaven and Earth in the Middle Ages*, 118–20; Gingerich, *The Eye of Heaven*, 145–46.

[112] Copernicus, *On the Revolutions of The Heavenly Spheres*, 26–27.

better than the widely discredited notion of a Flat Earth. Not only does this show that even in the sixteenth century, Lactantius and his thoughts on Earth's shape were well remembered, but dismissing both the theologian and his claims, even in a work dedicated to the Pope, was completely acceptable, even by the time of the turmoil of the early Protestant Reformation.[113]

Copernicus initially faced a rocky reception not only from Catholics, but also Protestants. None other than Martin Luther cited the Old Testament Book of Joshua, which depicted the Sun standing still in the sky, to justify the falsity of heliocentrism. However, in the following decades, Copernicanism would become acceptable by first Protestants, and then Catholics in the following century, notably reversing the trend of religious acceptance that the Flat Earth revival would experience in the nineteenth century.[114] However, the lack of seriousness given to Flat Earthism by Copernicus is a reflection of how little belief in a Flat Earth existed in late medieval Christendom. Even the fabulist travelogue of "Sir John Mandeville" from the late fourteenth century pays the Flat Earth no heed. In between fantastical descriptions of his visit to Prester John's court, the land of darkness surrounding the Garden of Eden, and the resting spot of Noah's Ark, Mandeville defends the spherical Earth from the belief of "simple men unlearned" that it is flat. At the end of chapter twenty, Mandeville describes a story he claimed to hear in his youth, of an Englishman who, by both sea and land, traveled east past India, and was ultimately surprised to return back to England. In defense of this story and the idea of the Earth's sphericity, Mandeville notes that the North Star would not be visible in the Southern Hemisphere due to the curvature of the Earth (though he does claim an equivalent south polar star, "Antarctic"); that gravity works by pulling objects toward the center of the Earth's sphere and not merely "down," and hence people in the Southern Hemisphere would not feel themselves to be upside down; and even gives a roughly accurate measurement of 20,425 miles (32,870 kilometers) for the Earth's circumference.[115]

113 Russell, *Inventing the Flat Earth*, 63–65; Gingerich, *The Eye of Heaven*, 277–78; Thomas S. Kuhn, *The Copernican Revolution: Planetary Astronomy in the Development of Western* Thought (Cambridge, MA: Harvard University Press, 1957), 106–08.

114 Kuhn, *The Copernican Revolution*, 185–228.

115 John Mandeville, *The Travels of Sir John Mandeville* (London: Macmillan and Co., 1900), 120–24; John Wyatt Greenlee and Anna Fore Waymack, "Thinking Globally:

Mandeville was widely read by even relatively common people, such as the sixteenth-century Italian miller and heretic Menocchio (1532–1599).[116] In terms of the spherical Earth, Mandeville's book is notable as one of the very few written sources of his time that depicted a journey—however conceptually—entirely around the circumference Earth, returning to one's point of origin from the opposite direction.[117] It is therefore not outrageous to claim that, for as embellished (to be generous) as much of Mandeville's depiction of the world was, he may have done more than any Church theologian or Renaissance scholar to popularize and explain the shape of the Earth to Western Christians. Mandeville biographer Giles Milton (b. 1966) has proposed that (the author writing as) Mandeville gained his measurements for the Earth's circumference and the idea of the voyage around it from Jean Buridan (c. 1301-c.1359), a philosopher at the University of Paris whose critiques of Aristotelian physics were an important predecessor to Copernicanism, and whose writings may have been studied by Copernicus during his 1501–1503 studies at the University of Padua.[118]

Also among the readership of *Mandeville's Travels* was a certain Genoese sailor named Cristoforo Colombo, better known in English as Christopher Columbus. Columbus had been present at the fall of Grenada at the start of 1492, an action which symbolized the eclipse of Islamic power in the Iberian Peninsula, but also heralded the acquisition of the voluminous Islamic libraries of the emirate by the Christian kingdom of Spain. The ocean voyage west to Asia that Columbus proposed to the victorious Spanish monarchs was likewise framed as an outgrowth of the medieval Crusades which had failed two centuries prior, by allowing the

Mandeville, Memory, and Mappaemundi," *The Medieval Globe*, Vol. 4, No. 2 (2018), 77–80; Russell, *Inventing the Flat Earth*, 15.

[116] Carlo Ginzburg, trans. John and Anne Tedeschi, *The Cheese and the Worms: The Cosmos of a Sixteenth-Century Miller* (Baltimore: Johns Hopkins University Press, 1992 [1976]), 41–44.

[117] Edson, *The World Map*, 108–09; Allegro, "The bottom of the universe," 73.

[118] Kuhn, *The Copernican Revolution*, 118–22; Giles Milton, *The Riddle and the Knight: In Search of Sir John Mandeville* (New York: Farrar, Straus and Giroux, 1996), 53–54.

Spanish to recoup their financial losses of the Grenada War in preparation for a new assault on Jerusalem.[119] Columbus not only brought a copy of *Mandeville's Travels* with him on his voyages—trying to fit the geography of the Caribbean into Mandeville's already-fantasized Asia—but according to Andrés Bernáldez (1450–1513), Archbishop of Seville and court historian to Ferdinand and Isabella, Mandeville's account of the English circumnavigator may have convinced both Columbus, and the Spanish monarchs, of the ability to sail from Europe to Asia.[120]

The Reconquista's physical transition of the Islamic and African centers of learning of al-Andalus to Christian dominion was symbolically exhibited by Columbus, as well. One of the other works which convinced him a voyage across the Atlantic to the Indies was possible were the writings of Abbasid astronomer al-Farghani (c. 800–870), whom Columbus was introduced to by the works of French cardinal Pierre d'Ailly. In a sign of things to come, however, Columbus fundamentally misunderstood al-Farghani's scholarship, believing he used Roman miles instead of the much longer Arabic miles, and that al-Farghani proved the world to be much smaller than it actually was.[121] The European conquest of the Americas, supported by the subsequent African slave trade, was therefore predicated on a basic intellectual error. Far from disproving a Flat Earth, Columbus used a flawed methodology more in line with those of the modern Flat Earth defenders. However, the state power represented by Columbus, along with the scientific progress it both symbolized and benefitted from, would allow the instantiation of his fantastical worldview into reality. If the colonial project relied on science as ideology, then perhaps it is only natural that some of those oppressed by that colonial system adopted a critique of science into their own anti-colonial ideologies. This is the subject of our next chapter.

[119] Gómez, *Tropics of Empire*, 385–86; Hamdani, "Islamic Background," 280–81; Fernández-Armesto, *Before Columbus*, 212–17; Fernández-Armesto, *Conquest of the Impossible*, 60–61, 64–65, 73.

[120] Greenlee and Waymack, "Thinking Globally," 101; Milton, *The Riddle and the Knight*, 215–18; Andrés Bernáldez, *Historia de los reyes católicos D. Fernando y Doña Isabel* (Seville: Impr. que fué de J. M. Geofrin, 1870), 357–58.

[121] Fernández-Armesto, *Conquest of the Impossible*, 37–38.

CHAPTER 2

African Science in the Diaspora

In 1725, the English naturalist Mark Catesby (1683–1749) was touring Britain's American colonies, as his sister was married to the Secretary of State of the Virginia Colony. Catesby, however, was at the Stono plantation near Charleston, South Carolina. Fourteen years later, Stono would become famous as the site of the largest slave revolt in British North America, a brutally suppressed rebellion led by literate, Catholic slaves taken from the Kingdom of Kongo in modern-day Angola. But in 1725, Stono was host to a different revolutionary development. As Catesby observed, the plantation's slaves—taken from the same Christianized, Portuguese-speaking Kongo state—had, while digging in a swamp, revealed several giant teeth. To their British owners, it was confirmation of the Book of Genesis, physical remnants of the Biblical giants killed in the Noachic flood. But the Kongolese slaves knew better—they correctly identified the remains as elephant teeth, recognizing them from their homeland.

Catesby agreed with the identification of the Kongolese slaves, as he had just seen African elephant teeth displayed in London. But this did not stop the prognosis from repeating. In 1782, African slaves digging in salt marshes in Virginia uncovered large bones, and again identified them as elephantine in contrast to the Biblical inclinations of the slaveowners. The bones were sent to the statesman Thomas Jefferson (1743–1826),

E. Guimont, *The Power of the Flat Earth Idea*,
Palgrave Studies in the History of Science and Technology,
https://doi.org/10.1007/978-3-032-23981-5_2

who at the time was writing his sole book, *Notes on the State of Virginia* (1785). Jefferson eventually matched the description of the large East Coast fossils with similar remains from Big Bone Lick, Kentucky, as well as those long known to Native Americans on the upper Missouri River and reported to French traders. In this, Jefferson was merely mostly correct; the Virginia, South Carolina, and Kentucky remains were of mastodons, while the Missouri ones were of Cretaceous dinosaurs. Because of this, one of the tasks then-President Jefferson gave to the explorers Meriwether Lewis (1774–1809) and William Clark (1770–1838) in 1804 was to search for the living creatures responsible for those bones. Unfortunately, both mastodons and dinosaurs proved to be long extinct, though it did set a precedent for Native American knowledge and labor being used to identify and extract fossil remains as settlers moved West. But this trend itself built on the initial identification of mastodon remains at Stono. It was there that the American science of paleontology was begun—a science owing its origin to the knowledge and labor of slaves brought over from the Kingdom of Kongo.[1]

Nor was the spread of knowledge from Africa to America limited to paleontology. Writing in 1946, the Black historian and civil rights pioneer W. E. B. Du Bois (1868–1963) argued, "What in our civilization is distinctly British or American? Nothing. Science was built on Africa and Religion on Asia."[2] The nautical knowledge of both free and enslaved Africans contributed greatly to American seafaring and diving in the Age of Sail.[3] The role of African knowledge in developing Western medicine and botany will be discussed in the next chapter. American astronomy

[1] Adrienne Mayor, *Fossil Legends of the First Americans* (Princeton: Princeton University Press, 2005), 56–61. For an overview of how paleoscience passes from unorthodox to accepted and back, see Richard Fallon and Edward Guimont, "Introduction: Conceptualising heterodox palaeoscience," *Interdisciplinary Science Reviews*, Vol. 49, No. 3–4 (Jul., 2024), 307–15.

[2] W. E. B. Du Bois, *The World and Africa: An Inquiry into the Part which Africa has Played in World History, New Enlarged Edition* (New York: International Publishers, 1965), 149.

[3] W. Jeffrey Bolster, *Black Jacks: African American Seamen in the Age of Sail* (Cambridge, MA: Harvard University Press, 1997), 44–67; Kevin Dawson, *Undercurrents of Power: Aquatic Culture in the African Diaspora* (Philadelphia: University of Pennsylvania Press, 2018), 82–118; Mary E. Hicks, *Captive Cosmopolitans: Black Mariners and the World of South Atlantic Slavery* (Chapel Hill: University of North Carolina Press, 2024), 35–74.

also owes a debt to African knowledge, through the figure of Benjamin Banneker (1731–1806).[4] Banneker was a free Black born near Baltimore, Maryland, who would later become a polymath, publishing an almanac, working as a surveyor, and studying mathematics and astronomy, among other subjects. In particular, Banneker is the first identified African-American recognized as having a dedicated interest in astronomy.[5] As a free Black, Banneker was a rarity at the time and place, but the time of his death was the start of a transition in status for Black Baltimoreans. While in 1790, there were only a few hundred free Blacks in the city, the number would swell to nearly 15,000 by 1830; that same year, the number of enslaved Blacks in the city would reach their peak number of 4100, declining thereafter.[6] Famed abolitionist Frederick Douglass (c. 1817–1895) would give his final public speech in the city in January 1894, making Baltimore the capstone for a certain type of nineteenth-century Black activism.[7] In 1791, Banneker took part in surveying the boundaries for what would become the new national capitol of Washington, D.C., a project meant to be based on geometrical precision and thus imbue mathematical logic onto the heart of the new republic.[8] Writing in 1876, Edward Wilmot Blyden (1832–1912), a West Indian-born Liberian education pioneer who was one of the few nineteenth-century Blacks to call for adoption of African culture over European culture, stated that

[4] Britt Rusert, *Fugitive Science: Empiricism and Freedom in Early African American Culture* (New York: New York University Press, 2017), 39–40, 191; Ronald LaMarr Sharps, *Black Folklorists in Pursuit of Equality: African American Identity and Cultural Politics, 1893–1943* (Lanham, MD: Lexington Books, 2022), 113.

[5] H. M. Oluseyi, "Contributions from the First Half-Century of African American Astronomers with a Focus on Solar Physics," *African Skies/Cieux Africains*, No. 11 (Jul., 2007), 42.

[6] Dexter J. Gabriel, "The British Emigration Scheme and the African American Emigration Movement to the Caribbean," in Ronald Angelo Johnson and Ousmane K. Power-Greene, eds., *In Search of Liberty: African American Internationalism in the Nineteenth-Century Atlantic World* (Atlanta: The University of Georgia Press, 2021), 168; Dexter J. Gabriel, *Jubilee's Experiment: The British West Indies and American Abolitionism* (Cambridge: Cambridge University Press, 2023), 264–68.

[7] Isabel Wilkerson, *The Warmth of Other Suns: The Epic Story of America's Great Migration* (New York: Random House, 2010), 40.

[8] Charles A. Cerami, *Benjamin Banneker: Surveyor, Astronomer, Publisher, Patriot* (New York: John Wiley & Sons, 2002), 129–46; Amir Alexander, *Proof!: How the World Became Geometrical* (New York: Farrar, Straus and Giroux, 2019), 227–57.

Banneker was "the most distinguished Negro produced by a Protestant country."[9]

After his death, Banneker's cabin burned to the ground in an apparent accident, destroying many of his personal papers and causing some details of his life to be obscured. As a result, his background is not quite clear. Sources agree that Banneker's father had been a slave taken directly from Guinea, an amorphously defined region in West Africa, who later took the name Robert after being baptized, and that Banneker's mother, Mary, had been a free person of African ancestry born in Maryland. However, there is a dispute whether Mary was of purely Black descent or not. In some accounts, Mary's mother was a white woman, Molly Welsh, who freed and married a slave named Bannaker [*sic*], Banaky, Banneky, or Bannaka, who had been taken directly from Africa and whose father had been a local king.[10] In either case, that both Banneker's father, Robert, and maternal grandfather, Bannaka,[11] were directly from Africa and are not challenged.

Efforts to learn more about Banneker's African ancestry have focused on searching for clues in his extant mathematical writing to determine what, if any, elements of traditional African mathematics he may have incorporated and circulated in his own work. Of these efforts, professor

[9] Edward Wilmot Blyden, *Christianity, Islam and the Negro Race* (Baltimore, MD: Black Classic Press, 1994 [1888]), 47; Robin D. G. Kelley, *Freedom Dreams: The Black Radical Imagination* (Boston: Beacon Press, 2002), 22–23.

[10] For a purely African background for Banneker, see William Cooper Nell, *The Colored Patriots of the American Revolution, With Sketches of Several Distinguished Colored Persons: To Which Is Added a Brief Survey of the Condition And Prospects of Colored Americans* (Boston: Robert F. Wallcut, 1855), 203–04; William Wells Brown, *The Black Man, His Antecedents, His Genius, and His Achievements* (New York: Thomas Hamilton, 1863), 51; Ellen E. Swartz, "Removing the Master Script: Benjamin Banneker "Re-Membered"," *Journal of Black Studies*, Vol. 44, No. 1 (Jan., 2013), 41. For a mixed ancestry for Banneker, see Martha Ellicott Tyson, *A Sketch of the Life of Benjamin Banneker; from notes taken in 1836* (Baltimore: Maryland Historical Society, 1854), 4–5; John Wesley Cromwell, *The Negro in American History* (Washington, D.C.: The American Negro Academy, 1914), 86; Cerami, *Benjamin Banneker*, 1–18; Ron Eglash, "The African Heritage of Benjamin Banneker," *Social Studies of Science*, Vol. 27, No. 2 (Apr., 1997), 308–09. For an agnostic take, see William J. Simmons, *Men of Mark: Eminent, Progressive and Rising* (Cleveland: George M. Rewell & Co., 1887), 344. For a broad overview of the controversy, see Sandra Perot, "Reconstructing Molly Welsh: Race, Memory, and the Story of Benjamin Banneker's Grandmother" (MA thesis, University of Massachusetts Amherst, 2008).

[11] I am choosing to use the spelling "Bannaka" for Banneker's grandfather to prevent confusion with his grandson's surname.

of information Ron Eglash (b. 1958) has argued that Robert likely originated from the area around the modern Ghana-Nigeria border, and that Bannaka may come from "the Wolof name *Banakas*, which came from the royalty of the Wolof kingdom of Walo—now in the St Louis region of Northern Senegal.... the name originated from the Arabic *Tanakas*, meaning 'belongs to the place', and was then 'West-Africanized', with the B consonant sound replacing the T."[12] Based on this, Eglash found evidence of Ifá and Fá divination numerology from Yoruba culture in some of Banneker's mathematical statements, when discussing the "hare and the hound" problem, which Banneker seemed to have associated with the same West African trickster stories that became Br'er Rabbit and Br'er Fox. Further, the quincrux, an important Wolof symbol, was often used in illustrations by Banneker, and appeared in dreams he recorded[13] (Fig. 2.1).

However, Eglash's Wolof ancestry is not the only proposed solution to Banneker's African background. In 2002, the economist Charles A. Cerami (1920–2010) published a biography of Banneker, where, in a passage without citations, he proposed that Bannaka's ancestry was Dogon, a people who live in what is now Mali.[14] Cerami's evidence was that Banneker allegedly claimed that his favorite star was Sirius and that he knew it was a double star long before the dim companion Sirius B was first recorded by Western astronomers in 1844. According to Cerami, the Dogon held Sirius in reverence and also knew of its binary status long before they had access to modern telescopes capable of resolving Sirius B.[15] Although Cerami does not cite his sources for the Dogon claims, he does not need to, for he drew on decades of speculation from the scientific fringe.

In a 1954, French anthropologists Marcel Griaule (1898–1956) and Germaine Dieterlen (1903–1999) published an article based on decades of research they had conducted on the cosmological views of the Dogon. Of particular relevance is their observation that the Dogon organized

[12] Eglash, "African Heritage," 309.

[13] Eglash, "African Heritage," 310–13.

[14] Isaac Samuel, "A complete history of Dogon country: Bandiagara from 1900 BC to 1900 AD: demystifying an ancient African society," *African History Extra*, March 5, 2023, accessed March 13, 2023, https://isaacsamuel.substack.com/p/a-complete-history-of-dogon-country.

[15] Cerami, *Benjamin Banneker*, 217–19.

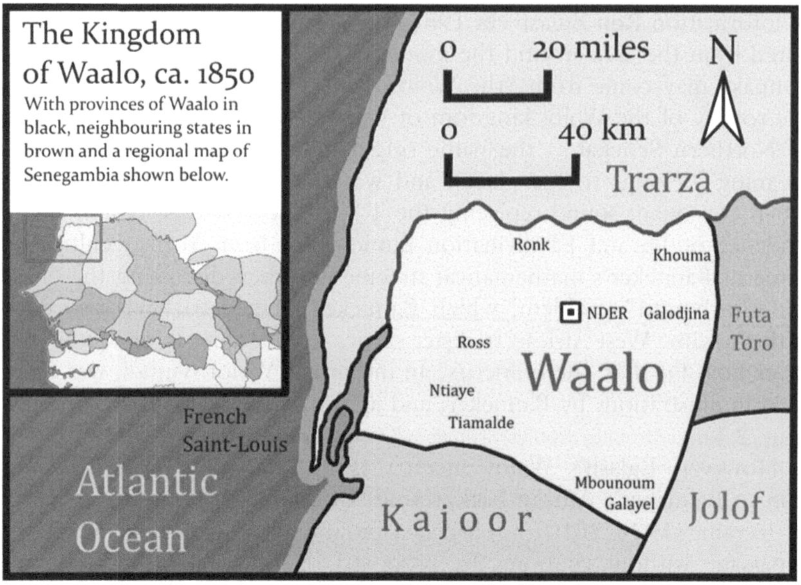

Fig. 2.1 Bannaka's potential homeland, Waalo. The Dogon are located near the center of the map insert (Wikimedia Commons)

their territory as much as possible in accordance with their belief that the world developed in the form of a spiral, and that their map of the world consisted of a human figure with its head pointing south.[16] In a smaller section of their article, Griaule and Dieterlen noted that the star Sirius played a major role in the Dogon religion, and that the Dogon religion appeared to incorporate knowledge of Sirius and its system that would have been impossible to know without modern telescopes. Since then, their methodology has undergone substantive criticism, with the suggestion that the Dogon may have learned about Sirius B from a team of French astronomers staying in Dogon territory—Mali being a French colony until 1960—to observe the 16 April 1893 solar eclipse. However,

[16] Marcel Griaule and Germaine Dieterlen, "The Dogon of the French Sudan," in Daryll Forde, ed., African Worlds: *Studies in the Cosmological Ideas and Social Values of African Peoples* (Oxford: James Currey Publishers, 1999 [1954]), 94–97, 103.

Griaule and Dieterlen explicitly did not attempt to offer any hypotheses for how the Dogon came by their knowledge.[17]

In 1965, American student Robert K. G. Temple (b. 1945) was shown Griaule and Dieterlen's article by the astrologer Arthur M. Young. In 1967, Temple read unidentified "futuristic essays" by British-born Sri Lanka-residing science fiction author Arthur C. Clarke (1917–2008), which in turn inspired Temple to look further into the Dogon beliefs, including having other scholarship by Griaule and Dieterlen translated into English for the first time. Temple also reached out to Clarke on the subject. Over the next several years, Clarke helped Temple with his research, passing along information on Sirius B and discussing Erich von Däniken's (1935-2026) 1968 ancient alien tome *Chariots of the Gods?*. Clarke also put Temple into contact with other researchers of what he termed "pet mysteries," such as British crystallographer Alan Lindsay Mackay (1926-2025) who studied the Minoan Phaistos Disc, and British historian of science Derek J. de Solla Price (1922–1983), who claimed that Babylonian mathematics could be found in New Guinea. The result of this collaboration with Clarke was what Temple called the Sirius Mystery, publishing his findings in a 1976 book by that title.[18] Griaule and Dieterlen had discussed the Nommo, amphibious ancestral spirits created by the Dogon sky god. Temple, in his book, argued that the Nommo were actually extraterrestrial visitors from Sirius, linking them not only to the development of Dogon culture, but to a wide range of Mesopotamian, Egyptian, and Greek beliefs.[19] Of note, Temple stated that to deny the obvious existence of extraterrestrial life would make one

> something of an intellectual freak equivalent to a believer in the Flat Earth Theory. I mention that theory because I once met a woman who appeared quite sane and yet who was a member of a cult who believe the Earth is flat. This was one of the more startling experiences anyone can have, and a salutary education to me. It taught me never to underestimate the power

[17] Griaule and Dieterlen, "The Dogon," 84–85; Walter E. A. van Beek, "Dogon Restudied: A Field Evaluation of the Work of Marcel Griaule," *Current Anthropology*, Vol. 32, No. 2 (Apr., 1991), 139–67; Isaac Samuel, "Star-gazing from the cliffs of Bandiagara; truths and mysteries of Dogon astronomy in the west African context," *African History Extra*, October 28, 2022, accessed March 13, 2023, https://www.patreon.com/posts/star-gazing-from-73897317.

[18] Robert K. G. Temple, *The Sirius Mystery* (New York: St. Martin's Press, 1976), 2–4.

[19] Griaule and Dieterlen, "The Dogon," 85–87; Temple, *The Sirius Mystery*, 203–27.

> of the human mind to believe what it wants to believe despite any amount of evidence.[20]

That a proponent of ancient aliens could come to that conclusion when faced with Flat Earthers is illustrative of a particular mental siloing among believers of heterodox notions, a siloing which as will be explored in the later chapters may not be as marked in the twenty-first century.

The Sirius Mystery was published the same year as Ivan Van Sertima's *They Came Before Columbus*, as discussed in the previous chapter. Ancient alien theorizing and Afrocentrist hyperdiffusionism both emerged from the same counterculture trends building in American culture since the 1960s. However, Temple's book reduced the achievements of Africans, while Van Sertima's built them up. In 1983, Van Sertima edited the book *Blacks in Science: Ancient and Modern*, which included an essay rejecting alien intervention with the Dogon, but instead attributing them with fantastic feats, such as originating in ancient Egypt and developing telescopes thousands of years before their development in the West.[21] Building off this synthesis, later Afrocentrists have continued to argue not only for fantastic Dogon astronomical knowledge, but that Dogon astronomy was gifted to young Banneker by Bannaka.[22]

But even if Benjamin Banneker was not constructing his telescopes using ancient Dogon glass-grinding techniques passed down from his royal grandfather, the fact that American science in the eighteenth and early nineteenth centuries drew from African knowledge makes a certain amount of sense. Consider that the South was the center of American science at the period, and Southern science was dominated by the elites whose livelihood came from their owning of slaves. Charleston in particular was not only the center of Antebellum slave owning society, as well as

[20] Temple, *The Sirius Mystery*, 14.

[21] Hunter Havelin Adams III, "African Observers of the Universe: The Sirius Question," in Ivan Van Sertima, ed., *Blacks in Science: Ancient and Modern* (New Brunswick, NJ: Transaction Books, 1983), 27–46.

[22] Swartz, "Removing the Master Script," 40–41; Robert Fikes Jr., "Careers of African Americans in Academic Astronomy," *The Journal of Blacks in Higher Education*, No. 29 (Autumn, 2000), 132; Jay Wright, *Transfigurations: Collected Poems* (Baton Rouge: LSU Press, 2000), 103–10; Vera M. Kutzinski, "The Descent of Nommo: Literacy as Method in Jay Wright's "Benjamin Banneker Helps to Build a City""," *Callaloo*, No. 19 (Autumn, 1983), 103–19.

the drive for secession, but the center of Antebellum Southern science.[23] Even prior to the Revolutionary War, the British slave trade in which Charleston was the largest North American port played a major role in the development of natural history and collecting from Africa and the Caribbean.[24] Particularly during his lifetime, the southern slaveowner Jefferson symbolized American science, while opposition to science was symbolized by those most opposed to him, New England Congregationalist clergy, who maintained their sect as a state religion as late as 1818 in Connecticut. As might not be a surprise, given the opposition to Galileo and Copernicus by the Church, astronomy was the epitome of the sciences criticized by the Congregationalist Yankees.[25]

This tension with astronomy made it all the more remarkable that, on the morning of 14 December 1807, multiple eyewitnesses reported loud thunder and a fall of stones from the sky in the town of Weston, Connecticut. Now known as the Weston meteorite, it was the first recorded meteorite fall in the Western hemisphere. Two professors from what was then the Congregationalist-affiliated Yale College, the chemist Benjamin Silliman (1779–1864) and the classicist James Luce Kingsley (1778–1852), investigated the Weston report, and announced that the stones had fallen from the celestial realm beyond the atmosphere. Reportedly, when hearing their conclusions, President Jefferson responded "that it was easier to believe that two Yankee Professors could lie than to admit that stones could fall from heaven." However, the word "reportedly" is the operative word here, as in reality this response from Jefferson seems to have originated with an 1874 speech by Silliman's son. Jefferson's actual response was far more tepid:

[23] Ronald L. Numbers and Janet S. Numbers, "Science in the Old South: A Reappraisal," *The Journal of Southern History*, Vol. 48, No. 2 (May, 1982), 163–84; Nancy Smith Midgette, "In Search of Professional Identity: Southern Scientists, 1883–1940," *The Journal of Southern History*, Vol. 54, No. 4 (Nov., 1988), 597–622; A. J. Angulo, "William Barton Rogers and the Southern Sieve: Revisiting Science, Slavery, and Higher Learning in the Old South," *History of Education Quarterly*, Vol. 45, No. 1 (Spring, 2005), 18–37; Cathryn J. Prince, *A Professor, A President, and A Meteor: The Birth of American Science* (New York: Prometheus Books, 2011), 73–86.

[24] Kathleen S. Murphy, *Captivity's Collections: Science, Natural History, and the British Transatlantic Slave Trade* (Chapel Hill, NC: University of North Carolina Press, 2023).

[25] Prince, *A Professor, A President, and A Meteor*, 87–100.

> A cautious mind will weigh well the opposition of the phenomenon to everything hitherto observed, the strength of the testimony by which it is supported, and the errors and misconceptions to which even our senses are liable. It may be very difficult to explain how the stone you possess came into the position in which it was found. But is it easier to explain how it got into the clouds from whence it is supposed to have fallen? The actual fact however is the thing to be established, and this I hope will be done by those whose situations and qualifications enable them to do it.[26]

Despite the mischaracterization of (the younger) Silliman's "misquote heard around the world," the fact remains that in the first decade of the nineteenth century, knowledge of the heavens in the United States was shaky enough that meteorites remained unexplained—and the southern Jefferson was wary of proposed explanations from New Englanders.[27]

Of course, skepticism over Yankee astronomy was not the only element of Southern science that might in turn be viewed with skepticism today. The somewhat related fields of Spiritualism and phrenology, both rooted in the use of Black slaves as subjects, were prominent in the South, from the pre-Civil War period through the late nineteenth century. Spiritualism in particular, with its crossover appeal to the new field of physics, seemed to allow for a blending of science and religion, much as Flat Earth thought claimed. But one of the most notorious innovations of Antebellum southern science was not Spiritualism, but the 1851 invention of drapetomania—a supposed mental illness which explained why slaves would try to flee their owners—by the Louisiana physician and future Confederate Army surgeon Samuel A. Cartwright (1793–1863).[28]

[26] Anna Berkes, "Who is the liar now?," *Thomas Jefferson's Monticello*, November 14, 2008, accessed September 1, 2022, https://www.monticello.org/research-education/blog/who-is-the-liar-now/.

[27] Prince, *A Professor, A President, and A Meteor*, 129–42; Joy Hakim, "A Founding Father of science," *Science Scope*, Vol. 37, No. 6 (Feb., 2014), 6–8.

[28] Alicia Puglionesi, *Common Phantoms: An American History of Psychic Science* (Palo Alto, CA: Stanford University Press, 2020), 159–200; Miranda Corcoran, "Science and the Supernatural: Spiritualism, Psychical Research and Pseudo-Science in the Nineteenth Century," in Clive Bloom, ed., *The Palgrave Handbook of Steam Age Gothic* (Houndsmills: Palgrave Macmillan, 2021), 512–13; Peter McCandless, "Mesmerism and Phrenology in Antebellum Charleston: "Enough of the Marvellous"," *The Journal of Southern History*, Vol. 58, No. 2 (May, 1992), 199–230; Jim Downs, *Maladies of Empire: How Colonialism, Slavery, and War Transformed Medicine* (Cambridge, MA: Harvard University Press, 2021), 137–66; Rachel E. Walker, *Beauty and the Brain: The Science of Human Nature in*

That what is now recognized as fringe science could be seen as mainstream science in Southern culture gives some context to how Flat Earth views might be seen as less controversial than today. But it is not the idea of a *Flat* Earth which initially received more widespread reception among racist proponents of American exceptionalism—but rather, a *Hollow* Earth.

The idea that the Earth is hollow, typically paired with the notion that the hollow interior can be accessed through enormous holes at the north and south poles, dates back at least to famous English physicist and astronomer Edmond Halley (1656–1742), of cometary namesake fame, who proposed the notion in 1692—the same year that on the other side of the Atlantic, the predecessors of the Yankee Congregationalists were placing suspected witches on trial in Salem, Massachusetts.[29] In the United States, however, Halley's notion would not become popularized until it was taken up in 1818 by John Cleves Symmes Jr. (1780–1829). Symmes' uncle was a prominent politician of the American Revolutionary period, and through him Symmes served as an officer in the US Army from 1802 until being honorably discharged at the end of 1815, having served throughout the War of 1812. From 1818 until his death—at which point his cause was taken up by his son Americus (1811–1896)—Symmes promoted the idea of a Hollow Earth accessible via the poles, advocating for polar exploration to enable the United States to claim and colonize the vast inner reaches of the globe.[30]

Early America (Chicago: University of Chicago Press, 2022); Courtney E. Thompson, *An Organ of Murder: Crime, Violence, and Phrenology in Nineteenth-Century America* (New Brunswick, NJ: Rutgers University Press, 2021); Wendy Gonaver, *The Peculiar Institution and the Making of Modern Psychiatry, 1840–1880* (Chapel Hill: University of North Carolina Press, 2019).

[29] David Standish, *Hollow Earth: The Long and Curious History of Imagining Strange Lands, Fantastical Creatures, Advanced Civilizations, and Marvelous Machines Below the Earth's Surface* (Boston: De Capo Press, 2006), 15–38; Richard Fallon, *Contesting Earth's History in Transatlantic Literary Culture, 1860–1935: Believers and Visionaries on the Borderlines of Geology and Palaeontology* (Oxford: Oxford University Press, 2025).

[30] Standish, *Hollow Earth*, 39–86; Robert J. Schadewald, "Is the World in Curious Shape?," *Isaac Asimov's Science Fiction Magazine*, Vol. 4, No. 12 (Dec., 1980), 101; L. Sprague de Camp and Willy Ley, *Lands Beyond: A Fascinating Expedition Into Unknown Lands* (New York: Rinehart & Company, 1952), 296–301; Irving Wallace, *The Square Pegs: Some Americans Who Dared to Be Different* (New York: Alfred A. Knopf, 1957), 223–42.

One of Symmes' most fervent adherents was the newspaper editor J. N. Reynolds (1799–1858). Reynolds began promoting the idea of an expedition to Antarctica to confirm the entrance to the Hollow Earth. However, while Symmes was seemingly a true believer in the Hollow Earth and viewed Antarctic advocacy as a way to confirm his geological theory, Reynolds appears to have been more interested in Antarctic exploration on its own, and using the interest in the Hollow Earth to promote that end. Reynolds' Hollow Earth advocacy was originally centered in Banneker's former home of Baltimore in 1826, and he quickly achieved the notice of another Baltimore resident, the horror author Edgar Allan Poe (1809–1849). Poe would become an advocate of the concept, including it in several of his works. In his 1835 short story "The Unparalleled Adventure of One Hans Pfaall," the title character flies to the Moon in a balloon, and during his ascent observes the entrance to the Hollow Earth located in the north pole. In Poe's sole novel, *The Narrative of Arthur Gordon Pym of Nantucket* (1837), the conclusion connects the Hollow Earth with early speculation about the Antarctic continent. Notably, one of Poe's early fans would be the French science fiction master Jules Verne (1828–1905), whose second novel, the 1864 *Journey to the Center of the Earth*, prominently featured a hollow interior of the Earth. Verne followed *Journey to the Center of the Earth* with his 1865 *From the Earth to the Moon*, which in addition to referencing Poe's "Hans Pfaall," also has its titular lunar voyage be carried out by the (fictitious) Baltimore Gun Club. Nor was the Hollow Earth the only alternative form of geography Verne's works would help spread; his 1869 *Twenty Thousand Leagues Under the Seas* would help popularize the modern notion of the sunken city of Atlantis. Notably, Verne's output of science fiction would also help popularize maps for public consumption in the late nineteenth century United States.[31]

[31] Standish, *Hollow Earth*, 87–142; Schadewald, "Curious Shape," 105; de Camp and Ley, *Lands Beyond*, 304–05; John Tresch, *The Reason for the Darkness of the Night: Edgar Allan Poe and the Forging of American Science* (New York: Farrar, Straus and Giroux, 2021), 95–103; Martin Brückner, "Popular Map Genres in American Literature," in *Literature and Cartography: Theories, Histories, Genres*, ed. Anders Engberg-Pedersen (Cambridge, MA.: The MIT Press, 2017), 346–49; Martin Brückner, "Maps, Movement, and American Literature," Mapping Movement in American History and Culture, 2016 (accessed May 24, 2017), https://mappingmovement.newberry.org/essay/maps-movement-and-american-literature.

Reynolds' lobbying efforts would nearly bear fruit in May 1828, when the US Congress passed a resolution urging then-President John Quincy Adams (1767–1848) to send an expedition to the South Pole to determine if there was an entrance to the Hollow Earth. However, that November, Adams lost re-election to Andrew Jackson (1767–1845), who would be inaugurated in March 1829 and remain as President until 1837. Jackson would ultimately approve an expedition in 1836, which would become the 1838–1842 United States Exploring Expedition, but would largely have the Hollow Earth aspects of the mission expunged in favor of general Pacific and Antarctic exploration and force projection for the expanding Republic. Ironically, Jackson's enemies would accuse him of ultimately endorsing the expedition because of an unsubstantiated rumor that he believed the Earth was flat.[32] The belief in the Hollow Earth by certain naval advocates would not vanish with the nineteenth-century US Navy; after World War I, the so-called World Ice Cosmogony would become popularized in Germany, especially among Nazi esoterics. A byproduct was the belief that we live on the inside skin of a Hollow Earth, with at least one German naval experiment in World War II being conducted to determine if the British navy could be detected by peering "across" the inner curvature of the Hollow Earth.[33]

Jackson's detractors attacking him by labeling him a Flat Earther in response to his approving an expedition originally influenced by Hollow Earthers is not the only aspect connecting the two theories. In a 2019 interview with journalist Tim Binnall, Mark Sargent (b. 1969), one of the most prominent Flat Earthers of the twenty-first century as reflected by his central role in the 2018 documentary *Behind the Curve*, stated that the Hollow Earth notion was his entry point (so to speak) to the Flat Earth.[34] And just as Symmes' interest in the Hollow Earth was promoting the continuation of American Manifest Destiny, it is of note that he named his

[32] Edward Pessen, *Jacksonian America: Society, Personality, and Politics* (Homewood, IL: Dorsey Press, 1969), 191.

[33] Gerard P. Kuiper, "German Astronomy During the War," *Popular Astronomy*, Vol. 54, No. 6 (Jun., 1946), 277–78; Eric Kurlander, *Hitler's Monsters: A Supernatural History of the Third Reich* (New Haven: Yale University Press, 2017), 131–61; Patrick Moore, *Can You Speak Venusian? A Guide to the Independent Thinkers* (New York: W. W. Norton & Company, 1972), 36–37, 66–72.

[34] Tim Binnall and Mark Sargent, "BoA Holiday Season 2019: Mark Sargent," *Binnall of America*, November 11, 2019 (accessed November 20, 2019), http://www.binnallofamerica.com/boaa111119.html.

son Americus Vespucius Symmes after Florentine sailor Amerigo Vespucci (1451–1512), who coined the term "New World" in 1501 and for whom the Americas were first named in 1507. Yet Vespucci was not the only Italian explorer of the Age of Discovery to relate with alternative shapes of the Earth in early nineteenth century America. No less than Christopher Columbus would play a role.

It should be noted that Vespucci's claims about his pioneering early role in the European exploration of the Americas have come to be viewed suspiciously by historians. However, the Age of Discovery was one where European fantasies bled into reality, the discovery of "new" lands allowing the embodiment of centuries-old fantasies by the new imperial geographers.[35] In this context, Vespucci's stories are no more outlandish than any other. Consider that when the volcanic Mount Hekla in Iceland erupted in 1104, less than two centuries after the island's settlement by the Norse, the destruction was so widespread that Christians across Europe believed that Hekla was the literal entrance to Hell (eclipsing the earlier belief that Sicily's Mount Etna held that distinction), a folk belief which persisted into the nineteenth century.[36] Within decades of the European discovery of the "New World," it was already being proposed that the Americas were none other than Atlantis (somehow not submerged) or various landmasses beyond the Atlantic mentioned in Classical texts by such authors as Seneca and the Pseudo-Aristotle.[37]

Throughout his expeditions, Columbus attempted to make several connections between mythical cartography and the geography he was encountering, including the belief that the Garden of Eden was near

[35] Felipe Fernández-Armesto, *Columbus and the Conquest of the Impossible* (New York: Saturday Review Press, 1974), 48; Evelina Gužauskytė, *Christopher Columbus's Naming in the Diarios of the Four Voyages (1492–1504): A Discourse of Negotiation* (Toronto: University of Toronto Press, 2014), 22; Renée Landell, "Naming 'Paradise': The Adamic Imagination, Colonial Toponyms, and Remembering the Indigenous Caribbean," Environmental History Now, April 15, 2021, accessed January 10, 2022, https://envhistnow.com/2021/04/15/naming-paradise-the-adamic-imagination-colonial-toponyms-and-remembering-the-indigenous-caribbean/.

[36] Sigurður Þórarinsson, trans. Jóhann Hannesson, *Hekla: A Notorious Volcano* (Reykjavík: Almenna bókafélagið, 1970), 5–8.

[37] Edward Brooke-Hitching, *The Phantom Atlas: The Greatest Myths, Lies and Blunders on Maps* (San Francisco: Chronicle Books, 2018), 24–29; James S. Romm, *The Edges of the Earth in Ancient Thought: Geography, Exploration, and Fiction* (Princeton: Princeton University Press, 1992), 215–22.

the Caribbean, perhaps along the Orinoco River.[38] As mentioned at the end of the previous chapter, not only did Columbus bring with him a copy of the book of John Mandeville, attempting to determine which areas of the Caribbean corresponded with which sections of Mandeville's fictionalized Asia, but such later European explorers and chroniclers as Walter Raleigh, Martin Frobisher, and Richard Hakluyt likewise used *Mandeville's Travels* as a basis for American cartography.[39] On his first voyage alone, Columbus named a group of islands after St. Ursula, who according to medieval legends was killed by Huns along with her 11,000 virgin handmaidens; thus did the islands lose their Taíno names and become the Virgin Islands.[40] The Virgin Islands themselves are part of the larger chain of islands termed the Antilles. The Caribbean islands were named in reference to the medieval legend of Antillia, an Atlantic island with seven cities on it which Christians fled to in order to escape the eighth-century Muslim conquest of Iberia.[41] Columbus believed the largest Caribbean island, Cuba, to be the realm of Prester John, a mythical Eastern Christian king whose stories emerged in the twelfth century and who Columbus hoped to create an anti-Muslim alliance with; in the next century, the Portuguese would in turn identify Prester John with the emperor of Ethiopia.[42] On his fourth and final voyage (1502–1504), Columbus came to believe that South America might not be Asia, but be the long-hypothesized Antipodes.[43]

[38] Ralph Bauer, *The Alchemy of Conquest: Science, Religion, and the Secrets of the New World* (Charlottesville: University of Virginia Press, 2019), 174–76; De Camp and Ley, *Lands Beyond*, 310–13.

[39] Giles Milton, *The Riddle and the Knight: In Search of Sir John Mandeville* (New York: Farrar, Straus and Giroux, 1996), 218–19.

[40] Fernández-Armesto, *Conquest of the Impossible*, 48.

[41] Gerald Roe Crone, "The Origin of the Name Antillia," *The Geographic Journal*, Vol. 91, No. 3 (Mar., 1938), 260–62; Abbas Hamdani, "An Islamic Background to the Voyages of Discovery," in Salma Khadra Jayyusi, ed., *The Legacy of Muslim Spain* (Leidein: E. J. Brill, 1992), 274–75; Brooke-Hitching, *Phantom Atlas*, 18–23.

[42] Nicolás Wey Gómez, *The Tropics of Empire: Why Columbus Sailed South to the Indies* (Cambridge, MA: The MIT Press, 2008), 374–80; Bauer, *The Alchemy of Conquest*, 174–76; Fernández-Armesto, *Conquest of the Impossible*, 129; Brooke-Hitching, *Phantom Atlas*, 194–97.

[43] Fernández-Armesto, *Before Columbus*, 248–52; Fernández-Armesto, *Conquest of the Impossible*, 164–66.

Outright legendary islands were also quite common in medieval European maps of the Atlantic. From Irish myths came Hy-Brasil, an island across the ocean which was bisected by a river. Despite this seeming connection, the legendary island was not the namesake of the modern country of Brazil. However, in line with the opening of the previous chapter, Afrocentric hyper diffusionists have proposed that the name of Brazil originated from Abu Bakr II naming it after the "Berber" tribe of the Barazil.[44] A more famous Irish connection to mythical Atlantic islands is the supposed island discovered by Saint Brendan the Navigator, a sixth-century Irish monk. In an effort to find the Blessed Island of antiquity, Brendan sailed west; that he could go west to find the Earthly Paradise that was traditionally in the east demonstrates that he recognized the sphericity of the Earth. In 1520, Ferdinand Magellan's crew named Samborombón Bay in modern Argentina after the saint, believing it to be the origin point for his wandering island. Approximately a decade earlier, the Spanish novel *Las sergas de Esplandián* invented a fictional island named California (its name likely derived from the caliph), which was itself instantiated into reality by its name being used on a map of the west coast of North America in 1541. The non-island nature of California was not decisively established until the 1770s.[45] As of writing, a similar process is now taking up a large amount of public discourse in the United States, with the second Donald Trump (b. 1946) presidential administration attempting to rename the Gulf of Mexico to the Gulf of America.

Indeed, legends of strange islands, especially in the Atlantic, were so common in medieval Europe that the Norse stories of Vinland were long taken as merely one example of them, rather than the only one which turned out to have an element of truth. But the 1960 Norse

[44] Evelyn Edson, *The World Map, 1300–1492: The Persistence of Tradition and Transformation* (Baltimore: Johns Hopkins University Press, 2007), 206–09; Hamdani, "Islamic Background," 275–77; Brooke-Hitching, *Phantom Atlas*, 130–33.

[45] Vincent H. Cassidy, "Other Fortunate Islands and Some that were Lost," *Terrae Incognitae: The Journal of the Society for the History of Discoveries*, Vol. 1, No. 1 (1969), 36; Felipe Fernández-Armesto, *Before Columbus: Exploration and Colonization from the Mediterranean to the Atlantic, 1229–1492* (Philadelphia: University of Pennsylvania Press, 1987), 247–48. Because of a separate account for the year 553 that Brendan "was seen ascending in a chariot into the sky this year," the saint has also been assigned the status of an early experiencer of the phenomenon that would today be called alien abduction: Jacques Vallee and Chris Aubeck, *Wonders in the Sky: Unexplained Aerial Objects from Antiquity to Modern Times and Their Impact on Human Culture, History, and Beliefs* (New York: Jeremy P. Tarcher/Penguin, 2009), 392–93.

finds at L'Anse aux Meadows were the exception to the geofiction of Europeans in their Age of Discovery. The idea of a "New World" was taken literally: a place where the mythologies, fantasies, and dreams of the Old World could be instantiated through the violent destruction not only of indigenous mythologies, fantasies, and dreams, but indigenous reality, and very nearly indigenous existence overall. It is only fitting that one of the core figures of the Age of Discovery, Columbus, would also play a role in building the myth of the Flat Earth—and that, much as Poe and Verne played a role in promoting the Hollow Earth, this would happen through the efforts of a fantasist.

Washington Irving (1783–1859) was one of the first American authors to win widespread acclaim. Irving is perhaps best known for his 1819 fantasy stories "Rip Van Winkle" and "The Legend of Sleepy Hollow." The latter in particular plays into the intersection of fiction and geography; in 1996, shortly after the Columbus Quincentenary, the town of North Tarrytown in New York voted to change its name to Sleepy Hollow and embrace the story that Irving wove around it (the town of Camptown, New Jersey had already voted to rename itself Irvingtown after the author in 1850). Irving had a longstanding fascination in the Spanish Reconquista and conquest of the Americas, an interest which would culminate in his 1842–1846 tenure as the US ambassador to Spain. From 1826 to 1829, Irving had lived in Spain, including a short period at the end living in the famous Alhambra. It was during this time that Irving was able to gain firsthand access to various records from the Reconquista and colonization of the Americas, from which several books sprung. Irving's books were ostensibly nonfiction history, but contained as much fantasy as his earlier short stories. In line with the above examples of Europeans translating myths to the New World, Irving's 1831 biography of Spanish explorer Juan Ponce de León (1474–1521) was responsible for the belief that the conquistador sought the Fountain of Youth in Florida, relocating it there from its more typical placement in

Africa.[46] However, by that point, Irving had made a far more reaching invention of geographic myth in one of his books.

One of the rising authors whom Irving served as a mentor to was Poe. Given Poe's role in promoting the Hollow Earth, it is only suitable that Irving helped promote the idea of Flat Earth—or more accurately, the myth of its belief. As mentioned in the previous chapter, the idea of a Flat Earth was extremely rare in medieval Catholicism. It was in the early nineteenth century that the Flat Error emerged—primarily due to Irving. In 1828, the same year that Verne was born and the US Congress authorized funding for a voyage to the Antarctic entrance of the Hollow Earth, Irving published his mammoth, three-volume *A History of the Life and Voyages of Christopher Columbus*, the most substantive biography of the navigator until the mid-twentieth century. It is in the first volume of his biography that Irving depicts the appearance of Columbus before the Catholic scholars of the University of Salamanca. Established in 1218, it is one of the oldest universities in the Christian world. In reality, the closest to this event actually happening was an informal gathering of geographers at the university in 1486, who raised objections to Columbus' ideas about the size—*not* shape—of the Earth. However, Irving altered this into Columbus arguing in 1492 before a group of Catholic priests—scholars mired in medieval Catholic Flat Earth belief—at the university for the sphericity of Earth. Although long, a direct quotation of this passage is worthwhile, given the enormous significance it has in establishing the belief of medieval Flat Earthism:

> Several of the objections opposed by this learned body have been handed down to us, and have provoked many a sneer at the expense of the University of Salamanca, but they are proofs not so much of the peculiar deficiency of that institution, as of the imperfect state of science at the time, and the manner in which knowledge, though rapidly extending, was still impeded in its progress by monastic bigotry. All subjects were still

[46] Washington Irving, *Voyages and Discoveries of the Companions of Columbus* (Philadelphia: Carey and Lea, 1831), 312–18; Douglas T. Peck, "Anatomy of an Historical Fantasy: The Ponce de León-Fountain of Youth Legend," *Revista de Historia de América*, No. 123 (Jan.–Dec., 1998), 79; John McGrath, "Sixteenth-Century Florida in the European Imagination," *The Florida Historical Quarterly*, Vol. 91, No. 3 (Winter, 2013), 402–04; Sebastian Major, "Episode #65 – Was There a Real Hunt for the Fountain of Youth?," *Our Fake History*, May 2, 2018, accessed September 16, 2022, https://ourfakehistory.com/index.php/season-3/episode-65-was-there-a-real-hunt-for-the-fountain-of-youth/.

contemplated through the obscure medium of those ages when the lights of antiquity were trampled out, and faith was left to fill the place of inquiry. Bewildered in a maze of religious controversy, mankind had retraced their steps, and receded from the boundary line of ancient knowledge. Thus, at the very threshold of the discussion, instead of geographical objections, Columbus was assailed with citations from the bible and the testament: the book of Genesis, the psalms of David, the orations of the Prophets, the epistles of the Apostles, and the gospels of the Evangelists. To these were added the expositions of various saints and reverend commentators: St. Chrysostom and St. Augustine, St. Jerome and St. Gregory, St. Basil and St. Ambrose, and Lactantius Firmianus, a redoubted champion of the faith. Doctrinal points were mixed up with philosophical discussions, and a mathematical demonstration was allowed no weight, if it appeared to clash with a text of scripture, or a commentary of one of the fathers. Thus the possibility of antipodes, in the southern hemisphere, an opinion so generally maintained by the wisest of the ancients, as to be pronounced by Pliny the great contest between the learned and the ignorant, became a stumbling block with some of the sages of Salamanca. Several of them stoutly contradicted this fundamental position of Columbus, supporting themselves by quotations from Lactantius and St. Augustine, who were considered in those days as almost evangelical authority. But, though these writers were men of consummate erudition, and two of the greatest luminaries of what has been called the golden age of ecclesiastical learning, yet their writings were calculated to perpetuate darkness in respect to the sciences.

The passage cited from Lactantius to confute Columbus, is in a strain of gross ridicule, unworthy of so grave a casuist. "Is there any one so foolish," he asks, "as to believe that there are antipodes with their feet opposite to ours; people who walk with their heels upward, and their heads hanging down That there is a part of the world in which all things are topsy turvy: where the trees grow with their branches downward, and where it rains, hails and snows upward? The idea of the roundness of the earth," he adds, "was the cause of inventing this fable of the antipodes, with their heels in the air; for these philosophers, having once erred, go on in their absurdities, defending one with another."

More grave objections were advanced on the authority of St. Augustine. He pronounces the doctrine of antipodes incompatible with the historical foundations of our faith; since, to assert that there were inhabited lands on the oppo site [sic] side of the globe, would be to maintain that there were nations not descended from Adam, it being impossible for them to

> have passed the intervening ocean. This would be, therefore, to discredit the bible, which expressly declares, that all men are descended from one common parent.
>
> Such were the unlooked-for prejudices which Columbus had to encounter at the very outset of his conference, and which certainly relish more of the convent than the university. To his simplest proposition, the spherical form of the earth, were opposed figurative texts of scripture. They observed that in the Psalms the heavens are said to be extended like a skin, that is, according to commentators, the curtain or covering of a tent, which among the ancient pastoral nations was formed of the skins of animals; and that St. Paul, in his Epistle to the Hebrews, compares the heaven to a tabernacle, or tent, extended over the earth, which they thence inferred must be flat.
>
> Columbus, who was a devoutly religious man, found that he was in danger of being convicted not merely of error, but of heterodoxy. Others more versed in science admitted the globular form of the earth, and the possibility of an opposite and habitable hemisphere; but they brought up the chimera of the ancients, and maintained that it would be impossible to arrive there, in consequence of the insupportable heat of the torrid zone. Even granting this could be passed, they observed that the circumference of the earth must be so great as to require at least three years to the voyage, and those who should undertake it must perish of hunger and thirst, from the impossibility of carrying provisions for so long a period. He was told, on the authority of Epicurus, that admitting the earth to be spherical, it was only in habitable in the northern hemisphere, and in that section only was canopied by the heavens; that the opposite half was a chaos, a gulf, or a mere waste of water.[47]

Here are some familiar concepts from Chapter 1: the torrid zone, the Antipodes, and Lactantius.

Indeed, the great irony of Irving's depiction of Columbus as the pioneer of the spherical Earth against the alleged Catholic doctrine of the Flat Earth is that, while not a Flat Earther himself—although, due to a 1498 miscalculation based on assuming the wrong star was Polaris, he did think the Earth to be pear-shaped, with the "nipple" located in the Garden of Eden—Columbus anticipated their methods. Influenced

[47] Washington Irving, *A History of the Life and Voyages of Christopher Columbus, Vol. I* (New-York, G. & C. Carvill, 1828), 75–78.

by the incorrect sizing of the Earth by Arab scholar Ibn Rushd (1126–1198, often Latinized as Averroes) and Pierre d'Ailly, along with the latter's astrological and apocalyptic ideas, Columbus ignored the existing (and generally correct) scientific orthodoxy on the Earth's size, because promoting a smaller Earth—with a correspondingly shorter distance to sail—made it easier for him to promote his case to the Catholic Monarchs Ferdinand and Isabella and receive his funding from them.[48] If not a Flat Earther, Columbus can perhaps be called a "Small Earther."

At this point, it would be instructive to mention that a number of modern Flat Earthers of the late twentieth and early twenty-first centuries have turned to various indigenous American cosmologies to show that Flat Earth belief was present in the Western Hemisphere prior to the arrival of Europeans. In this, Flat Earth believers coexist with various cryptozoologists, New Age spiritualists, and ancient alien theorists who also view indigenous beliefs as a form of "purer" evidence that shore up their own views, a sort of conspiracy theory take on the noble savage trope.[49] It is also true that various indigenous American cultures had worldviews that could be attributed to Flat Earth interpretations.

For example, a number of Northeastern Woodlands cultures in what is now the northeastern United States and southeastern Canada have a creation story involving Turtle Island, the notion that the inhabited world resides upon the back of an enormous turtle swimming in the ocean.[50] Among the Plains Indians and Pueblo peoples, a common belief is that our world is located in between an upper and lower world, and

[48] Jeffrey Burton Russell, *Inventing the Flat Earth: Columbus and Modern Historians* (New York: Praeger, 1991), 5–11, 51–56; Rudolph Simek, trans. Angela Hall, *Heaven and Earth in the Middle Ages: The Physical World before Columbus* (Woodbridge: The Boydell Press, 1996), 24–38; Gómez, *The Tropics of Empire*, 107–58; Bauer, *The Alchemy of Conquest*, 162–63, 174–76; Hamdani, "Islamic Background," 279; Fernández-Armesto, *Conquest of the Impossible*, 36–41; De Camp and Ley, *Lands Beyond*, 289–92, 310–13.

[49] Kenneth L. Feder, *Archaeological Oddities: A Field Guide to Forty Claims of Lost Civilizations, Ancient Visitors, and Other Strange Sites in North America* (London: Rowman & Littlefield, 2019), 214–20; John G. Abizaid, *The Enlightenment of the World, Second Edition* (Boston: John G. Abizaid, 1912), 20; Mark Wilding, "Circle Or Sphere? Inside The UK's First 'Flat Earth' Conference," *Esquire*, October 9, 2018, accessed July 27, 2020, https://www.esquire.com/uk/culture/a22725244/inside-the-uks-first-flat-earth-conference/.

[50] William N. Fenton, "This Island, the World on the Turtle's Back," *The Journal of American Folklore*, Vol. 75, No. 298 (Oct.–Dec., 1962), 283–300; Jay Miller, "Why the World is on the Back of a Turtle," *Man*, New Series, Vol. 9, No. 2 (Jun., 1974), 306–08.

that the region they inhabit is at the center, oriented along the cardinal directions.[51] The Cherokee creation myth starts with Galunlati, the Sky World, over the arch of the Sky-Vault which domed over our world, Elohi, which at the time was only water. When the animals who lived in the Sky World became too crowded, they sought for a new place to live on the water beneath the dome, but nothing existed until Water Beetle dove beneath the surface, bringing up mud which spread to create the solid world beneath the Sky World dome.[52] The *Popol Vuh*, the sacred text of the K'iche' Maya people centered in what is now Guatemala, explicitly describes both the Earth and the Sky as having four sides and four corners.[53] The Ch'orti' Maya, also located in modern Guatemala, likewise had a cosmology incorporating a square Earth and square universe oriented to the cardinal directions.[54]

It can therefore be asserted that at least some pre-Columbian cultures of the Americas had a cosmology that could be described as Flat Earther, and that some of these were colonized by the Europeans very early on—the Maya were conquered by the Spanish followers of Columbus around thirty years after his first voyage. The fact that the Spanish conquest included severe repression of the Maya religion and introduction of Catholicism has led to criticism over how accurate the recording of pre-Columbian religion is. For example, the *Popol Vuh* was transcribed into book form from earlier hieroglyphic codices in the mid-1550s, and which in turn was copied and translated into Spanish by the Dominican friar Francisco Ximénez (1666–1729) in 1701. None of the pre-Ximénez versions of the text remain, making it uncertain how much Ximénez altered (intentionally or otherwise) his translation.[55] The four corners of

[51] Jerry H. Gill, *Native American Worldviews: An Introduction* (Amherst, NY: Humanity Books, 2002), 67–87.

[52] Christopher B. Teuton, *Deep Waters: The Textual Continuum in American Indian Literature* (Lincoln, NE: University of Nebraska Press, 2010), xi–xii.

[53] Allen J. Christenson, trans., *Popol Vuh: The Sacred Book of the Maya* (Norman, OK: University of Oklahoma Press, 2003), 54–55.

[54] John M. Watanabe, "In the World of the Sun: A Cognitive Model of Mayan Cosmology," *Man*, New Series, Vol. 18, No. 4 (Dec., 1983), 721–23.

[55] Christenson, *Popol Vuh*, 15–31; Douglas T. Peck, "Re-Examination of Spanish Colonial Period Documents Related to Prehistoric Maya History and Mythology," *Revista de Historia de América*, No. 136 (Jan.–Dec., 2005), 30–33. For more detail on Maya astronomy and cosmology, see Gerardo Aldana, *The Apotheosis of Janaab' Pakal: Science, History, and Religion at Classic Maya Palenque* (Boulder: University Press of Colorado,

the *Popol Vuh*'s world would seem to correspond to the four corners of the Earth from the Book of Revelations which influenced Cosmas Indicopleustes, as mentioned in Chapter 1. Likewise, the Lenape legend of Turtle Island was first written in 1678 by Jasper Danckaerts (1639–c.1704), leader of a colony of radical Dutch Calvinists in Maryland.[56]

But even this is significant, in that the Spanish Catholics and Dutch Protestants were not adopting or correlating the Flat Earth views of the indigenous peoples they had conquered. To the degree that they noticed the various indigenous concepts of a Flat Earth at all, the colonizers merely recorded. Unlike their twentieth and twenty-first century Christian followers, the sixteenth and seventeenth century Christian colonizers saw no confirmation of their cosmological views with the Maya or Northeastern Woodlands peoples. This is despite the evidence of incorporation of Maya cartographic tropes into at least some Spanish-produced maps, resulting in such hybridized cartographic projects as the 1579 Alfaro Map of Tabasco.[57] But despite the incorporation of Maya cartography, there is no indication of a Flat Earth. Indeed, the first Spanish-language text on the Americas, the 1519 *Suma de Geografía* of Martín Fernández de Enciso (c. 1470–1528), opens with a description of the spherical Earth.[58] By the end of the century, the globe would be included as a symbol of power in portraits of both Iberian and English monarchs.[59]

It is clear that the idea that the Spanish Catholics who authorized the voyage of Columbus believed in a Flat Earth can be put to rest. The Flat Error therefore emerged from Irving's false depiction of Columbus' debate with the officials at Salamanca. But what was the reason for Irving altering the historical facts in that way, given he had access to many firsthand Spanish records from the fifteenth and sixteenth centuries?

2007), 127–80. In 2024, my wedding ceremony was held directly over the crypt in which Ximénez's remains are interred, at the Santo Domingo Church in Antigua, Guatemala.

56 Miller, "Back of a Turtle," 306.

57 Amara L. Solari, "The Relación Geográfica Map of Tabasco: Hybrid Cartography and Integrative Knowledge Systems in Sixteenth-Century New Spain," *Terrae Incognitae: The Journal of the Society for the History of Discoveries*, Vol. 41, No. 1 (2009), 38–58.

58 Ricardo Padrón, *The Spacious Word: Cartography, Literature, and Empire in Early Modern Spain* (Chicago: The University of Chicago Press, 2004), 84–87.

59 Kristen Lippincott, "Power and Politics: The Use of the Globe in Renaissance Portraiture," *Globe Studies*, No. 49/50 (2002), 121–38; Eleanor Stephenson, "Maps, instruments, and maritime power," in Victoria Avery and Jake Subryan Richards, eds., *Black Atlantic: Power, People, Resistance* (London: Philip Wilson Publishers, 2023), 88–90.

One possible explanation is that anti-Catholicism, rampant in the United States of the early- to mid-nineteenth century, reached a fever pitch in the 1830s–1840s when combined with racial panic over large-scale immigration of Irish and Germans and the annexation of one third of Mexico.[60] By portraying Columbus as the proponent of the scientifically accurate spherical Earth against his opponents (recast from geographers to theologians) who were steeped in the Catholicism of the so-called "Dark Ages," Columbus was transmuted from problematic Catholic of southern European origin and employ, to the predecessor of both Martin Luther and Nicolaus Copernicus. Through the historical fiction of Irving, the discovery of the Americas—and ultimately the establishment of the United States—could become the foundational moment in the establishment of modern science and the inception of the Protestant Reformation.

The anti-Catholic impulses driving the Flat Error were reflected in another aspect of American society in the early nineteenth century, as well. In 1837, Danish antiquarian Carl Christian Rafn (1795–1864) published his book *Antiquitates Americanæ*, which argued that the Vinland described by the Norse in several obscure medieval sagas was not simply another mythical island like Antillia or Hy-Brasil, but North America. Rafn's basic idea would eventually be proved correct with the 1960 uncovering of the L'Anse aux Meadows site in Newfoundland, but almost every detail Rafn used to buttress his hypothesis was wrong. This included misidentifying various Native American and colonial artifacts, such as the Newport Tower, Dighton Rock, and Fall River Skeleton in Armor—none of which he viewed in person—as Norse constructs, locating Vinland in New England. Rafn's arguments were extremely popular in New England, where locals believed that the austere, hardscrabble lives of their ancestors reflected the lives of Norse settlers. But across the United States, Rafn's ideas became widely accepted, because they allowed for an alternate origin story. No longer would the "discovery" of the Americas be at the hands of problematically Catholic Southern

[60] Will Sommer, *Trust the Plan: The Rise of QAnon and the Conspiracy That Unhinged America* (New York: HarperCollins, 2023), 105; Colin Dickey, *Under the Eye of Power: How Fear of Secret Societies Shapes American Democracy* (New York: Viking, 2023), 90–111; Luke Ritter, "Nativism and the Roman Catholic Plot to Destroy America," in Luke Ritter, ed., *American Conspiracism: An Interdisciplinary Exploration* (London: Routledge, 2024), 177–96.

Europeans. Instead, its "discoverers" were Norse: northern European, with a democratic republican tradition (the *thing*), and implicitly Protestant. The Norse craze and the Flat Error were two sides of the same pseudohistorical narrative.[61] A pop culture example of this is the 1928 film *The Viking*. The first feature-length Technicolor film, it depicted a fictionalized account of Leif Erikson's life. During his voyage to Vinland, Erikson's pagan crew almost mutiny, based on the belief they will sail over the edge of the Earth; ironically, it is Erikson's rational Christian knowledge of the Earth's sphericity which calms his crew. At the end of the movie, the Vinland colony established by Erikson is explicitly identified with the Newport Tower of Rhode Island, one of Rafn's signature claims; the final scene of the movie is a transition from the Vinland set to contemporary footage of the Newport Tower as "God Bless America" plays.[62]

It is not a coincidence that before the late twentieth-century re-evaluation, the few examples of rejections of the anti-Catholic Flat Error largely came from Catholic scholars at Catholic universities.[63] Interestingly, Flat Earthers also rejected the Flat Error. Charles Johnson, President of the International Flat Earth Research Society, wrote several articles in his newsletter arguing that Columbus actually proved the world *flat*, and in 1980 said to Robert Schadewald,

[61] Annette Kolodny, *In Search of First Contact: The Vikings of Vinland, the Peoples of the Dawnland, and the Anglo-American Anxiety of Discovery* (Durham: Duke University Press, 2012), 213–55; Brian Regal, *The Battle over America's Origin Story: Legends, Amateurs, and Professional Historiographers* (London: Palgrave Macmillan, 2022), 103–40; Brooke-Hitching, *Phantom Atlas*, 170–73; Fernández-Armesto, *Columbus and the Conquest of the Impossible*, 210–12.

[62] *The Viking*, directed by Roy William Neill. Beverly Hills, CA: Metro-Goldwyn-Mayer, 1928; Kenneth L. Feder, *Encyclopedia of Dubious Archaeology: From Atlantis to the Walam Olum* (Santa Barbara, CA: ABC-CLIO, 2010), 193–94; Feder, *Archaeological Oddities*, 112–18. Notably, one of the Native Americans was played by "Iron Eyes Cody" (né Espera Oscar de Corti) in one of his first acting roles; an Italian actor, he claimed to be Native American, and famously played the "Crying Indian" in the 1971 Keep America Beautiful public service announcement.

[63] Francis S. Betten, "The Knowledge of the Sphericity of the Earth during the Earlier Middle Ages," *The Catholic Historical Review*, Vol. 9, No. 1 (Apr., 1923), 74–90; Charles W. Jones, "The Flat Earth," *Thought: Fordham University Quarterly*, Vol. 9, No. 2 (Sep., 1934), 296–307.

> There was a dispute out on the ship, but it was because Columbus was a flat-earther. The others believed the earth to be a ball, and they just knew that they were falling over the edge and couldn't get back. Columbus had to put them in irons and beat them until he convinced them they weren't going over any curve, and they could return. He finally calmed them down.[64]

By the middle of the century, recognition was also coming from those of a different type of speculative occupation. Suitably, given the role of fiction authors in popularizing both the Hollow Earth and Flat Error, the science fiction author L. Sprague de Camp (1907–2000) and the science popularizer Willy Ley (1906–1969) published their *Lands Beyond: A Fascinating Expedition into Unknown Lands* in 1952, critiquing both notions among many others. Their tenth chapter, "The Shape of the Earth," begins with a rejection of the Flat Error and noting that the Salamanca debate was between Columbus and the court philosophers over the size of the Earth, not its shape.[65] Despite that, they follow it up with the statement that "Nearly all primitives and barbarians, unless they have been told otherwise by civilized men, believe that the earth is the shape it looks: a flat disk under a hemispherical blue dome"—incorrect, as the descriptions of non-European cosmologies across this book show.[66]

The same year, American science fiction author Philip José Farmer (1918–2009) published his short story "Sail On! Sail On!," a clockpunk alternate history which depicts Columbus' crew actually sailing over the edge of the Flat Earth and into orbit beyond. However, they are able to communicate back using radios (invented by the sainted Roger Bacon and utilizing the luminiferous aether), and may be picked up by the "flying sausages" which are either Ottoman balloons or UFOs.[67] Farmer's story

[64] Charles K. Johnson, "Columbus Proved World Flat," *Flat Earth News* (Dec., 1978), 3; Charles K. Johnson, "Columbus Proved World Flat," *Flat Earth News* (Sep., 1979), 2; Charles K. Johnson, "Columbus Proved World Flat Didn't Go Over Edge of Ball," *Flat Earth News*, No. 79 (Sep., 1990), 1; Charles K. Johnson, "Columbus Proved Earth Flat," *Flat Earth News*, No. 83 (Sep., 1992), 1; Robert J. Schadewald, "The Flat-out Truth: Earth Orbits? Moon Landings? A Fraud! Says This Prophet," Science Digest, Vol. 88, No. 1 (Jul., 1980), 61.

[65] De Camp and Ley, *Lands Beyond*, 280–81.

[66] De Camp and Ley, *Lands Beyond*, 281.

[67] Philip José Farmer, "Sail On! Sail On!," *Startling Stories*, Vol. 28, No. 2 (Dec., 1952), 55–61.

brings to mind Alfred, Lord Tennyson's 1833 poem "Ulysses," at the end of which the titular hero states that "my purpose holds / To sail beyond the sunset, and the baths / Of all the western stars, until I die."[68] In 1987, science fiction grandmaster Robert Heinlein (1907–1988), one of whose inspirations was Farmer, would use one of the more Flat Error-based lines from Tennyson's poem for the title of his final novel, *To Sail Beyond the Sunset*.[69]

By 1952, it seemed like the notion that Columbus was the pioneer of the spherical Earth over Catholic Flat Earth orthodoxy was so ridiculous it could only be taken seriously in such an absurd science fiction satire. In science fiction, it occasionally reappears. In the 1989 film *Star Trek V: The Final Frontier*—released just a few years before the Columbus Quincentenary—Spock's half-brother Sybok, having turned away from Vulcan logic to embrace emotion, hijacks the USS *Enterprise* to try to find the planet at the center of the galaxy where God lives. Justifying such an insane premise to Captain Kirk, Sybok argues, "The people of your world once believed the world was flat. Columbus proved it was round." The prevalence of the Flat Error into the twenty-third century can perhaps be attributed to Sybok not knowing human history, or as the danger of embracing emotion over logic.[70] Of interest is that the character of Sybok was inspired by 1980s televangelists like Jim Bakker (b. 1940), showing a connection to the Protestant origins of the Flat Error.[71]

The Flat Error appears in more subtle forms in science fiction, as well. One of the modern classics of the genre is Larry Niven's (b. 1938) 1970 novel *Ringworld*, about a massive ring-shaped space habitat built to encircle a distant star. When four explorers from our portion of the galaxy reach Ringworld, they discover that a thousand years earlier, the Ringworld inhabitants had suffered a massive civilizational collapse—a literal fall of civilization, as their flying cities crashed to the ring floor.

[68] Alfred Lord Tennyson, "Ulysses," in *Alfred Lord Tennyson: Selected Poems*, ed. Christopher Ricks (London: Penguin Books, 2007), 50.

[69] Robert A. Heinlein, *To Sail Beyond the Sunset: The Life and Loves of Maureen Johnson (Being the Memoirs of a Somewhat Irregular Lady)* (New York: Putnam Books, 1987), ix.

[70] *Star Trek V: The Final Frontier*, directed by William Shatner. Los Angeles, CA: Paramount Pictures, 1989. This is also referenced in the 2007 film *The Man from Earth*, written by *Star Trek* writer Jerome Bixby.

[71] William Shatner and Chris Kreski, *Star Trek Movie Memories* (New York: HarperCollins, 1995), 278–79.

Amongst the natives living in this "dark age," the explorers encounter one named Seeker, so called because he is on a quest to walk until he reaches the base of the "Arch." From the point of view of the post-fall Ringworld natives, their world is a flat surface, with the ring as an arch going overhead; logically, it must intersect with the ground at the fore and aft. In *Ringworld*, there is therefore a Flat Error suitable for a world whose shape is indeed dramatically different from Earth's, with accompanying "Firmament" (the Arch), "dark age" (which, appropriately, the Ringworld natives do not consider themselves to be living in) and "age of discovery" tropes as well[72] (Fig. 2.2).

Of course, it is not only in the medium of science fiction that the Flat Error lingers on. In 2017, James J. Allegro, an historian of colonial America, claimed that Flat Earth belief was more common than thought

Fig. 2.2 The 'Arch' of the Ringworld, illustrated by Tim Russell[73]

[72] Larry Niven, *Ringworld* (New York: Ballantine Books, 1970), 310.

[73] Tim Russell, "Ringworld Ocean," The Universe of Larry Niven, n.d., accessed March 13, 2023, https://larryniven.net/files/ringworld_ocean-r.jpg.

until Magellan's voyage in 1522. However, Allegro's was a rare example of such a claim, and quickly shut down in turn by Philipp Nothaft, Oxford historian of astronomy in early modern Europe.[74] More substantive was the French philosopher of science Bruno Latour (1947–2002) identifying Europeans as "the Moderns," with the idea of American professor of Transatlantic studies Ralph Bauer that the term Age of Discovery applies to European "discoveries" of both the Americas and science. Spanish literature scholar Ricardo Padrón has likewise noted that the explorers and scholars of the Spanish Empire invented "America" at the same time they invented the modern map, both creations being intimately linked with each other. The conservative American historian Daniel Boorstin (1914–2004), who served as Librarian of the United States Congress 1975–1987—and who used his position to promote the Flat Error in the 1980s—saw Columbus as the first of the Moderns. In contrast, Mexican historian Edmundo O'Gorman (1906–1995) saw Columbus as the last of the Ancients, with Vespucci as the first Modern given his promotion of the continent that now bears his name as a separate entity from the oikoumene of old.[75]

For all of the anti-Catholic sentiment that the Flat Error was tied to and which was so prominent in early American politics, there was one of the Thirteen Colonies that was explicitly founded as a refuge for English Catholics: Maryland. The colony (and later state's) largest city, Baltimore, was named after the colony's founder and proprietor, Cecil Calvert, Second Baron Baltimore (1605–1675), a Catholic member of the House of Lords. Maryland would remain in the Baltimore family until they were overthrown in 1689 by local Protestants in the aftermath of the Glorious Revolution. Under the new regime, Catholicism in Maryland

[74] James J. Allegro, "The bottom of the universe: Flat earth science in the Age of Encounter," *History of Science*, Vol. 55, No. 1 (2017), 61–85; C. Philipp E. Nothaft, "Zaccaria Lilio and the shape of the earth: A brief response to Allegro's "Flat earth science"," *History of Science*, Vol. 55, No. 4 (2017), 490–98.

[75] Bruno Latour, trans. Catherine Porter, *An Inquiry into Modes of Existence: An Anthropology of the Moderns* (Cambridge, MA: Harvard University Press, 2018), 2–23; Padrón, *The Spacious Word*, 1–44; Bauer, *The Alchemy of Conquest*, 1–46. For Boorstin's role in the Flat Error, see Daniel Boorstin, *The Discoverers: A History of Man's Search to Know His World and Himself* (New York: Random House, 1983), 107–15; David Hutchings and James C. Ungureanu, *Of Popes and Unicorns: Science, Christianity, and How the Conflict Thesis Fooled the World* (Oxford: Oxford University Press, 2022), 58–60; Russell, *Inventing the Flat Earth*, 3–4.

was outlawed, only becoming legal again following the religious toleration established by the United States Constitution two centuries later. Given this, it is somewhere between ironic and appropriate that in the late nineteenth century, Baltimore—the former Catholic refuge, home of Benjamin Banneker, center of Reynolds' Hollow Earth advocacy, and setting for Jules Verne's space program—was a major center for Flat Earth thought in the United States.

As mentioned in the introduction, the major organizational Flat Earth movement of the century, the Universal Zetetic Society, was founded in December 1883 and based in London, created by the disciples of the founder of modern-day Flat Earth thought, Samuel Rowbotham. One of Rowbotham's followers was a printer in Kent named William Carpenter (1830–1896). A vegetarian and a Spiritualist, by 1864 Carpenter had begun to print Flat Earth pamphlets. In 1879, he emigrated to Baltimore, where he continued to promote the Flat Earth until his death. Taking the moniker of "Professor" Carpenter, he had a reputation for hanging out at Johns Hopkins University, accosting such visiting figures as the astronomy popularizer Garrett P. Serviss (1851–1929), and after 1885, handing out copies of his pamphlet, *One Hundred Proofs the Earth Is Not a Globe*.[76]

Among the *One Hundred Proofs* are examples drawn from both local Baltimore news, as well as the imperial focus of the Universal Zetetic Society (as will be explored more in Chapter 5). Take for example proofs 4 and 40, with their usage of rivers and canals in Egypt, linking the British Empire and one of the most important locations of the Old Testament:

> There are rivers that flow for hundreds of miles towards the level of the sea without falling more than a few feet—notably, the Nile, which, in a thousand miles, falls but a foot. A level expanse of this extent is quite incompatible with the idea of the Earth's "convexity." It is, therefore, a reasonable proof that Earth is not a globe.... The Suez Canal, which joins the Red Sea with the Mediterranean, is about one hundred miles long; it forms a straight and level surface of water from one end to the other; and

[76] Robert J. Schadewald, *The Plane Truth: A History of the Flat Earth Movement* (Wendy S. Schadewald, 2015), 91–100; Christine Garwood, *Flat Earth: The History of an Infamous Idea* (New York: Thomas Dunne Books, 2007), 162; John Michell, *Eccentric Lives and Peculiar Notions* (London: Thames and Hudson, 1984), 26–27; Kelly Weill, *Off the Edge: Flat Earthers, Conspiracy Culture, and Why People Will Believe Anything* (Chapel Hill: Algonquin Books, 2022), 25–26.

> no "allowance" for any supposed "curvature" was made in its construction. It is a clear proof that the Earth is not a globe.[77]

The Suez Canal would remain popular among Flat Earthers into the late twentieth century, who would see its construction as essentially a large-scale replication of the Bedford Level experiment of Samual Rowbotham.[78]

Proof 66 relied on the false assertion that Ptolemy was a Flat Earther, a notion disproven in the prior chapter:

> It is often said that the predictions of eclipses prove astronomers to be right in their theories. But it is not seen that this proves too much. It is well known that Ptolemy predicted eclipses for six-hundred years, on the basis of a plane Earth, with as much accuracy as they are predicted by modern observers. If, then, the predictions prove the truth of the particular theories current at the time, they just as well prove one side of the question as the other, and enable us to lay claim to a proof that the Earth is not a globe.[79]

Proof 78 takes aim at the HMS *Challenger* expedition of 1872–1876, which laid the foundations of the modern science of oceanography via its extension of British imperial power; ironically, the expedition was promoted by the English zoologist of no relation, William *Benjamin* Carpenter (1813–1885):

> "Yes, but we can circumnavigate the South easily enough," is often said—by those who don't know. The British Ship Challenger recently completed the circuit of the Southern region—indirectly, to be sure—but she was three years about it, and traversed nearly 69,000 miles [111,044 kilometers]—a stretch long enough to have taken her six times round on the globular hypothesis. This is a proof that Earth is not a globe.[80]

[77] William Carpenter, *One Hundred Proofs that the Earth is Not a Globe, Fifth Edition* (Baltimore: William Carpenter, 1886), 6–7, 13.

[78] Charles K. Johnson, "Proof in Egypt," *Flat Earth News*, No. 83 (Sep., 1992), 1–2.

[79] Carpenter, *One Hundred Proofs*, 18.

[80] Carpenter, *One Hundred Proofs*, 20. Through misdated fossil teeth recovered from the sea floor by the *Challenger*, the expedition was also responsible for creating the urban legend that the prehistoric giant shark megalodon is still alive, demonstrating that the

But the most prominent topic in Carpenter's rhetorical repertoire were the midcentury British pioneers of balloon aviation, particularly James Glaisher (1809–1903) and Henry Tracey Coxwell (1819–1900). Between 1862 and 1866, the pair made several ascents together to conduct meteorological studies of the atmosphere. Their 5 September 1862 third flight resulted in them setting a record for human ascent, reaching an estimated eleven kilometers.

Because of their record setting, the two aeronauts achieved a cultural status similar to those of astronauts a century later. Aeronauts' perception of the horizon from such a height still resembling a straight line were of great interest to Flat Earthers—a marked contrast to how astronauts would be perceived by the Flat Earthers of the twentieth century, as will be discussed in Chapter 6.[81] For Carpenter, the experiences of aeronauts like Glaisher and Coxwell formed the basis for proofs 1, 6, 54, and 64:

> The aeronaut can see for himself that Earth is a Plane. The appearance presented to him, even at the highest elevation he has ever attained, is that of a concave surface—this being exactly what is to be expected of a surface that is truly level, since it is the nature of level surfaces to appear to rise to a level with the eye of the observer. This is ocular demonstration and proof that Earth is not a globe.... If we stand on the sands of the seashore and watch a ship approach us, we shall find that she will apparently "rise"—to the extent of her own height, nothing more. If we stand upon an eminence, the same law operates still; and it is but the law of perspective, which causes objects, as they approach us, to appear to increase in size until we see them, close to us, the size they are in fact. That there is no other "rise" than the one spoken of is plain from the fact that, no matter how high we ascend above the level of the sea, the horizon rises on and still on as we rise, so that it is always on a level with the eye, though it

same expedition which created the science of oceanography also contributed to pseudoscience, in this case the field of cryptozoology. Edward Guimont, "The Megalodon: A Monster of the New Mythology," *M/C Journal*, Vol. 24, No. 5, October 6, 2021, accessed January 12, 2022, https://journal.media-culture.org.au/index.php/mcjournal/article/view/2793.

[81] Samuel Rowbotham, *Zetetic Astronomy: Earth Not a Globe, Third Edition* (London: Simpkin, Marshall, and Co., 1881), 36–40; S. L. Kotar and J. E. Gessler, *Ballooning: A History, 1782–1900* (Jefferson, NC: McFarland & Company, 2010), 210–11; Lacy Davis, ed., John Holman, *On the Wings of Ezekiel: The Story of Reverend Burrell Cannon's Remarkable Airship* (Pittsburg, TX: Pittsburg/Camp County Museum Association, 2002), 3; Schadewald, *The Plane Truth*, 25.

> be two-hundred miles away, as seen by Mr. J. Glaisher, of England, from Mr. Coxwell's balloon.... The aeronaut is able to start in his balloon and remain for hours in the air, at an elevation of several miles, and come down again in the same county or parish from which he ascended. Now, unless the Earth drag the balloon along with it in its nineteen-miles-a-second motion, it must be left far behind, in space: but, since balloons have never been known thus to be left, it is a proof that the Earth, does not move, and, therefore, a proof that the Earth is not a globe.... Mr. Glaisher, in speaking of his balloon ascends, says: "The horizon always appeared on a level with the car." Now, since we may search amongst the laws of optics in vain for any principle that would cause the surface of a globe to turn its face upwards instead of downwards, it is a clear proof that the Earth is not a globe.[82]

Carpenter was far from the only Flat Earther to use Glaisher's observation as evidence. In his 1893 pamphlet *The Shape of the World*, A. E. Skellam cites several balloon aeronauts, but particularly Glaisher's report from his 11 October 1862 ascent.[83] When Glaisher died on 7 February 1903, he received an obituary in the UZS journal *The Earth*, titled "He Only Saw the Plane." In it, author H. H. D'Arcy Adams mentions Glaisher and Coxwell's record-breaking 1862 ascent and their perception of the horizon as linear, and makes the interesting argument that

> Without doubt, had Mr. Glaisher been nearer 20 than 66 years of age, and had he possessed the telescopic eyes of a South African Boer, or a Bedouin, he would have seen nearly as far again as he did; in fact his view was only limited by the strength of his vision, as the other conditions recorded were favourable.[84]

In other words, a Boer (Dutch settler of southern Africa) or Bedouin (indigenous inhabitant of northern Africa) would have been able to see the flatness of the Earth just as well as Glaisher reported, a statement with interesting imperial connotations—especially considering the end of the Second Boer War the year prior—as will be unpacked in Chapter 5. Adams' obituary is also curious as he states that Glaisher was 66 years

[82] Carpenter, *One Hundred Proofs*, 6, 7, 16–17, 18.

[83] A. E. Skellam, *The Shape of the World* (London: Harrison & Sons, 1893), 2.

[84] H. H. D'Arcy Adams, "He Only Saw the Plane," *The Earth*, Vol. III, Nos. 33 & 34 (Apr./May, 1903), 179.

old in 1862, when he had actually only been 53. This brings to mind the 2019 film *The Aeronauts*, which portrays a highly fictionalized version of the flight; Coxwell is omitted entirely (replaced by a fictional character played by Felicity Jones) while the 53-year-old Glaisher is portrayed by the 36-year-old actor Eddie Redmayne.[85]

Flat Earthers were not the only conspiracy theorists to take an interest in balloons. Almost as soon as the modern UFO era began in 1947, weather balloons were a favorite explanation for flying saucers, leading to "weather balloon" becoming a pop culture meme. Later in the twentieth century, weather balloons were also a favorite source for chemtrail believers, who use (questionable) observational data from them to try to claim that humidity in the upper atmosphere is too low for contrails to persist and that airplanes are actually spraying nefarious chemicals.[86] Nor were balloons the end of Zetetic interest in flying objects. In 1909, UZS president Lady Blount designed an airship based on the flapping of a bird's wings. Samuel Shenton (1903–1971), who would become the leading Flat Earther in Britain in the midcentury, was led into Flat Earth thought by both designing similarly esoteric airships from 1919 to 1939 while misunderstanding how the rotation of the Earth impacted airplane flights. Charles Johnson, who took over from Shenton as the leading Flat Earther of the late twentieth century, himself worked as an airplane mechanic.[87] These figures, along with modern Flat Earth misconceptions about air and space travel, will be discussed more in Chapters 4 and 6.

But even into the twentieth century, Zetetics did not lose their interest in balloons. The 1918 pamphlet *Is the Earth a Globe Whirling in Space?*, initially written by Karl A. Smith but revised and enlarged by Lady Blount's cofounder Albert Smith, has a section titled "Aeronauts" (Fig. 2.3). It notes that

> Mr. Glaisher, of the Royal Observatory, Greenwich, says:—"The horizon always appears *on a level with* the car."—Mr. Glaisher's report in *Leisure Hours*, Oct. 11th, 1862. The following diagram (Fig. 5) illustrates the phenomena observed by these, and other aeronauts. The horizon A, B, is always on a level with the eye *at any altitude*; and the earth A, C,

[85] *The Aeronauts*. Directed by Tom Harper. Burbank, CA: Mandeville Films, 2019.

[86] Mick West, *Escaping the Rabbit Hole: How to Debunk Conspiracy Theories Using Facts, Logic, and Respect* (New York: Skyhorse, 2018), 125.

[87] Moore, *Venusian*, 18–19; Weill, *Off the Edge*, 35, 57–58, 68.

> B, seems like a great basin beneath the balloon. This is what should be observed in accordance with the laws of perspective, at an elevation above a plane surface. But if the earth were a globe the horizon would gradually fall away from the observer, and would naturally dip downwards more and more as he ascended; so that the supposed curvature of the earth's surface should be distinctly visible at great altitudes, if it existed. As no dip of the horizon is seen, and no curvature observed anywhere, we are bound to conclude that the earth is not a globe; but, that as already proved by observations and experiments, it is a vast extended plane.[88]

However, the Smiths do not rely entirely on Glaisher's account. They also state that

> It will be interesting to hear what is the view of such regarding the shape of the world. To describe this I cannot do better than refer you to Mr. Elliott, an American aeronaut, who, in a letter giving an account of his ascension from Baltimore, U.S.A., thus speaks of the appearance of the earth from an elevated balloon:—"I don't know that I ever hinted heretofore that the aeronaut may well be the most sceptical man about the rotundity of the earth. Philosophy imposes the truth upon us; but the view of the earth from the elevation of a balloon is that of an immense terrestrial basin, the deeper part of which is that directly under one's feet. As we ascend the earth beneath us seems to recede—actually to sink away—while the horizon gradually and gracefully lifts a diversified slope stretching away farther and farther to a line that, at the highest elevation, seems to close with the sky. Thus upon a clear day the aeronaut feels as if suspended at about an equal distance between the vast blue oceanic concave above, and an equally expanded terrestrial BASIN below".[89]

The "Mr. Elliott" is George Elliott (1800–1855), a pioneering American balloonist from Baltimore, who flew across Maryland, Pennsylvania, and Virginia for decades. Like Carpenter, he took the moniker "Professor." In 1834, when forced to cancel a planned ascension in Philadelphia, the assembled crowd attacked and destroyed his balloon. However, this did not put an end to his career. In the final year of his life, he resided in Richmond, Virginia; it is therefore apt that this aeronaut cited by a Flat

[88] Karl A. Smith and Albert Smith, *Is the Earth a Globe Whirling in Space?, Second Edition* (Abington: Albert Smith, 1918), 24.

[89] Smith and Smith, *Is the Earth a Globe*, 23.

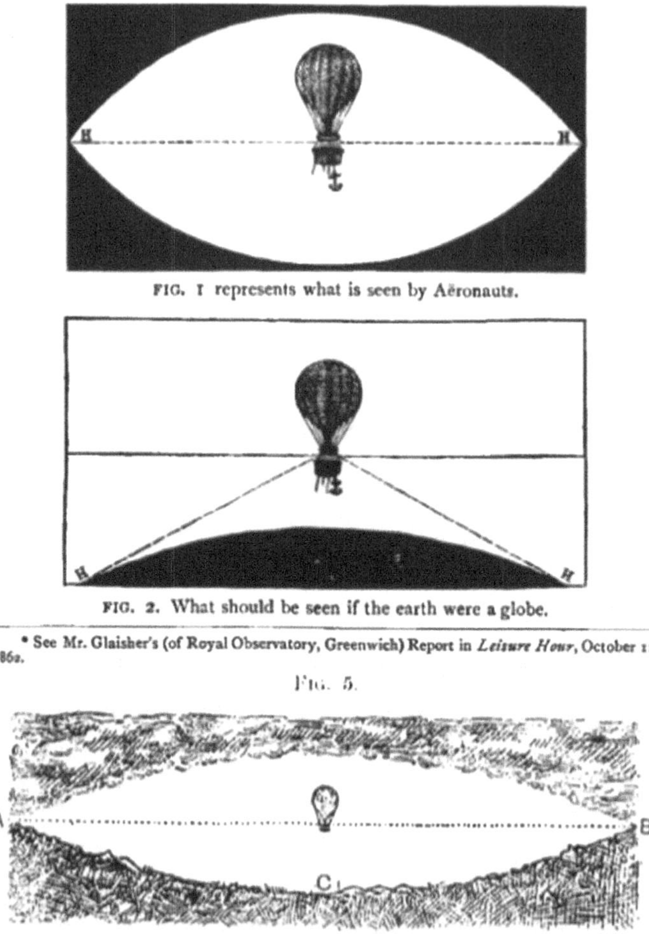

Fig. 2.3 Two usages of illustrations of the horizon by aeronaut James Glaisher by Flat Earthers. The top is from the 1893 pamphlet *The Shape of the World* by A. E. Skellam, and the bottom is from the 1918 pamphlet *Is the Earth a Globe Whirling in Space?* by Karl A. Smith and Albert Smith[90]

[90] Skellam, *The Shape of the World*, 2; Smith and Smith, *Is the Earth a Globe*, 24.

Earther began his life in Baltimore, and ended it in Richmond, the two leading cities in the American Flat Earth movement.[91]

While Richmond's role in the Flat Earth will be discussed next chapter, for now it should be noted that Baltimore's role in Flat Earth belief went beyond Carpenter's adoptive home, or Elliott's ballooning. On 17 July 1873, a New York Zetetic Society (NYZS)—the first organized "Flat Earth society" on Earth—had been established by Dr. George Davey, President of the Ocean Express Steam Navigation Company. While Davey's Flat Earth belief was seemingly genuine, he was also highly interested in publicity, and one of the NYZS' first actions was attempting to raise funds to support the attempt by famous aeronaut "Professor" John Wise (1808–1879) to cross the Atlantic in a balloon.[92] Other notable members of the NYZS council included Hans Powell (1843–1885), an Irish-born medical doctor who had served in the Union (Northern) Army during the US Civil War (1861–1865) and who had just recently been elected Surgeon-General of the Grand Army of the Republic (GAR), the Union veterans' organization. Notably, the GAR as an institution, and white veterans overall, tended to be more supportive of Black political rights than other northern whites, especially in the years immediately after the war.[93] Another doctor among the board was George Henry Colton Salter (1826–1895), who had also served in the Union Army at the start of the Civil War, before being appointed as a United States consul to China by President Abraham Lincoln (1809–1865), continuing in that service after the end of the conflict. Notably, Carpenter's proof 67 involved the sixth-century Grand Canal in China: "Seven-hundred miles is said to be the length of the great Canal, in China. Certain it is that, when this canal was formed, no 'allowance' was made for 'curvature.' Yet the canal is a fact without it. This is a Chinese proof that the Earth is not a globe."[94]

But the most relevant member of the NYZS council was the Reverend John Nelson McJilton (1806–1875). An Episcopal minister, McJilton was appointed rector of Baltimore's St. James African Episcopal Church from

[91] Kotar and Gessler, *Ballooning*, 148–49.

[92] Schadewald, *The Plane Truth*, 92–93; Kotar and Gessler, *Ballooning*, 240.

[93] Stephen A. Goldman, *One More War to Fight: Union Veterans' Battle for Equality through Reconstruction, Jim Crow, and the Lost Cause* (Lanham, MD: Rowman and Littlefield, 2023), 73–106.

[94] Carpenter, *One Hundred Proofs*, 19.

1841 to 1848, despite being white; it was common in Southern states before Emancipation for Black churches to legally be placed under white trustees.[95] The next year, McJilton—who had been a schoolteacher since 1835—was elected Treasurer to the Baltimore Board of School Commissioners, a position which despite the title essentially gave him leadership of the body. McJilton had been a strong advocate for the Union during the war. The year after Union victory, after years of urging on his part, the formal position of Superintendent of Public Instruction in Baltimore was established, and McJilton was elected its inaugural officeholder.[96]

At the start of the nineteenth century, Maryland had 125,222 Blacks, of whom 19,587 were free. By the start of the secession crisis in 1860, it had 171,131 Blacks, of whom 83,942 were free. The proportion of free to slave continued to rise until the Civil War, at which point slavery in the state collapsed more rapidly than anywhere else in the country.[97] Immediately upon Emancipation, education was a primary objective of the freed slaves as a signifier for freedom, but progress was slow. By 1875 only about half of the South's overall children attended school and by 1880, 70% of Southern Blacks remained illiterate.[98] However, with its role as a major center of pre-war free Black population in the South, Baltimore took a leading role in post-war Black education. The Baltimore Association for the Moral and Educational Improvement of the Colored People was one of the groups amalgamated into the American Freedmen's Union Commission at the end of the war, advocating for Black education throughout the South.[99] Before the Civil War, there were at least six

[95] Eric Foner, *Reconstruction: America's Unfinished Revolution, 1863–1877* (New York: Harper & Row, 1988), 88.

[96] W. Bird Terwilliger, "John Nelson McJilton. Humorist, Divine, Educator," *Maryland Historical Magazine*, No. 32 (Dec., 1937), 301–31; Frederick N. Rasmussen, "Judge campaigns to restore reputation of city's first schools superintendent," *The Baltimore Sun*, August 28, 2010. Accessed January 29, 2021. https://www.baltimoresun.com/maryland/bs-xpm-2010-08-28-bs-md-backstory-john-mcjilton-20100828-story.html.

[97] W. E. B. Du Bois, *Black Reconstruction in America, 1860–1880* (New York: Touchstone, 1995 [1935]), 563–64; Gabriel, "The British Emigration Scheme," 167–68; Foner, *Reconstruction*, 39–40.

[98] Heather Cox Richardson, *The Death of Reconstruction: Race, Labor, and Politics in the Post-Civil War North, 1865–1901* (Cambridge, MA: Harvard University Press, 2001), 138.

[99] Robert C. Morris, *Reading, 'Riting, and Reconstruction: The Education of Freedmen in the South, 1861–1870* (Chicago: The University of Chicago Press, 1976), 44.

private schools for free Blacks in Baltimore; after 1865, Northern philanthropists funded additional schools for freedmen. Also in 1865, the state government exerted control over education for the first time, with taxes on Blacks being apportioned to Black schools. From 1868 to 1872, state law required a ten cent tax per 100 dollars, which was the limit of funding for Black education. In 1872, the state appropriated an extra $50,000 for Black schools ($1.3 million in 2025), and then $100,000 ($3.2 million in 2025) in 1878 from the general funding for white schools, which remained in effect until 1888.[100]

It was in this environment that McJilton became the formal head of Baltimore's public education the year after the war's end. Without a formal policy in place, McJilton established two schools for Blacks in the city in the 1866–1867 period. This led to confrontation with the public when McJilton doubled down in favor of the establishment of publicly funded city schools. In July 1867, the pro-Democratic *Baltimore Sun* newspaper not only claimed the military administration of the Reconstruction South was an enormous drain on the national budget ($110 million; $2.4 billion in 2025), particularly in the funding of Black education, but that the control of the military government by Republicans was setting up a one-party state where those who spoke views against the government consensus would be subjected to arrests and fines—a claim that sounds very similar to American conspiratorial conservative attacks (including by Flat Earthers) on public school boards from 2020 on.[101]

While Maryland was not one of the states under Reconstruction occupation, the newspaper signified the shifting of the winds in the state's public opinion. Ultimately, McJilton's advocacy in favor of public Black education cost him his job, and he was removed from his position on 31 January 1868. McJilton relocated to New York City, where he became rector of Madison Street Protestant Episcopal Chapel, but retired within a few months due to ill health. It was in that poor state of health that he joined the NYZS council in 1873. The Society did not long survive McJilton's death in 1875, and there was no subsequent organized Flat Earth movement in the United States at the time of Carpenter's arrival

[100] Du Bois, *Black Reconstruction*, 566.

[101] Richardson, *The Death of Reconstruction*, 58.

in the country at the end of the decade.[102] McJilton's image began to be rehabilitated in the early twenty-first century, but said rehabilitation has ignored his Flat Earth views, perhaps embarrassed that a prominent advocate of Black education also believed the Earth was flat.[103] But there is a potential explanation for that contradiction—a way to square that circle, so to speak.

In her 2022 book on Flat Earth conspiracy culture, Kelly Weill argues that McJilton and his other prominent companions were mocking Rowbotham by agreeing to join the NYZS.[104] With respect to Weill, there seems to be more to it than that. As mentioned above, Powell and Salter were Union Army veterans, and McJilton a supporter of the Union and defender of Reconstruction policies. The British Zetetic movement that Carpenter came from included several Afro-Caribbean leaders and promoted pro-Black policies in southern Africa, as will be discussed over the next two chapters. But there was an even more prominent and local figure in the American Flat Earth movement who connected the idea that the Earth was flat with the defense of the emancipated slaves of the American South—a man who personified the American Flat Earth idea in the late nineteenth century, even if by the twenty-first century he has largely been overlooked in favor of Carpenter. That man was John Jasper, a Baptist preacher and former slave in Richmond, Virginia.

[102] Terwilliger, "John Nelson McJilton," 310–11; Rasmussen, "Judge;" Schadewald, *The Plane Truth*, 93.

[103] Erica L. Green, "Baltimore educators recognize official who started black schools," *The Baltimore Sun*, November 12, 2010. Accessed July 31, 2020. https://www.baltimoresun.com/news/bs-xpm-2010-11-12-bs-ci-first-superintendent-recognitio20101112-story.html.

[104] Weill, *Off the Edge*, 25.

CHAPTER 3

"The Unmatched Negro Philosopher"

John Jasper (1812–1901) became nationally famous for preaching a sermon titled "The Sun Do Move and the Earth Am Flat," which used the Old Testament as evidence to support the notion that the Earth was flat and orbited by the Sun. First given to his Richmond, Virginia congregation in early March 1878, Jasper would preach the sermon an additional 252 times during his life, the final time shortly before his death at the start of the twentieth century on 30 March 1901. These included sermons where notable individuals from both white Southern and Northern society, as well as Europeans, visited his largely Black Sixth Mount Zion Baptist Church in Richmond. He also engaged in national tours to the North, preaching the sermon to primarily white audiences, all while charging for attendance and soliciting donations. Through this income stream and network of connections, Jasper turned his church into a center of Black political activism in Richmond. In 1954, Black historian Rayford Logan (1897–1982) coined the term "nadir of American race relations" to refer to the period of Black oppression from the 1877 end of Reconstruction to 1901.[1] This almost exactly coincides to the period when Jasper preached "The Sun Do Move." That at such a period of

[1] Rayford Logan, *The Negro in American Life and Thought: The Nadir, 1877–1901* (New York: Dial Press, 1954), 3–11.

E. Guimont, *The Power of the Flat Earth Idea*, Palgrave Studies in the History of Science and Technology, https://doi.org/10.1007/978-3-032-23981-5_3

race relations in the country, a former slave preacher was able to achieve such prominence by promoting Flat Earth views speaks to the ideological power of the belief.

There are two main sources of biographical information on Jasper. The first substantive biography of him—with the suitably substantive title *The Life of Rev. John Jasper, Pastor of Sixth Mt. Zion Baptist Church, Richmond, Va., from His Birth to the Present Time, with His Theory on the Rotation of the Sun*—was published in 1884 by Edwin Archer Randolph. Randolph had been born a free Black in Virginia in 1854. In 1880, he became both the first Black graduate of Yale Law School and the first Black lawyer in Connecticut, before returning to Virginia in 1881 to take up a political career and write for the Black-owned *Richmond Planet* newspaper. It was on his return to Richmond that Randolph befriended Jasper, and based his biography on firsthand information gained from Jasper, as well as others in Richmond who knew him.[2]

The second major biography—*John Jasper: The Unmatched Negro Philosopher and Preacher*—was published in 1908 by William E. Hatcher (1834–1912), a white Baptist preacher in Richmond who also knew Jasper and called him a friend. Hatcher makes his perspective clear from the start of his work, in his initial description of Jasper:

> The man in question was a negro, and if you cannot appreciate greatness in a black skin you would do well to turn your thoughts into some other channel.... But do not be afraid that you are to be fooled into the fanatical camp. This story comes from the pen of a Virginian who claims no exemption from Southern prejudices and feels no call to sound the praises of the negro race. Indeed, he never intended to write what is contained within the covers of this book. It grew up spontaneously and most of the contents were written before the book was thought of.[3]

[2] Edwin Archer Randolph, *The Life of Rev. John Jasper, Pastor of Sixth Mt. Zion Baptist Church, Richmond, Va., from His Birth to the Present Time, with His Theory on the Rotation of the Sun* (Richmond: R. T. Hill & Co., 1884), v-vi. In 1882, an anonymously-written two-page biography of Jasper was included in a pamphlet printing of his sermon from New York City: John Jasper, *"The Sun Do Move!" The Celebrated Theory of the Sun's Rotation Around the Earth* (New York: Brentano's Literary Emporium, 1882), 14–15.

[3] William E. Hatcher, *John Jasper: The Unmatched Negro Philosopher and Preacher* (New York: Fleming H. Revell Company, 1908), 7–8.

Indeed, the significance of Hatcher's sympathy to Jasper is put into perspective by the fact that Hatcher had published a defense of lynching in 1894, which brought about a condemnation from Randolph's *Richmond Planet*.[4] Both Randolph and Hatcher's works are generally sympathetic to Jasper, showing that his appeal could cross to both educated Black and white Southern elites. Perhaps not surprisingly, Randolph's biography is generally more accurate and detailed, but Hatcher's includes a longer perspective given its publication date.[5]

Randolph states that Jasper was born as a slave in Fluvanna County, Virginia, on 4 July 1812 (Hatcher ignores all earlier aspects of Jasper's life). His parents were Philip and Tina, slaves of the Peachy estate, and Jasper was the last of their astounding twenty-four children. Philip was a Baptist preacher of great renown in the local slave community, but did not pass any of this on to his son, as he died two months before John's birth. John became a "mamma's baby;" Tina was also of relatively high status amongst the estate slaves, being the leader of the field workers, and then taken into the house, while also serving as a healer whose skill was supposedly beyond many white doctors. She survived to become a free woman, dying in 1867.[6]

Jasper later saw his birth date as auspicious. 4 July was both the anniversary of the 1776 Declaration of Independence and the simultaneous 1826 deaths of Founding Fathers Thomas Jefferson (b. 1743) and John Adams (b. 1735); Jasper would later view himself as converting to Christianity on 4 July; and he was born at the start of the War of 1812 which was interpreted as the second war of independence against Britain.[7]

[4] "Rev. Dr. Hatcher's Surprising Assertions," *Richmond Planet* (Jun. 23, 1894), 2; https://encyclopediavirginia.org/entries/rev-dr-hatchers-surprising-assertions-richmond-planet-june-23-1894/ (accessed 24 December 2022).

[5] Christopher Z. Hobson, "The Lord Is a Man of War: John Jasper, Covenant, and Apocalypse," *African American Review*, Vol. 44, No. 4 (Winter, 2011), 619; Thomas Strange, "A black and white sermon: John Jasper's capitalization on white exploitation in late nineteenth-century Virginia," *American Nineteenth Century History*, Vol. 18, No. 3 (2017), 276.

[6] Randolph, *The Life of Rev. John Jasper*, 3–5, 25–26; Howard H. Harlan, *John Jasper: A Case History in Leadership* (Charlottesville: University of Virginia, 1936), 16.

[7] Randolph, *The Life of Rev. John Jasper*, 1–3, 9–10; Isaac James, *"The Sun Do Move:" The Story of the Life of John Jasper* (Richmond: Hunter's Letter Shop, 1954), 8; Jasper, *"The Sun Do Move!"*, 14; William J. Simmons, *Men of Mark: Eminent, Progressive and Rising* (Cleveland: George M. Rewell & Co., 1887), 1064.

Randolph recounts the impact of an 1831 meteor shower on the young Jasper, which occurred four years after Jasper had been

> brought back to Richmond and hired out to Mr. Samuel Cosby, to work in the [tobacco] factory, then situated on the corner of Sixteenth and Cary streets. He worked in this factory about six or seven years, during which time stars fell. The falling of the stars, as it was so termed in this part of the world, excited the colored people very much. I remember, when I was a little fellow, I used to hear the old people talking about the time when the stars fell, and I always thought I would liked to have seen the sight, grand and beautiful as it must have appeared. Mr. Jasper, who has always made a study of the planetary system, using no instruments at all save the natural eye, was the first, of course, among his co-laborers in the factory, to discover the falling of the stars; he called the attention of the other hands to the fact. This was in the night season, about two hours before the break of day, though the slaves were at work (it was nothing unusual for the slaves to be at their work two or three hours before daylight); they watched the falling stars until nigh the rising of the sun. The next year [1832] there was a great epidemic in the land of Virginia, known as the Cholera year. This disease was so disastrous among the colored people, that it caused a very large number of them, as they used to say, "to sit out to pray," believing that the last day was not far off. While their views were by some considered not at all intelligent, but rather superstitious, a great good was the result and many were added unto the church.[8]

From a young age, Jasper was therefore aware of how celestial events could be interpreted as either apocalyptic or liberatory for the Black population. In this, he was far from alone among the Antebellum country.

Nat Turner (1800–1831), a slave in southern Virginia, saw the solar eclipse of 12 February 1831 as a sign to start his uprising.[9] As a young man in 1833, Abraham Lincoln saw the Leonid meteor shower, whose 12 November appearance that year was notable for the number and intensity

[8] Randolph, *The Life of Rev. John Jasper*, 6–7. For details on the Richmond tobacco industry whose labor Jasper helped enrich, see Richard C. Wade, *Slavery in the Cities: The South 1820–1860* (New York: Oxford University Press, 1964), 33–35, and Selden Richardson, *Built by Blacks: African American Architecture and Neighborhoods in Richmond* (Charleston, SC: The History Press, 2008), 49. For details on the Richmond cholera outbreak of 1832, see Wade, 135–36.

[9] Gloria D. Rall, "Follow the Drinking Gourd," *The Planetarian*, Vol. 23, No. 3 (Sep., 1994), 11.

of meteors falling. Lincoln's companions thought they represented the end of the world but thirty years later during the Civil War, the president thought their regularity meant the nation would be preserved. The escaped slave and abolitionist Harriet Tubman (1822–1913) also saw the Leonids as a child with other slaves who thought they were a sign of the end of the world, though she herself did not. Tubman's fellow escaped slave and abolitionist Frederick Douglass likewise did not associate the 1833 Leonids with the end of the world, but with the Second Coming. Writing in February 1859, free Black physician Martin Delany (1812–1885) compared the fugitive bodies of comets to the fugitive bodies of escaped slaves. The Great Comet of 1860, visible since fall 1859, was linked to the Harper's Ferry raid of October 1859 by figures such as Walt Whitman, Henry David Thoreau, and Herman Melville. In July 1860, Charles Grady, a slave in Mississippi, saw a meteor shower and assumed it was a portent of the coming Civil War. The Great Comet of 1861, visible in July, three months after the war's start, was seen as a symbol of the outbreak of fighting.[10]

Writing in 1887, the emancipated slave, theologian, and biographer of Blacks William J. Simmons (1849–1890) went so far as to say of Jasper's encounter,

> Had John Jasper been an educated man he would surpass Herschel, Kepler, Halley, Encke, Biela or Faye, for he is of an astronomical turn of mind, and his soul is never happier than when he revels in the beauty of the starry heavens. The great event known as the falling of the stars occurred while he was at work at this place [Cosby's tobacco factory], and John was the first among his fellows to see this magnificent sight. While to others it was a source of terror, to him it was a scene of joy.[11]

Being compared to such astronomers as Sir John Herschel (1792–1871, whose exploits will be discussed in chapter 5), Johannes Kepler (1571–1630, theorist of the laws of planetary motion), Halley (1656–1742, namesake of the famous comet), Johann Franz Encke (1791–1865, comet discoverer), Wilhelm von Biela (1782–1856, another comet discoverer),

[10] Thulani Davis, *The Emancipation Circuit: Black Activism Forging a Culture of Freedom* (Raleigh, NC: Duke University Press, 2022), 20, 343; Rusert, *Fugitive Science*, 164.

[11] Simmons, *Men of Mark*, 1065.

and Hervé Faye (1814–1902, yet another comet discoverer) was a higher honor indeed.

Being placed in such astronomical company was especially an honor considering that after the Civil War, the network of Black colleges established in the South did not promote astronomy, as this was considered one of the spurious "peripheral sciences" seen at odds with the practical industrial education promoted by Booker T. Washington (1856–1915).[12] But prior to the war, slaves were well familiar with some tenets of astronomy, particularly the use of the Big Dipper constellation and the North Star, Polaris, it pointed to. Using this guide, slaves could head north—a method which only works on a spherical Earth. Such was this association that Douglass named his 1847–1851 newspaper *The North Star*.[13] In the twentieth century, the slave use of the Big Dipper and Polaris became associated with the spiritual "Follow the Drinking Gourd," with the notion that the "drinking gourd" represented the Big Dipper and was meant as a covert way of teaching slaves practical astronomy their owners would have forbidden. In this conception, which was popularized and spread by white Texan folklorist H. B. Parks (1879–1958) in 1928, slaves were taught the song by an itinerant white man named Peg-Leg Joe, who wandered from plantation to plantation.[14]

In 2008, English professor James B. Kelley challenged this interpretation, arguing that much of it was invented by Parks and embellished by others later as a way to promote astronomy to Black students. However, Kelley acknowledges that the song may still have served "as a synonym (but not a code term) for the Big Dipper" and reflecting the desire of slaves to escape to the North, as opposed to secret instructions on how to do so. Kelley proposed that the song's reference to quail reflects their presence in the Book of Exodus of the Old Testament, a passage relevant to slaves seeking freedom—see for example Tubman's nickname

[12] Fikes, "Careers of African Americans in Academic Astronomy," 132.

[13] Rall, "Follow the Drinking Gourd," 11; Rusert, *Fugitive Science*, 123–26, 168–69; John Wesley Cromwell, *The Negro in American History* (Washington, D.C.: The American Negro Academy, 1914), 38–39.

[14] H. B. Parks, "Follow the Drinking Gourd," *Publications of the Texas Folk-Lore Society*, No. 7 (1928), 81–84; James B. Kelley, "Song, Story, or History: Resisting Claims of a Coded Message in the African American Spiritual "Follow the Drinking Gourd"," *The Journal of Popular Culture*, Vol. 41, No. 2 (Apr., 2008), 265; Rall, "Follow the Drinking Gourd," 10–11. This Parks is not to be confused with the Black Episcopal clergyman Henry Blanton Parks (1856–1936).

of Moses. Most intriguingly, Kelley suggested that Peg-Leg Joe may reflect the figure of Papa Legba, a West African deity adopted by voudon practitioners in the Americas. "Follow the Drinking Gourd" may therefore serve as a combination of astronomical knowledge of the global Earth, West African knowledge, and Old Testament allegory, a mix that will be extremely potent to the Flat Earth conceptualization that Jasper established.[15]

For his part, there is no evidence that Jasper knew "Follow the Drinking Gourd," and he certainly did not escape along the Underground Railroad, remaining a slave in Virginia. While "walking through [Richmond's] Capitol Square, on the Fourth day of July, in the year 1839, in the midst of the great crowds of white people, while they were celebrating this memorable and historic day," Jasper became unsettled and convinced he was a great sinner, which led to him to proclaiming his faith and being proclaimed a Baptist minister in 1840.[16] It was through preaching in public that Jasper first began to gain the respect of white people, who defended him from "some white ruffians" accosting him during his sermon. Jasper's street preaching would thereafter often attract more whites than Blacks.[17]

Within a few years, Jasper was selected as the delegate of the Black Baptist churches in Richmond to the Foreign Mission Board, a Southern Baptist organization established in 1845. The Foreign Mission Board was headquartered in the city of Petersburg, Virginia, forty kilometers from Richmond, and Jasper annually was selected to take the donations from Richmond's Black Baptists to the Foreign Mission Board annual meeting. The donations that Jasper took were specifically earmarked for missionary work in Africa.[18] From the available record, the collection of missionary funds for African missions appears to be Jasper's only direct connection to Africa itself, an early model of African development promoted by Edward

[15] Kelley, "Song, Story, or History," 274–75.

[16] Randolph, *The Life of Rev. John Jasper*, 9–10; Harlan, *John Jasper*, 17–18.

[17] Randolph, *The Life of Rev. John Jasper*, 14–15.

[18] Randolph, *The Life of Rev. John Jasper*, 18. Since 1997, the Foreign Mission Board has been known as the International Mission Board. For more on the Foreign Mission Board and Richmond Blacks, see Floyd J. Miller, *The Search for a Black Nationality: Black Emigration and Colonization, 1787–1863* (Urbana, IL: University of Illinois Press, 1975), 68–69.

Blyden.[19] As with his preaching, the ability to travel to other cities with the possession of large sums of money was a degree of freedom that many slaves did not experience (even within the looser system of urban slavery, in contrast to the life of rural plantation slaves), and which stemmed from Jasper's skill as a preacher.[20]

Starting approximately 1859/60, Jasper's owner hired him out to work at a mill located along the James River in Richmond. Jasper continued to work there throughout the Civil War, during which he also preached to wounded Confederate soldiers in a military hospital in the city. On 3 April 1865, Richmond was captured by Union forces, and Jasper became a free—albeit impoverished—man.[21] Retreating Confederate soldiers set fire to their former capitol on 2 April, causing Richmond to become enveloped in flames that still raged when President Lincoln toured the conquered city on 4 April, meeting with thousands of freed slaves. There is no evidence that Jasper was among those who viewed the Great Emancipator, but these were the conditions of Jasper's introduction to freedom—a would-be national capitol that was burning so heavily, Union soldiers compared it to an erupting volcano.[22]

After several months earning a living by cleaning bricks from the city streets, Jasper spent time at Baptist churches in Petersburg and Weldon, North Carolina from July 1865 to September 1867.[23] In the aftermath of the Civil War, Emancipation, and the implementation of Reconstruction, the composition of Southern churches was rapidly changing. Before the war, mixed-race but segregated denominations were common. After 1865, Black churches rapidly multiplied with the departure of Blacks from mixed congregations, stemming from a new Black desire for autonomy and unwillingness of whites to treat Blacks equally. Richmond, which had only four Black churches in 1860—the first from only 1823—was

[19] Blyden, *Christianity, Islam and the Negro Race*, 30–53; Kelley, *Freedom Dreams*, 22–23.

[20] Wade, *Slavery in the Cities*, 80–82, 241–46.

[21] Randolph, *The Life of Rev. John Jasper*, 27–28; Hatcher, *The Unmatched Negro Philosopher*, 97–100.

[22] Brenda Wineapple, *Ecstatic Nation: Confidence, Crisis, and Compromise, 1848–1877* (New York: HarperCollins, 2013), 360–63.

[23] Randolph, *The Life of Rev. John Jasper*, 28–30.

no exception.[24] Not only were the number of Black churches expanding rapidly, but their leadership was changing as well. All four of those Black churches had white pastors until 1866, when former slave James Henry Holmes (1826–1900) took over leadership of the First African Baptist Church, formed in 1841 from the Black congregation of the city's 1780 First Baptist Church.[25]

Because Black churches were one of the few institutions of Black autonomy before and after Reconstruction, Black preachers had unique authority in Southern Black society, even in contrast to preachers' roles in Southern white society.[26] Jasper would soon join that elite position. On 1 September 1867, Jasper was called to lead one of the new Black churches, an outgrowth from First African Baptist's expanding congregation. Jasper's new Sixth Mount Zion Baptist Church was initially located in a small wood shack that had been used as a government stable, but the increasing size of his congregation required them to move several times. The church relocated to its current position in 1890, in the city's Jackson Ward. Jackson Ward was popularly known as "Africa" due to its large Black population, which enabled Blacks to control its local electoral politics.[27]

The origin of Jasper's Flat Earth sermon came from a dispute between two of his parishioners in early 1878. A Black man named Lester Woodson, a sexton of the church, and an unidentified second man got into an argument which Randolph categorizes as "a hot conversation about the rotation of the sun, or about the meaning of the third verse of the fifteenth chapter of Exodus, 'The Lord is a man of war, the Lord is his name.'" Hatcher states (quoting testimony from another of Jasper's flock, "a truly excellent woman" aged seventy named Virginia Adams) that "Dey got inter a spute es to wheddur de sun went 'roun' de wurl' ur not." Already a difference, though slight, is evident in the origins of the

[24] Foner, *Reconstruction*, 88–90; Wade, *Slavery in the Cities*, 166–67; Richardson, *Built by Blacks*, 144–48.

[25] Albert J. Raboteau, *Slave Religion: The "Invisible Institution" in the Antebellum South* (New York: Oxford University Press, 1978), 196–97; Federal Writers' Project, *The Negro in Virginia* (New York: Hastings House Publishers, 1940), 250.

[26] Katharine L. Dvorak, *An African-American Exodus: The Segregation of the Southern Churches* (Brooklyn: Carlson Publishing, 1991), 34–39.

[27] Randolph, *The Life of Rev. John Jasper*, 30–33; Hatcher, *The Unmatched Negro Philosopher*, 58–64; Richardson, *Built by Blacks*, 148–49.

sermon. Another is present in who the second parishioner in the argument was. Randolph states that it was an unidentified white man, while Hatcher provides no details. Later historians using Sixth Mount Zion records identified him as a white man named Frank Combs. The white Hatcher may have elected to keep that detail out due to embarrassment of Flat Earth belief not being limited to the Black population. But the account shows that Jasper's renown was such that he still managed to attract whites to his Black-led church; the white and Black parishioners debated theological matters as equals; and Flat Earth thought preexisted in the congregation—indeed, Jasper absorbed it from them. Woodson and Combs asked Jasper to mediate the dispute, and the result was the first preaching of the sermon that was originally called "The Lord is the God of War," but became universally known as "The Sun Do Move."[28]

The key arguments Jasper cites for "The Sun Do Move" are taken from a variety of Biblical sources, mostly Old Testament. The only New Testament source is the Book of Revelation, chapter seven, verse one, which states that four angels stand on the four corners of the Earth, requiring it to be flat and square if taken literally. Revelations, with its apocalyptic imagery, had become central to Southern Black Christianity before and after Reconstruction.[29] Its prominence in Black theology may have given a literal interpretation of its geography more weight to both Jasper and his audience. From the Book of Jeremiah, chapter 31, verse 37, comes the idea that "heaven above can be measured, and the foundations of the earth searched out beneath." Jasper's longest evidential source was taken from the Book of Joshua, chapter ten, verses twelve through fourteen—notably, the same passages that Martin Luther used to critique Copernican heliocentrism three centuries earlier:

> Then spake Joshua to the Lord, in the day when the Lord delivered up the Amorites before the Children of Israel, Sun, stand thou still on Gibeon, and thou moon in the valley of Ajalon. And the sun stood still, and the moon stayed, until the people avenged themselves upon their enemies. And

[28] Randolph, *The Life of Rev. John Jasper*, 34–35; Hatcher, *The Unmatched Negro Philosopher*, 76–78; Nancy B. Miner, "De Sun Do Move: The Story of John Jasper," *Fluvanna History: A Biannual Publication of the Fluvanna County Historical Society*, No. 73 (2003), 25; Strange, "A black and white sermon," 273, 286n1.

[29] Dvorak, *An African-American Exodus*, 25–34.

> there was no day like that before it or after it, that the Lord hearkened unto the voice of a man; for the Lord fought for Israel.

Jasper's other examples are drawn from literal readings of the Books of Exodus, Malachi, Isaiah, Ecclesiastes, Judges, and Psalms. Most of these are merely statements about the Sun rising above and setting below the Earth. Jasper also raises practical concerns to reinforce his religious arguments, such as the idea that it is impossible to chart the distance between the Sun and the Earth because no tape measure would be long enough and the prospective cosmic prospector could not take enough food to last such a journey.[30] This seems to speak to the fact that Jasper was not aware of contemporary science fiction by the likes of Poe, Verne, or Congregationalist minister and abolitionist Edward Everett Hale (1822–1909), as mentioned above.[31]

Jasper's arguments therefore are that the Bible says the Earth is a flat square, that the Sun passes above and below the Flat Earth, and that in particular the Book of Joshua serves as an example through its depiction of God stopping the Sun overhead to help the Israelites defeat the Gibeonites. Ironically, in his 1615 letter to Christina of Lorraine (1565–1637), Grand Duchess of Tuscany, Galileo argued that the Copernican heliocentric system was the only way the stopping of the Sun in the Book of Joshua could be explained as described.[32] Jasper was also not the only nineteenth-century Flat Earther to draw from Joshua. In 1890, Charles A. L. Totten (1851–1908), an advocate of British Israelism (discussed in the next chapter), published a work that he claimed was a scientific endorsement of the Sun's ability to halt as described in Joshua. At the time, Totten was a military officer assigned to Yale University to teach military science as a program analogous to the modern Reserve Officers' Training Corps. Despite not being a member of the Yale faculty proper, Totten and his supporters depicted him as such, giving him academic credentialism much as Carpenter and Rowbotham styled themselves as "Professors."

[30] Randolph, *The Life of Rev. John Jasper*, 47–51; Hatcher, *The Unmatched Negro Philosopher*, 133–49.

[31] Hale served as Chaplain of the US Senate from 1903 until his death. His 1869–1870 serialized novel "The Brick Moon" was the first depiction of both a space station and communications satellite, and the longest work of any kind about a satellite until 1952.

[32] Jean Dietz Moss, "Galileo's Letter to Christina: Some Rhetorical Considerations," *Renaissance Quarterly*, Vol. 36, No. 4 (Winter, 1983), 547–76.

Totten's work was endorsed by Carpenter's Flat Earth companions in Britain as scientific evidence of Biblical truth, particularly that the Sun orbits the (flat) Earth.[33]

Jasper's usage of those particular Old Testament sources fits in with the evolving usage of the Bible for Black political ends after the Civil War, and particularly after the failure of Reconstruction. As mentioned above, during the period of slavery, the Book of Exodus—also the model for the Flat Earth of Cosmas Indicopleustes, as described in chapter one—tended to be highlighted, with God delivering the Israelites from Egyptian slavery into the Promised Land seen as a metaphor for the deliverance of Black slaves from the South to the Promised Land of the North (or Canada, Mexico, Haiti, Liberia, or Guiana).[34] However, Jasper's sermon came at a moment of change in race relations in both the state and the country. As a result of the disputed presidential election of 1876, Democrats allowed Republican candidate Rutherford B. Hayes (1822–1893) to become President, in exchange for him then announcing the end of Reconstruction in the South on assuming the office. Jasper's sermon would have been given not long after Hayes' 4 March 1877 inauguration. While the Federal military occupation of Virginia had ended in 1870 and the state had been in Democratic hands since, Hayes' ending of Reconstruction spurred the rise of the Readjuster Party, a moderate biracial party which held power in Virginia from 1879–1883. Jasper—a supporter of the Radical Republicans, as were the overriding majority of Blacks at the time—created "The Sun Do Move" in between those two events.[35] Spurred by the fear that a Democrat would win the Presidency in the 1880 election and work with the Democratic majorities in Congress to further erode Black civil rights,

[33] Charles A. L. Totten, *Joshua's Long Day and the Dial of Ahaz: A Scientific Vindication and "A Midnight Cry"* (New Haven: The Our Race Publishing Company, 1890); Albert Smith, "Professor Totten and Joshua's Long Day," *The Earth not a Globe Review*, No. 7 (May, 1894), 145–49; Michael Barkun, *Religion and the Racist Right: The Origins of the Christian Identity Movement* (Chapel Hill: University of North Carolina Press, 1997), 18–21; Ian Richard Netton, *Islam, Christianity and the Realms of the Miraculous: A Comparative Exploration* (Edinburgh: Edinburgh University Press, 2020), 158–60; Ronald Fritze, *Invented Knowledge: False History, Fake Science, and Pseudo-Religions* (London: Reaktion Books, 2009), 115–16.

[34] Kelley, *Freedom Dreams*, 16–17; Hobson, "The Lord Is a Man of War," 622–26; Strange, "A black and white sermon," 275; Davis, *The Emancipation Circuit*, 9.

[35] Randolph, *The Life of Rev. John Jasper*, 105; Hobson, "The Lord Is a Man of War," 621–22; Davis, *The Emancipation Circuit*, 104–05.

the period of 1877–1879 saw a revival of migration proposals amongst Southern Blacks. While some of them were interested in Liberia, the primary wave that emerged in 1879 was called the Kansas Fever Exodus. Its participants, termed Exodusters, were the first widescale migration of Blacks out of the South.[36]

One of the Exodusters was William Saunders Crowdy (1847–1908), an escaped slave from Maryland and Union Army veteran. In 1893, Crowdy became one of the founders of what are today known as the Black Hebrew Israelites, another alternative interpretation of Old Testament Israelite narratives among post-Emancipation Blacks which argued that Africans had been the original Israelites and modern Jews are interlopers who had stolen the identity with their Kingdom of Israel.[37] Another Black Hebrew Israelite was the self-styled Prophet Frank S. Cherry (1875–1963), who in 1915 founded the Church of the Living God, the Pillar Ground of Truth for All Nations in Philadelphia. However, Cherry claimed he had established his church in 1886 in Chattanooga, Tennessee as a way to establish primacy over the claims of Crowder's 1896 establishment of his Church of God and Saints of Christ in Kansas, widely recognized as the first Black Hebrew Israelite congregation. The details of Cherry's life are difficult to ascertain for certain as he gave contradictory, if not outright fabricated, accounts to different people, and early biographers of him further complicated matters by not leaving clear citations of their sources. This includes the facts that Cherry was seemingly never resident in Tennessee, and was many years younger than he later claimed to be. But Cherry does seem to have been self-educated, experienced as both a sailor and railroad worker, genuinely fluent in both Yiddish and Hebrew, and motivated by extreme hatred of the Southern Jim Crow system.[38]

[36] Nell Irvin Painter, *Exodusters: Black Migration to Kansas After Reconstruction* (New York: W. W. Norton & Company, 1976), 106–07, 184–211; Foner, *Reconstruction*, 598–600; Isabel Wilkerson, *The Warmth of Other Suns: The Epic Story of America's Great Migration* (New York: Random House, 2010), 161.

[37] Jacob S. Dorman, *Chosen People: The Rise of American Black Israelite Religions* (Oxford: Oxford University Press, 2013), 23–55; Elly M. Wynia, *The Church of God and Saints of Christ: The Rise of Black Jews* (New York: Garland Publishing, 1994), 19–28. For a general overview of alternative Black religions in the wider American context of the early twentieth century, see Barry Hankins, *Jesus and Gin: Evangelicalism, the Roaring Twenties and Today's Culture Wars* (New York: Palgrave Macmillan, 2010), 149–70.

[38] Wynia, *Church of God and Saints*, 15–17; Dorman, *Chosen People*, 91; Arthur Huff Fauset, *Black Gods of the Metropolis: Negro Religious Cults of the Urban North*

Cherry believed not only that the Earth was only six thousand years old, but that every two thousand years a major transition occurred. The first had been the Noachic flood, the second Christ, and the third would be Christ's Second Coming in the year 2000. To Cherry's followers—who had a large overlap with support for Pan-Africanist Marcus Garvey (1887–1940)—the Black Israelites had been driven from Palestine to West Africa by the Roman Empire, and sold into slavery in the Americas as punishment for disobeying the commandments laid out by God in Deuteronomy 28. The Second Coming would restore the Blacks to their proper status as the true Hebrews, as part of a process which would include them returning to their homeland in Israel. Cherry's cosmology was that the Earth was flat and square—likely drawing from the description of the "four corners of the world" from Revelation 7:1, and popularized among Flat Earthers by an 1893 map by hotelier and quack healer Orlando Ferguson (1846–1911) of South Dakota—and that there were two layers of Heaven overhead, with God residing in the uppermost.[39] These beliefs lasted in Cherry's congregation well into the Space Age, causing them to become skeptical of NASA, a development among Flat Earthers that will be discussed further in the last two chapters. In addition to his chronological falsehood for when he established his church in contrast to Crowder's, Cherry's use of Flat Earth being drawn from the same Book of Revelation he based an apocalyptic-utopian vision of Southern Blacks' liberation and return to their "homeland" is a clear sign of the utility of Flat Earth in gaining support from at least a subset of the Black community, linking with Biblical passages associated with liberation theology.

While the Exodusters and Black Hebrew Israelites show that the association with the Book of Exodus and the Israelites was still present by the late 1870s, it was shifting. The end of Reconstruction meant it was more vital than ever that Blacks take active part in preserving the vestiges of their freedom, now that Federal protection was gone. In Exodus, the

(Philadelphia: University of Pennsylvania Press, 1971 [1944]), 32–36; James E. Landing, *Black Judaism: Story of an American Movement* (Durham, NC: Carolina Academic Press, 2002), 340–43; Merrill Singer, "The Southern Origin of Black Judaism," in Hans A. Baer and Yvonne Jones, eds., *African Americans in the South: Issues of Race, Class, and Gender* (Athens, GA: University of Georgia Press, 1992), 128–29; TheMatrixQ, "F.S. Cherry church 1886 founding year doubtful," *?verything!*, March 27, 2013, accessed June 8, 2023, https://farpointe.wordpress.com/2013/03/27/f-s-cherry-church-1886-founding-year-doubtful/.

[39] Singer, "Southern Origin," 129.

Israelites are guided out of Egypt by Moses and his divine intervention. But in the Book of Joshua, while God helps by stopping the Sun, the Israelites play a much more active role in securing their military victory. That Jasper used Joshua over Exodus as his primary source demonstrates that "The Sun Do Move" was part of this more activist approach to theology.[40]

Sixth Mount Zion Church was renovated in 1925, almost a quarter century after Jasper's death. The next year, a John Jasper Memorial Room and Museum was established on its ground floor, providing a historical interpretation to his life more accessible to most of Richmond's Blacks than academic publications.[41] But renewed scholarly attention to Black history in the United States would only explode with the New Deal (1933–1939), the term for the various Federal programs instituted by the presidential administration (1933–1945) of Franklin D. Roosevelt (1882–1945) to combat the Great Depression. Some of the New Deal's intellectual programs built off earlier efforts—one such being the Phelps Stokes Fund. Established in 1911, its goal was to promote education and social improvement of Native Americans and Blacks in both the United States and what then were the African colonies, primarily of Britain. This included funding historical and anthropological research on Black society in various Southern universities, including the University of Virginia.[42]

One beneficiary was Howard Harper Harlan, a white student who came to the University of Virginia in 1931 as an undergraduate, receiving his BA in 1933. He remained there for his graduate work in sociology, earning an MA in 1934 and PhD in 1941, before being appointed at the University of Alabama until 1946, and at Birmingham Southern College from then until 1963. In his graduate student capacity, Harlan was a Phelps Stokes Fellow from 1933–1935, and published two research papers through the Publications of the University of Virginia Phelps-Stokes Fellowship Papers program based on the research that was funded by the Phelps Stokes Fund. The first, number 13 in the series, was a

[40] Allen Dwight Callahan, *The Talking Book: African Americans and the Bible* (New Haven: Yale University Press, 2006), 192–93; Cleophus J. LaRue, *The Heart of Black Preaching* (Louisville, KY: Westminster John Knox Press, 2000), 31–36; Hobson, "The Lord Is a Man of War," 628.

[41] Richardson, *Built by Blacks*, 149.

[42] Eric S. Yellin, "The (White) Search for (Black) Order: The Phelps-Stokes Fund's First Twenty Years, 1911–1931," *The Historian*, Vol. 65, No. 2 (Winter, 2002), 319–52.

1935 study of Zion Town, a Black neighborhood in Richmond. The second, paper 14 in the series, was his 1936 *John Jasper: A Case History in Leadership*.[43]

Harlan's thesis was that the "Negro preacher [w]as a racial leader," but never a "nationalistic" leader, focusing on their own small communities instead of larger groups of Blacks. To this end,

> The story of John Jasper may be taken as typical. He was one of the last of these early Negro preachers who rendered the type of leadership we have designated as "primary". That he was one of the last of these leaders makes him no less typical, for during the greater part of his life the Church was the only social institution of any importance that belonged to the Negro, and specialized leadership could develop only after the growth of other, specialized organs.[44]

Harlan's sources were the Randolph and Hatcher biographies of Jasper, plus "numerous conversations with persons living in Richmond who knew and remember the famous preacher."[45] However, Harlan seemed to make the most use of Hatcher's biography. Perhaps as Hatcher had been a life-long resident of Richmond, Harlan viewed the white Baptist as more of an authentic regional voice than the Black Connecticut-educated lawyer.

Harlan's publication does contain some details not found in either Randolph's or Hatcher's publications, presumably meaning they were authentic stories in the Richmond community of the 1930s. Harlan does not specify whether his interview subjects were Black or white, but either way, they were going off memories that were a third of a century old at the youngest. Perhaps the most notable example is when Harlan discusses Jasper's parentage and birth, stating of his preacher father Philip, "John was the last of twenty-four children, and when his father died, two months before the boy was born, he prophesied that this child would become

[43] Harlan, *John Jasper*, 1; Howard H. Harlan, *Zion Town: A Study in Human Ecology* (Charlottesville: University of Virginia, 1935), 5–6; "Higher Degrees in Sociology Conferred in 1941," *American Journal of Sociology*, Vol. 48, No. 1 (Jun., 1942), 88; "Current Items," *American Sociological Review*, Vol. 8, No. 5 (Oct., 1943), 604; Lauranett L. Lee and Shelby M. Driskill, *"Knowledge of This Cannot Be Hidden": A Report on the Westham Burying Ground* (Richmond: University of Richmond, 2019), 17n30; Richardson, *Built by Blacks*, 133.

[44] Harlan, *John Jasper*, 15–16.

[45] Harlan, *John Jasper*, 16n5.

a great preacher, and admonished his wife to exercise great care in his rearing."[46] Philip's prophecy is not found in either Randolph or Hatcher, and therefore seems to have been something which developed amongst the Richmond community in the years after Hatcher's 1908 biography.

Harlan gives to Jasper a particular motivation for creating a church, that being a means of providing leadership to "liberated members of his race drifting aimlessly, without anchors, nothing to cling to, no way to employ their constructive energies" following Emancipation.[47] The only reference to "The Sun Do Move" is in a single paragraph, where it is only identified as a geocentrist sermon based on Biblical passages, and used to shore up Jasper's role as an excellent Biblical scholar—in service of Harlan's claim of unimpeachable Biblical knowledge being the source of Jasper's community leadership.[48] Harlan's Jasper urges his parishioners to read everything they can possibly get their hands on, and notes that through the fundraising via his sermons he was able to remodel his original church for $6000 ($200,000 in 2025) and build a new one in 1887.[49] Harlan concludes by arguing that Jasper.

> was more than a great preacher. As has been noted previously, he was an educator before there were any schools for his people. He drew from his own rich experience and from the knowledge he had obtained and gave freely what instruction he could. He established a Sunday School in the church which was a powerful educational organ for the Negro youth. He was an organizer, a builder. He was a man of great and varied interests, but all were to one end,--to benefit his people.[50]

But, Harlan hastens to add, "What he thought of as his people were those of his church, his community. He never achieved the wider view. He was never interested in the larger race problems, never even interested himself in the problems and policies of his denomination."[51]

[46] Harlan, *John Jasper*, 16.

[47] Harlan, *John Jasper*, 22–23.

[48] Harlan, *John Jasper*, 22.

[49] Harlan, *John Jasper*, 21, 24.

[50] Harlan, *John Jasper*, 25.

[51] Harlan, *John Jasper*, 25–26.

This demonstrates the degree to which Harlan favored the Hatcher biography—emphasizing Jasper's alleged non-political nature—to Randolph's, which mentions his support for African missionary efforts and Radical Republicanism. Harlan's overall view of Jasper—emphasizing local community over a national or racial community, and supporting basic literacy and fundraising over advanced education or political campaigning—also is reminiscent of the racial uplift doctrine endorsed by Booker T. Washington at his famous 1895 Atlanta Compromise speech. Even by Washington's death in 1915, much less in 1936, this was an outdated concept among Black activists.[52] Notably Washington heavily criticized Black Baptist preachers in the South for lacking formal education and being more interested in making money than enlightening their flocks, criticisms that could certainly be aimed at Harlan's depiction of Jasper.[53]

Very little information is available on Harlan beyond the details of his professional career related above, so speculation on his motivations and personal views must be taken with a large grain of salt. However, Harlan's use of Jasper to seemingly promote an outdated, apolitical view of Black community organization occurred several years into the New Deal, which even in segregated Southern states promoted increased political activity among Blacks. It was the actions of Roosevelt during his first term in office (1933–1937) which almost entirely realigned Blacks toward the Democratic Party, away from their support of Republicans stemming from Lincoln in the 1860s.[54] In Virginia alone, where almost 70% of the white electorate had voted for Roosevelt in 1932 (and over 70% would vote to re-elect him in 1936), Blacks set up the Hampton County Civic League, the Petersburg League of Negro Voters, the Portsmouth Civic and Welfare Club, and the Newport News Young Men's Democratic Club as political movements during the New Deal, while Black college students

[52] Booker T. Washington, *The Future of the American Negro* (Boston: Small, Maynard & Company, 1900), 218–44; David Sehat, "The Civilizing Mission of Booker T. Washington," *The Journal of Southern History*, Vol. 73, No. 2 (May, 2007), 323–62.

[53] Booker T. Washington, "The Colored Ministry: Its Defects and Needs," *Christian Union*, Vol. 42, No. 7 (Aug. 14, 1890), 190–200.

[54] Nancy Joan Weiss, *Farewell to the Party of Lincoln: Black Politics in the Age of FDR* (Princeton, NJ: Princeton University Press, 1983), 209–35; Lauren Rebecca Skarloff, *Black Culture in the New Deal: The Quest for Civil Rights in the Roosevelt Era* (Chapel Hill: University of North Carolina Press, 2009), 15–32.

were as enthusiastic for Roosevelt and the New Deal as white students were.[55]

Culturally, one of the most important organizations of the New Deal was the Federal Writers' Project (FWP), established in 1935 to employ out of work writers of both fiction and nonfiction. Henry G. Alsberg (1881–1970), the initial director of the FWP from 1936–1939, appointed Sterling Allen Brown (1901–1989) of the Howard University English Department to be National Editor of Negro Affairs. Alsberg hired two hundred Black authors and put pressure on state directors in the South to hire more. The most famous Black cultural product of the FWP was the Slave Narrative Collection undertaken from 1936 to 1938, but the FWP also produced numerous Black-centric biographies, bibliographies, and state studies, and the FWP was the first publisher of such major Black authors as Ralph Ellison (1913–1994).[56]

Despite Alsberg's efforts, the state guides produced by the FWP and its state affiliates only sparsely included Black topics and history. As such, Brown began a dedicated book on Black history, which would eventually be published in 1940 as *The Negro in Virginia*, the state chosen because it was the first colony in what became the United States to import African slaves. The primary supervisor of the project was Hampton Institute professor Roscoe E. Lewis, who led the Negro Unit of the Virginia Writers' Project (VWP) and was known for building productive relationships with white colleagues. The project also benefitted from the fact that the state director of the VWP, white former suffragist Eudora Ramsay Richardson (1891–1973), shared Alsberg's commitment to working with Black writers and carrying out Black studies. Brown's project worked out of the Hampton Institute and used sources dating back to 1619. Upon its publication, *The Negro in Virginia* was the first FWP book fully dedicated to Blacks and was well received by both Black and white reviewers. It was followed by similar but less intensive books on Blacks in Illinois

[55] Weiss, *Farewell to the Party of Lincoln*, 232–33.

[56] Weiss, *Farewell to the Party of Lincoln*, 239–40; Skarloff, *Black Culture in the New Deal*, 81–122; Jerre Mangione, *The Dream and the Deal: The Federal Writers' Project, 1935–1943* (Boston: Little, Brown and Company, 1972), 258; Harvard Sitkoff, *A New Deal for Blacks: The Emergence of Civil Rights as a National Issue, Volume I: The Depression Decade* (Oxford: Oxford University Press, 1978), 72; Ronald LaMarr Sharps, *Black Folklorists in Pursuit of Equality: African American Identity and Cultural Politics, 1893–1943* (Lanham, MD: Lexington Books, 2022), 291–301.

and New York, as well as a shorter, nationally oriented *Cavalcade of the American Negro*.[57]

Jasper is present in *The Negro in Virginia* even before the first page. The inside of the front cover is "A Historical Map of the Negro in Virginia," in which the thirtieth entry in the key is "John Jasper—'De Sun Do Move.'" The particular image is one of six set in Richmond, and depicts Jasper preaching to a crowd with the Sun overhead (Fig. 3.1).[58]

Jasper's first appearance in the text is in the chapter on slave preachers. In discussing the toleration for itinerant slave preachers, it is noted that "Tom Stokes of Henrico County, who thinks he is getting near ninety, recalls John Jasper as such a preacher; John Jasper would go from 'place to place preachin' funerals fo' slaves. Sometimes dem slaves been daid an' buried a year or mo'. Den one Sunday ole Jasper would preach one big sermon over dem all.'"[59] Interestingly, Tom Stokes does not appear to have been one of the former slaves interviewed for the FWP's Slave Narrative Collection.

But Jasper is later discussed properly for two pages, a significant amount considering the book covers over three centuries of Black Virginian history in 352 pages. It emphasizes how "The Sun Do Move" was not only extremely popular with both wealthy whites and Blacks, it helped Jasper pay off the $17,050 ($364,000 in 2025) he had put into renovating the Sixth Mount Zion church, and that the sermon had turned him into a "Richmond institution" and one of the most prominent preachers in Virginia.[60] Jasper is later noted as having one of the largest number of white attendees to the funeral of a prominent Richmond Black.[61] Interestingly for a Black-authored book on Virginia Blacks, the white Hatcher's biography of Jasper is cited as one of the sources used, but not the Black Randolph's.[62] Again, this is perhaps due to Hatcher's deeper Virginia connection.

[57] Skarloff, *Black Culture in the New Deal*, 83, 109; Mangione, *The Dream and the Deal*, 259–60; Jeutonne P. Brewer, *The Federal Writers' Project: A Bibliography* (Metuchen, NJ: The Scarecrow Press, 1994), 109; Sharps, *Black Folklorists in Pursuit of Equality*, 299–300.

[58] *The Negro in Virginia*, i.

[59] *The Negro in Virginia*, 77.

[60] *The Negro in Virginia*, 250–51.

[61] *The Negro in Virginia*, 298.

[62] *The Negro in Virginia*, 360.

Fig. 3.1 "A Historical Map of the Negro in Virginia" from the inside cover of *The Negro in Virginia*, depicting Jasper preaching "The Sun Do Move" in Richmond as the thirtieth image (lower center)[63]

Awareness of the politically organizing role of Jasper's sermon has waxed and waned among scholars. In his 1972 study of American slavery, Marxist historian Eugene Genovese (1930–2012) used Jasper's sermon, which he categorized as merely geocentric, as an example of the power

[63] *The Negro in Virginia*, i.

that sermons gave certain Black preachers in the South, even over white audiences.[64] In a 2006 study of African American interpretations of the Bible, Black religious scholar Allen Dwight Callahan (b. 1957) pointed out that Jasper was one of a number of Blacks who saw their resistance authorized by the Bible, but described Jasper as simply resistant to science in favor of geocentrism.[65] Likewise, in historian Eric Foner's (b. 1943) landmark 1988 study of Reconstruction, the sole reference to Jasper (misspelled "Jaspar") is as an example of the "the anti-intellectualism of ex-slave preachers" for whom "Educated members of the free black elite often found the freedmen's religious practices excessively emotional, and were appalled by." As with Callahan, Foner only describes Jasper's views as anti-heliocentric.[66] A 1992 overview of Black Southern religion identified Jasper merely as "a popular African American preacher from Virginia, who was regularly asked to give his sermon on Joshua to both slaves and whites"—not accurate by the time Jasper began his sermon in 1878, and not giving any indication of what its contents were.[67] Ironically, even geocentrists could get Jasper's name wrong. In his 2013 book promoting geocentrism, the Traditionalist Catholic astronomer Gerardus Bouw, mentioned in chapter 1, included two versions of Jasper's sermon, one in dialect and one in modern English, but misidentified both sermon and author as "'The Sun Does Move' by James Jasper."[68] (Fig. 3.2).

In contrast, Harlem Renaissance figures Arna Bontemps (1902–1973)—who had worked as a supervisor on the Illinois section of the FWP and its *Cavalcade of the American Negro*—and Langston Hughes (1901–1967) included "The Sun Do Move" in a 1958 edited volume of "Negro folklore," with the former praising Jasper in his introduction for being one of the most successful "of the old-time Negro preacher."[69]

[64] Eugene D. Genovese, *Roll, Jordan, Roll: The World the Slaves Made* (New York: Vintage Books, 1972), 269–70.

[65] Callahan, *The Talking Book*, 242–43.

[66] Foner, *Reconstruction*, 101.

[67] Singer, "Southern Origin," 126.

[68] Gerardus D. Bouw, *Geocentricity: Christianity in the Woodshed* (Miamitown, OH: Daystar Publishing, 2013), 719–39.

[69] John Jasper, "De Sun Do Move," in Langston Hughes and Arna Bontemps, eds., *The Book of Negro Folklore* (New York: Dodd, Mead & Company, 1958), 225–33; Arna Bontemps, "Introduction," in Langston Hughes and Arna Bontemps, eds., *The Book of Negro Folklore* (New York: Dodd, Mead & Company, 1958), xii. For Bontemps' work on

Fig. 3.2 The cover of a pamphlet copy of Jasper's sermon, published in New York City in 1882 and referring only to the geocentrism, and not Flat Earthism, of Jasper's theology[70]

The start of the Civil Rights Movement saw a number of new publications about Jasper come out, highlighting his appeal to both Blacks and whites.

the FWP, see Skarloff, *Black Culture in the New Deal*, 109 and Mangione, *The Dream and the Deal*, 127–28, 257.

[70] Jasper, *"The Sun Do Move!"*.

The implication given the context is clearly that Jasper could serve as a model for both the religious leaders of the movement and for integration as a whole.[71] In 1957, Sixth Mount Zion was targeted for demolition in order to make room for the development of the I-95 highway. The congregants staged a protest which was ultimately successful, causing the highway to curve around the church, sparing it the fate of the other historical buildings in the surrounding Black neighborhood. This rare win by a Southern Black community against the Federal government spoke to the organizational and inspirational power of Jasper's legacy, but also resulted in the highway cutting the church off from the homes of its congregation.[72]

In 1960, future Malcolm X biographer and *Roots* author Alex Haley (1921–1992) pitched a biographical article of Jasper to *Reader's Digest*; nine years later, he stated that Jasper was one of his favorite Black orators.[73] The same year, the Hampton Institute would go on to reprint *The Negro in Virginia* with its Jasper chapter.[74] At the turn of the millennium, Baptist minister and professor of Black theology Cleophus LaRue (b. 1953) included Jasper first in his study of distinctive Black preaching, arguing that "The Sun Do Move" was not meant to be an attack on established science but rather to reveal "the mighty acts of God on behalf of the marginalized and powerless."[75] In his 2021 study of Black Christianity, Black historian Henry Louis Gates, Jr. (b. 1950) identified Jasper as one of the early members of the activist preacher tradition which would inspire Martin Luther King, Jr. (1929–1968). In an inversion of Foner

[71] Elizabeth Dabney Coleman, "Black Boanerges: John Jasper's Conviction that "De Sun Do Move" Brought Him Worldwide Fame," *Virginia Cavalcade*, Vol. 4, No. 3 (Winter, 1954), 22; James, "*The Sun Do Move*", 40.

[72] Richardson, *Built by Blacks*, 149–50; Wilson, "Sixth Mount Zion."

[73] Alex Haley, "Foreward," in Marcus H. Boulware, *The Oratory of Negro Leaders: 1900–1968* (Westport, CT: Negro Universities Press, 1969), xiv; Alex Haley, "Manuscript by Alex Haley, John Jasper Annotated Typescript and Holograph Notes," Broward County Library Digital Archives, 1960; accessed June 26, 2020, https://digitalarchives.broward.org/digital/collection/alexhaley/id/1984.

[74] *Federal Writers' Project: A Bibliography*, 109.

[75] LaRue, *The Heart of Black Preaching*, 31–36, 131–37.

and Callahan, Gates noted that while Jasper's sermon was unscientific, scientific accuracy was not the point in its appeal.[76]

Paradoxically, a testament to the power of Jasper as an emerging icon of civil rights was his appropriation by *opponents* of civil rights. In 1954, Isaac James (1914–1992), a Black preacher who had been assistant minister at Sixth Mount Zion, published a laudatory biography of Jasper which drew upon not only the records of Sixth Mount Zion, Jasper's descendants, and Carter Godwin Woodson (1875–1950), a prominent Black historian, but also prominent white Virginia politician William M. Tuck (1896–1983), who had served as governor from 1946–1950 and was beginning a 1953–1969 tenure in the US House where he would be a strong opponent of civil rights. James' biography largely avoids politics and emphasizes Jasper's ability to bring white and Black audiences together peacefully through his sermons.[77] The same year, Elizabeth Dabney Coleman—white daughter of a Virginia legislator and member of one of the First Families of Virginia—also published an article on Jasper claiming he was against political engagement for his flock, an error seemingly deliberate, presumably either to widen Jasper's appeal to her white audience or dissuade Black readers from action.[78] But perhaps the most evident example of this was a 29 March 1965 speech in the US Senate by A. Willis Robertson (1887–1971) white senator from Virginia (1946–1966) and father of future conservative televangelist Pat Robertson (1930–2023); Robertson cited Jasper and "The Sun do Move" and its ability to bring multi-racial crowds together as an example of the "brotherly love" existent in the South that made the proposed Voting Rights Act of 1965 unnecessary and divisive.[79] The Flat Earth turn to conservative political aims will be discussed in chapters 6 and 7, but for now it is clear that Jasper had become such a strong rallying cry for at least some Black activists that opponents of Civil Rights needed to defuse him by adopting him for their own ends.

At this point, it would be appropriate to mention other Black religious beliefs from the nineteenth and early twentieth centuries that parallel

[76] Henry Louis Gates, Jr., *The Black Church: This Is Our Story, This Is Our Song* (New York: Penguin Press, 2021), 6, 86–87.

[77] James, *"The Sun Do Move"*, 6, 40, 43.

[78] Coleman, "Black Boanerges," 22.

[79] *Congressional Record: Proceedings and Debates of the 89th Congress Volume 111—Part 5* (Washington, D.C.: United States Government Printing Office, 1965), 6227.

white Flat Earth thought. During slavery, there was a belief among slaves that native-born Africans had the ability to fly and could sometimes use that to escape from slavery.[80] After Emancipation, this developed into an interest in flying machines, particularly among more esoteric Black religious figures. For example, on 5 July, 1941, the Harlem preacher Father Divine (likely born George Baker, growing up the son of freed slaves in Baltimore) announced that the concept of a transparent airship had been revealed to him, which would help Blacks escape from racial persecution.[81] The Old Testament Book of Ezekiel, in which the titular prophet encounters angels descended from Heaven surrounded by objects described as wheels within wheels, inspired Black composer William L. Dawson (1899–1990) to pen the spiritual "Ezekiel Saw the Wheel," but was also inspirational to whites. In 1901, the Texas Baptist preacher Burrell Cannon (1848–1922) designed the "Ezekiel Airship," a prototype airplane he claimed was based on the vision of the prophet, with the wheels within wheels interpreted as propellers.[82] Of interest is that one of the few known details of Cannon's life prior to his embarking on the airship project was that he served as a missionary in Africa; one wonders if his interest in Ezekiel stemmed from any interaction with Black religious leaders during that time.[83] Erich von Däniken's 1968 *Chariots of the Gods*? would popularize (though did not invent) the notion that Ezekiel had actually seen ancient alien visitors and their UFOs; the prophet's vision even provides the source for von Däniken's title. Within years, Black religious movements like the Nation of Islam would begin to incorporate ideas of UFOs and ancient aliens into their belief structure, creating an anti-colonial inversion of the ancient alien concept by making ancient Africans the creators of the UFOs. These ideas would be broadly popularized by works such as the Black musician George Clinton's (b.

[80] Callahan, *The Talking Book*, 178; Caesar Grant, "All God's Chillen Had Wings," in Langston Hughes and Arna Bontemps, eds., *The Book of Negro Folklore* (New York: Dodd, Mead & Company, 1958), 62–65.

[81] Father Divine, "A Transparent Airplane That Can Take on Any Color," in Langston Hughes and Arna Bontemps, eds., *The Book of Negro Folklore* (New York: Dodd, Mead & Company, 1958), 275–76.

[82] Allen Dwight Callahan, "Perspectives for a Study of African American Religion: From the Valley of Dry Bones," *Nova Religio: The Journal of Alternative and Emergent Religions*, Vol. 7, No. 1 (Jul., 2003), 55–57; "The Ezekiel Airship," *Scientific American*, Vol. 85, No. 15 (Oct. 12, 1901), 228; Davis, *On the Wings of Ezekiel*, 32–34.

[83] Davis, *On the Wings of Ezekiel*, 19.

1941) albums *Mothership Connection* (1975) and *The Awesome Power of a Fully Operational Mothership* (1996).[84]

While ancient UFOs (whether of extraterrestrial or African origin) do not directly connect to Flat Earth ideas, their usage in Black religious belief stems from the same nineteenth century interpretations of Old Testament beliefs for political ends that Jasper's sermon emerged from. Further, whether Father Divine's invisible plane or George Clinton's Mothership, the Black religious community continued an interest in alternative modes of air and space travel, just as the Flat Earth community's nineteenth interest in balloons evolved into a similar interest in speculative airship designs in the early twentieth century. While interest in fanciful airplane ideas does not mean a belief in the Flat Earth, it is an area of overlap, and therefore a potential route by which Black members of these religious communities could be exposed to the idea of a Flat Earth, even outside of Jasper's explicit—and explicitly political—sermons on the topic.

But besides being an example of an emerging liberation theology, Jasper's Flat Earth sermon also reflected a wider shift in Black political goals. The 1865 convention of freed Blacks in Virginia was mainly of elites (like Randolph) who had been born free, were literate, and had stable employment. Black education was established in Richmond at the end of the war in April 1865. According to statistics of the Freedmen's Bureau, there were 90 schools for Blacks in Virginia in 1865, employing 195 teachers and serving 12,898 students. Of the fourteen states (and Washington, D.C.) covered by the Freedmen's Bureau, only Louisiana and the district comprised of Kentucky, Tennessee, and northern Alabama had larger numbers. By start of June 1865, there were 74 Black teachers and 3224 Black students in eastern Virginia, including Richmond. Black elites pushed the idea that education was the key to Black advancement after Emancipation. And it is true that Black education was extremely popular among Blacks; Jasper himself spoke in favor of the hiring of Black teachers and building Black schools in Richmond, and there is little evidence that Black Christian leaders pursued any sort of anti-intellectualism.[85]

[84] Dorman, *Chosen People*, 150–51; Callahan, *The Talking Book*, 178–79; Kelley, *Freedom Dreams*, 29–35; Colin Dickey, *The Unidentified: Mythical Monsters, Alien Encounters, and Our Obsession with the Unexplained* (New York: Viking, 2020), 178–79.

[85] Davis, *The Emancipation Circuit*, 101; Dvorak, *An African-American Exodus*, 116; Randolph, *The Life of Rev. John Jasper*, 105–06.

However, very quickly, the leadership role of those pre-war Black elites began to decline. In contrast to the Blacks elected to the 1865 Virginia convention, the 24 Black men elected to the 1867–1868 state constitutional convention largely lacked traditional education but had access to widespread informational networks among Black communities, were itinerant and had partial histories of stable income, and had direct experiences with survival under slavery. Jasper's sermon, anti-intellectual as it was, reflected that shift away from the leadership of pre-war Black elites in favor of new leaders emerging from the emancipated, and who valued passionate racial solidarity—such as offered in Jasper's parable—over logical, intellectual programs.[86] Of course, not all advocates for political progress needed to turn to Flat Earth views. Nineteenth-century liberal reformers and socialist revolutionaries alike saw planetary cosmology as a pattern for terrestrial political change. Writing in February 1859, Martin Delany—an educated free Black from Virginia—argued that planetary revolution above explicitly formed a basis for slave revolution below.[87] But in Jasper's case, Flat Earth theology can be seen as liberation theology for a Black community navigating the end of Reconstruction and the best ways to maintain both their rights and unity in that unsteady era.

In this light, it makes sense why the community of Flat Earthers surrounding the New York Zetetic Society tended to be Northern veterans of the Civil War, Republican political operatives, and advocates for Black education and rights. The association of Reconstruction with Flat Earth views has proven so durable that in some cases Reconstruction officials have been retroactively connected to Flat Earth views. A 2019 textbook on geometry notes, after erroneously stating that Jasper and William Carpenter were friends, that

[86] Steven Hahn, *A Nation under Our Feet: Black Political Struggles in the Rural South from Slavery to the Great Migration* (Cambridge, MA: Harvard University Press, 2003), 276–79; Davis, *The Emancipation Circuit*, 108; Strange, "A black and white sermon," 284–85; Morris, *Reading,'Riting, and Reconstruction*, 22, 43; Foner, *Reconstruction*, 97, 101.

[87] Simon Schaffer, "The nebular hypothesis and the science of progress," in *History, Humanity, and Evolution: Essays for John C. Greene*, ed. James R. Moore (Cambridge: Cambridge University Press, 1989), 142–53; Edward Guimont, "As Above, So Below: The Role of Astronomical Evolution, Imperialism, and Religion in the Long Nineteenth Century," in *Evolutionary Theories and Religious Traditions: National, Transnational, and Global Perspectives, 1800–1920*, eds. Bernard V. Lightman and Sarah A. Qidwai (Pittsburgh: University of Pittsburgh Press, 2023), 84–85; Rusert, *Fugitive Science*, 164.

> Joseph Woods Holden, an American politician and the son of William Woods Holden, an American politician and the thirty-eighth and fortieth Governor of North Carolina – appointed by the United States president, Andrew Johnson, in 1865 and eventually elected in 1868 (leader of the state's Republican Party during the Reconstruction Era) – was another proponent who delivered numerous lectures on the flatness of the Earth.[88]

While both Holdens were real, the textbook appears to conflate Joseph W. Holden (1844–1875), Speaker of the North Carolina House of Representatives from 1868–1870 and mayor of the state capitol of Raleigh from 1874 until his death, with Maine mill owner and Flat Earther Joseph W. Holden (1815–1900). Adding to the confusion was that on his death, the Maine Holden was prominent enough that a North Carolina newspaper not only reported on his death, but endorsed his views.[89] Interestingly, the Maine Holden's gravestone states that he "discovered that the Earth is flat and stationary, and that the sun and moon do move," terminology which seem to indicate the influence of Jasper's sermon.[90]

If Richmond's white community were scornful of a Flat Earth sermon being a foundation for Black community action, that could even be seen as a useful tool for maintaining Black unity in the city. But it is not that simple—consider, for example, the role of the white Frank Combs in the origins of the sermon. Recall also that even after the end of the Civil War, the white community in the South was extremely economically and socially segregated. In 1875, only about half of the South's children attended school, and in 1880 nearly a quarter of Southern whites over the age of ten were illiterate.[91] Even among educated whites, scientific illiteracy remained high throughout the end of the rest of the century.[92]

[88] Layla S. Mayboudi, *Geometry Creation and Import With COMSOL Multiphysics* (Dulles, VA: Mercury Learning and Information, 2019), 52–53.

[89] The assertion that Jasper and Carpenter were friends, and the knowledge of Joseph Holden, indicate that Mayboudi's source was Garwood, *Flat Earth*, 163–66.

[90] Jean and David Hankins, "Otisfield, Maine: A Short History," Otisfield Historical Society, June 17, 1999, accessed January 1, 2023, https://sites.rootsweb.com/~mecotisf/bicenhs.htm.

[91] Robert A. Margo, *Race and Schooling in the South, 1880–1950: An Economic History* (Chicago: The University of Chicago Press, 1990), 6–8; Richardson, *The Death of Reconstruction*, 138.

[92] Jeremi Suri, *Civil War by Other Means: America's Long and Unfinished Fight for Democracy* (New York: PublicAffairs, 2022), 253–54.

As an example, the story mocking Jasper's sermon in the 19 March 1878 edition of the *Richmond Whig* newspaper misspelled Copernicus as "Capernicus."[93] Nor were misspellings the only example of the lack of astronomical knowledge. For much of the nineteenth century, the mainstream explanation for how light waves flowed through space was the belief in luminiferous aether. Its existence was disproven in 1887 by the Michelson–Morley experiment, but—predicting modern Flat Earthism—the scientific evidence did not halt public belief.[94] Eventually, in 1919, Einstein's general theory of relativity would be confirmed through observations of the Sun during an eclipse, allowing for an alternative method of light propagation to finally replace aether. Einstein's 1919 confirmation also allowed for the final debunking of the existence of Vulcan, a supposed planet circling the Sun within the orbit of Mercury. Among the promoters of Vulcan was John Hower Tice (1809–1883), a Southern white (and for a time head of the St. Louis, Missouri public school system) who claimed he could predict the weather using observations of this non-existent planet.[95]

Outside of these inaccurate but widely accepted views of astronomy, the prevalence of Spiritualism and phrenology among mainstream Antebellum Southern science has already been discussed. The later nineteenth century also saw the spread of Christian Science, mesmerism, and hypnotism among mainstream American society. While the US Geological Survey was established in 1879, it did not denounce the practice of dowsing until 1917.[96] Spiritualism continued to be popular among postwar Southern whites, and from 1927 to 1965 the botanist Joseph Banks Rhine (1895–1980) turned Duke University in North Carolina into a world center of parapsychology research. However, post-Civil War Spiritualism also drew in Black practitioners, who claimed contact with the ghosts of slaves, Haitian revolutionaries, and United States Colored

[93] Randolph, *The Life of Rev. John Jasper*, 52.

[94] Daniel Kennefick, *No Shadow of a Doubt: The 1919 Eclipse That Confirmed Einstein's Theory of Relativity* (Princeton, NJ: Princeton University Press, 2019).

[95] John H. Tice, "The Supposed Planet Vulcan," *Scientific American*, Vol. 35, No. 25 (Dec. 16, 1876), 389; Thomas Levenson, *The Hunt for Vulcan: …And How Albert Einstein Destroyed a Planet, Discovered Relativity, and Deciphered the Universe* (New York: Random House, 2015).

[96] Arthur Jackson Ellis, *The Divining Rod: A History of Water Witching* (Washington, D.C.: Government Printing Office, 1917).

Troops killed in the Civil War.[97] There is even a quasi-Spiritualist connection to Jasper; in the twenty-first century, the noises of radiators in Sixth Mount Zion have led some parishioners to believe that it is haunted by Jasper's ghost, and the belief that his spirit could return to lay vengeance upon parishioners if they tore down his church dates back to at least 1925.[98]

Worth noting is that Jasper was not the only figure of national renown who paired what would now be considered pseudoscience with progressive political activism after the Civil War. Ignatius Loyola Donnelly (1831–1901) served as the second Lieutenant Governor of Minnesota from 1860 to 1863, and as a member of the US House of Representatives from then until 1869. A Radical Republican at the time, he advanced the cause of Black and women's suffrage on both a state and national level, and advocated for increased funding of the Freedman's Bureau and for a National Bureau of Education to grant education to all citizens regardless of race. Losing re-election in 1868 due to intra-party squabbling, Donnelly returned to Minnesota, becoming active in state politics and eventually helping establish the People's Party in 1892, unsuccessfully running for both Vice President and Governor of Minnesota under its auspices.[99]

Outside of his progressive political activism, Donnelly was also what might be called a pseudoscientist and conspiracy theorist. One of his first forays into politics was an 1852 attack on Horace Greeley based on phrenology.[100] His 1888 book *The Great Cryptogram* pioneered the

[97] Tiya Miles, *Tales from the Haunted South: Dark Tourism and Memories of Slavery from the Civil War Era* (Chapel Hill: University of North Carolina Press, 2015); Puglionesi, *Common Phantoms*, 110.

[98] Jeffrey P. Moran, "The Scopes Trial and Southern Fundamentalism in Black and White: Race, Region, and Religion," *The Journal of Southern History*, Vol. 70, No. 1 (Feb., 2004), 105; Brad Wilson, "Excellence in Video: "Sixth Mount Zion: The Sun Do Move but this Church Does Not"," YouTube, February 5, 2018, accessed October 14, 2023, https://www.youtube.com/watch?v=Bw9O0EbZ-bE.

[99] John D. Hicks, "The Political Career of Ignatius Donnelly," *The Mississippi Valley Historical Review*, Vol. 8, No. 1/2 (Jun.-Sep., 1921), 80–132; Martin Ridge, *Ignatius Donnelly: The Portrait of a Politician* (Chicago: The University of Chicago Press, 1962), 100–02, 211–26; David D. Anderson, *Ignatius Donnelly* (Boston: Twayne Publishers, 1980), 25–31; Justin McHenry, *Lemuria: A True Story of a Fake Place* (Port Townsend, WA: Feral House, 2024), 47–62.

[100] Ridge, *Donnelly*, 10; Anderson, *Donnelly*, 21.

claim that William Shakespeare's plays had actually been written by Francis Bacon, and his 1891 novel *Doctor Huguet* was a plea for racial tolerance based on a white man and a Black southerner magically exchanging bodies.[101] However, he is best known for his ostensibly-nonfiction works *Atlantis: The Antediluvian World* (1882) and *Ragnarok: The Age of Fire and Gravel* (1883), where he respectively established the modern conception of Atlantis as an ancient super-civilization in the Atlantic—a trope termed ATAC (ancient, technologically advanced civilization) by archaeologist Stephanie Halmhofer[102]—whose refugees established the Maya, Egyptian, and Sumerian civilizations; and the idea that a comet impacted Earth causing widespread ecological and cultural destruction approximately twelve thousand years ago. Inspired by the science fiction novels of Jules Verne, Donnelly's allegedly-nonfiction works, particularly *Atlantis*, drew widespread interest. As with Jasper, Donnelly's pseudoscientific musings led him embarking on speaking tours of the US and UK, and corresponding with such figures as Mark Twain, Charles Darwin, and William Gladstone.[103] As with Jasper, Donnelly paired his conspiratorial views with his populist political activism, including of Black educational and political rights, only to have them be adopted in the twentieth and twenty-first centuries by figures such as Immanuel Velikovsky (1895–1979) and Graham Hancock (b. 1950) with less progressive political views. Even their deaths were in proximity; Donnelly died on 1 January 1901, several hours into the dawn of the twentieth century and less than three months before Jasper's own 30 March death.[104]

[101] Ridge, *Donnelly*, 227–44, 289–90; Anderson, *Donnelly*, 55–67, 81–91; John R. Bovee, "Doctor Huguet: Donnelly on Being Black," *Minnesota History*, Vol. 41, No. 6 (Summer, 1969), 286–94.

[102] Stephanie Halmhofer, "Manufacturing History: Atlantis, Aryans, and the Use of Pseudoarchaeology by the Far-Right," in Christopher T. Conner; Matthew N. Hannah and Nicholas J. MacMurray, eds., *Conspiracy Theories and Extremism in New Times* (Lanham, MD: Rowman & Littlefield, 2024), 62–63.

[103] Ridge, *Donnelly*, 197–244; Anderson, *Donnelly*, 33–54; Donald H. DeMeules, "Ignatius Donnelly: A Don Quixote in the World of Science," *Minnesota History*, Vol. 37, No. 6 (Jun., 1961), 229–34; Davis, *The Emancipation Circuit*, 321; John Michell, *Eccentric Lives and Peculiar Notions* (London: Thames and Hudson, 1984), 22–26; Garwood, Flat Earth, 201–11; Kenneth L. Feder, *Encyclopedia of Dubious Archaeology: From Atlantis to the Walam Olum* (Santa Barbara, CA: ABC-CLIO, 2010), 28–29, 33–34.

[104] Anderson, *Donnelly*, 110.

Jasper was therefore not unique in post-Reconstruction America in connecting pseudoscience to progressive politics. Donnelly was inarguably better known, both at the time and since. But while Donnelly's scientific and political beliefs were both sincere, they emerged from different streams, being linked only via the figure advocating them. Jasper's Flat Earth sentiments—a geographical view not held by Donnelly—stemmed from the same religious sentiments that were the core of his political philosophy. And while Jasper may not have been as popular as Donnelly, that does not mean that he was unpopular—far from it.

Consider that "The Sun Do Move" was not one of Jasper's best sermons, either rhetorically or oratorically. It was also one of his longest, at ninety minutes. After his initial March 1878 sermon, Jasper only repeated it by request. Despite all these seeming negatives, the requests still came—252 of them, and not only from his own parishioners.[105] Those who came to hear Jasper preach "The Sun Do Move" at Sixth Mount Zion were delegates from the South Carolina Press Association and the American Bankers' Association; a governor of Pennsylvania; and the English poet Matthew Arnold (1822–1888).[106] Within the state, Jasper spoke to the Virginia General Assembly (which contained Black delegates until 1890), which had adjourned specifically to hear him; to Richmond's Jewish community; and to the headquarters of the Virginia post of the Grand Army of the Republic. Others in the white community who heard him included Edward W. Hudgins (1882–1958), who would go on to serve as Chief Justice of the Supreme Court of Virginia from 1947–1958, and John Fulmer Bright (1877–1953), who would serve as Richmond's segregationist Democratic mayor from 1924–1940.[107]

An early description of Jasper's speaking appeal was provided in the 1880 book *My Southern Home: Or, the South and Its People* by William Wells Brown (c. 1814–1884). Born into slavery in Kentucky, he escaped

[105] Miner, "De Sun Do Move," 25.

[106] In a piece of curious coincidence, in his unfinished final novel *The Mystery of Edwin Drood* (1870), novelist Charles Dickens (1812–70) based a character named John Jasper on Arnold. Pete Orford, "Picturing John Jasper," *Cloisterham Tales*, July 31, 2013; accessed July 30, 2020, https://cloisterhamtales.wordpress.com/2013/07/31/picturing-john-jasper/.

[107] Randolph, *The Life of Rev. John Jasper*, 36, 77; Hatcher, *The Unmatched Negro Philosopher*, 107; Strange, "A black and white sermon," 275–76; James, *The Sun Do Move*, 24, 40; Miner, "De Sun Do Move," 25; Coleman, "Black Boanerges," 20–21.

in 1834 and relocated to Boston, where he became an abolitionist activist. In 1853, he became the first African American novelist with the publication of *Clotel*, about a fictional mulatto daughter of Thomas Jefferson. Brown was also a self-proclaimed medical doctor, something in common with a number of Flat Earthers.[108] In his book, written for a white Northern audience, Brown recounts in his twenty-second chapter how

> I went to hear the somewhat celebrated negro preacher, Rev. John Jasper. The occasion was one of considerable note, he having preached, and by request, a sermon to prove that the "Sun do move," and now he was to give it at the solicitation of forty-five members of the Legislature, who were present as hearers.
> Those who wanted the sermon repeated were all whites, a number of whom did it for the fun that they expected to enjoy, while quite a respectable portion, old fogy in opinion, felt that the preacher was right.
> On reaching the church, I found twelve carriages and two omnibuses, besides a number of smaller vehicles, lining the street, half an hour before the opening of the doors. However, the whites who had come in these conveyances had been admitted by the side doors, while the streets were crowded with blacks and a poorer class of whites.
> By special favor I was permitted to enter before the throng came rushing in. Members of the Legislature were assigned the best seats, indeed, the entire centre of the house was occupied by whites, who, I was informed, were from amongst the F. F. Vs [First Families of Virginia]. The church seats one thousand, but it is safe to say that twelve hundred were present at that time.
> Rev. John Jasper is a deep black, tall, and slim, with long arms and somewhat round-shouldered, and sixty-five years old. He has preached in Richmond for the last forty-five years, and is considered a very good man. He is a fluent speaker, well versed in Scriptures, and possesses a large amount of wit. The members of Jasper's church are mainly freedmen, a large number of whom are from the country, commonly called "corn-field niggers."
> The more educated class of the colored people, I found, did not patronize Jasper. They consider him behind the times, and called him "old fogy." Jasper looked proudly upon his audience, and well he might, for he had before him some of the first men and women of Virginia's capital. But these people had not come to be instructed, they had really come for a good laugh and were not disappointed....

[108] Simmons, *Men of Mark*, 447–50.

The church was now completely jammed, and it was said that two thousand people sought admission in vain. Jasper's text was "God is a God of War." The preacher, though wrong in his conclusions, was happy in his quotations, fresh in his memory, and eloquently impressed his views upon his hearers.

He said, "If the sun does not move, why did Joshua command it to 'stand still?' Was Joshua wrong? If so, I had rather be wrong with Joshua than to be right with the modern philosophers. If this earth moves, the chimneys would be falling, tumbling in upon the roofs of the houses, the mountains and hills would be changing and levelling down, the rivers would be emptying out. You and I would be standing on our heads. Look at that mountain standing out yonder; it stood there fifty years ago when I was a boy. Would it be standing there if the earth was running round as they tell us?"

"No, blessed God," cried a Sister. Then the laugh came, and Jasper stood a moment with his arms folded. He continued: "The sun rises in the east and sets in the west; do you think any one can make me believe that the earth can run around the world in a single day so as to give the sun a chance to set in the west?"

"No, siree, that doctrine don't go down with Jasper."

At this point the preacher paused for breath, and I heard an elderly white in an adjoining pew, say, in a somewhat solemn tone—"Jasper is right, the sun moves."

Taking up his bandanna, and wiping away the copious perspiration that flowed down his dusky cheeks, the preacher opened a note which had just been laid upon the desk, read it, and continued,—"A question is here asked me, and one that I am glad to answer, because a large number of my people, as well as others, can't see how the children of Israel were able to cross the Red Sea in safety, while Pharaoh and his hosts were drowned. I have told you again and again that everything was possible with God. But that don't seem to satisfy you.

"Those who doubt these things that you read in Holy Writ are like the infidel,—won't believe unless you can see the cause. Well, let me tell you. The infidel says that when the children of Israel crossed the Red Sea, it was in winter, and the sea was frozen over. This is a mistake, or an intentional misrepresentation." Here the preacher gave vivid accounts of the sufferings and flight of the children of Israel, whose case he likened to the colored people of the South. The preacher wound up with an eloquent appeal to his congregation not to be led astray by "these new-fangled notions."[109]

[109] William Wells Brown, *My Southern Home: Or, the South and Its People* (Boston: A. G. Brown & Co., Publishers, 1880), 201–07.

The information provided by Brown is instructive. This was extremely early in Jasper's Flat Earth career—likely within the first year given the publication date—and already he was attracting hundreds of Virginia's most socially, economically, and politically connected white residents. Most, but not all, of the whites wanted to hear the sermon to make fun of it, but some sided with Jasper. Brown notes that at least one of the First Family members he was seated next to seemed to have been converted from skepticism by Jasper's sermon. Outside of the wealthy white visitors, Brown notes that Jasper's congregation, while mostly poor Blacks, also contained poor whites, while educated Blacks avoided attending Jasper, seemingly even before his turn to the Flat Earth. It is also worth noting that due to their curiosity over hearing the Flat Earth sermon, those First Family members were exposed to a religious argument on how Southern Blacks should fight for their freedom just as ancient Israelites had.

One of the benefits of Jasper's wide audience was that, between speaking fees and donations—even white visitors who came to mock his views typically left donations—Jasper was able to provide enough money to keep not only himself but the church financially solvent and free from the need to seek support, fiscal or otherwise, from white organizations. An example of this is that in 1884, Jasper's personal wealth was over five thousand dollars (approximately $161,000 in 2025) in, and in 1890 he was able to fund a $22,000 refurbishment of the church (approximately $763,000 in 2025). This is particularly striking considering he had been penniless and homeless upon Emancipation in 1865.[110]

Jasper's sermons did not merely attract visitors from out of state. In 1883, he undertook a tour of Washington, D. C., Baltimore, Philadelphia, and several cities in New Jersey. In doing so, Jasper's tour anticipated the destination of Blacks leaving the South in the Great Migration forty years later (ironically, one of the most popular histories of the Great Migration is *The Warmth of Other Suns*).[111] In Washington, Jasper preached at Lincoln Hall to a crowd including members of Congress and several other white dignitaries, who reportedly received him well.[112] During his

[110] Randolph, *The Life of Rev. John Jasper*, 28; Strange, "A black and white sermon," 282–83; Miner, "De Sun Do Move," 19–20.

[111] Wilkerson, *The Warmth of Other Suns*, 527–38.

[112] Randolph, *The Life of Rev. John Jasper*, 36; Simmons, *Men of Mark*, 1070–71.

stay in Baltimore, Jasper received a letter of interest from Bridgeport, Connecticut, dated 13 March 1883:

> Rev. Mr. Jasper, Richmond, Va.:
> Respected Sir,–From your active and important labors, may I request a few moments to settle a discussion regarding what you teach as to the earth. You are reported as saying, and fortifying your position by Scripture texts, that–
> I. The earth is square and immovable.
> II. The sun rises in the east and sets in the west; and,
> III. The world is flat.
> These were the beliefs of Pious Bede, of holy memory, one of the most devout of the early English writers, and whose writings, although composed in the seventh century, are still authoritative upon the subjects of which they treat.
> Now, can I make bold to ask you to briefly tell me your belief upon the shape of the earth, how the sun returns from the east to the west, and the Scripture-texts upon which you found your belief (and to explain what protects people from falling over the edge of the earth and the waters from flowing over).
> Enclosed find stamps (6 cents).
> I have the honor to remain,
> Your obedient servant,
> GEO. L. PORTER, M. D.[113]

Being compared to the Venerable Bede was certainly no small matter for a Black preacher in the late nineteenth-century United States, although as mentioned in the first chapter, Bede was not a Flat Earther. But the authorship of this letter is more interesting than its contents. "Geo. L. Porter, M.D.," was Dr. George Loring Porter (1838–1919), who had begun his medical career as a surgeon in the Union Army. For a brief period in 1862, he was a prisoner of war, tending to both Union and Confederate wounded in a Presbyterian church in Strasburg, Virginia, which had been temporarily converted into a makeshift hospital. In 1864, Porter was assigned to be Assistant Surgeon at the Washington Arsenal, a position which he held until 1867. A year into his appointment, he supervised the burial of John Wilkes Booth's body, as well as pronounced several other of the Lincoln conspirators dead after their executions. In

[113] Randolph, *The Life of Rev. John Jasper*, 78–79.

1868, disillusioned by his reassignment to the western frontier and its wars against Native Americans, he resigned his commission and relocated to Bridgeport with his family. In Bridgeport, Porter became interested in a new field of medical research: the potential use of electricity to re-animate the dead. In 1880, Porter was on hand for the hanging of a murderer; immediately afterwards, he attempted to revive the corpse using galvanic currents. Although he was unsuccessful, he believed that he had successfully induced respiration, and that had he also resumed the heartbeat, he would have succeeded in bringing the dead murderer to life.[114]

Porter's experiments gave him national, if brief, notoriety, although it is unclear if Jasper would have recognized the name of the doctor who had attempted to copy the miracle of Lazarus. If Jasper was familiar with Porter, he might have associated him with the legend of the Night Doctors, a myth started and circulated by Southern whites to try and scare Black Southerners in approximately the 1880s–1920s to prevent their embarking on the Great Migration to northern cities. Hooded figures, they would capture Blacks at night, vivisecting them or subjecting them to invasive surgical procedures. The myth became self-perpetuating in Black communities in both the North and South, and may have contributed to twentieth-century alien abduction folklore; Jasper would certainly have been familiar with the notion in the later decades of his life.[115]

Porter would attempt to resurrect a corpse in Bridgeport again in 1888, although by then he would ultimately be opposed by local clergy. By that time, he had a collaborator: Dr. George Poe (1847–1914), son of Edgar Allan Poe's second cousin and whose inspiration was an early handwritten copy of the poem "The Raven" (1845) he owned. In turn, Porter and Poe's experiments may have gone on to influence horror author H. P. Lovecraft (1890–1937), himself inspired by Edgar Allan, to write his seminal proto-zombie story "Herbert West—Reanimator" (1922).[116] But outside of his letter to Jasper, there is another potential thread, however minor, connection Porter to Flat Earth thought. The

[114] Michael J. Bielawa, *Wicked Bridgeport* (Charleston: The History Press, 2012), 38–51.

[115] Gladys-Marie Fry, *Night Riders in Black Folk History* (Knoxville: University of Tennessee Press, 1975), 170–211. For the potential alien abduction connection, see my forthcoming article "Early UFO Culture and New England Folklore."

[116] Michael J. Bielawa, "American Frankensteins: George Porter and George Poe, and Their Attempts to Reanimate the Dead in New England," in *Lovecraftian Proceedings*

Grand Army of the Republic seventh national encampment was held on 14–15 May, 1873 in New Haven, Connecticut. It was at that meeting that Hans Powell was elected Surgeon-General of the organization. Bridgeport is located around two-thirds of the way to New Haven from where Powell resided in New York City.[117] It is easy to imagine that Powell may have met with his fellow surgeon—both of whom had received their medical training in Philadelphia—in Bridgeport en route to the muster, if they did not in fact meet in New Haven. Considering that Powell was only a few months from joining the NYZS council, one wonders whether the topic of the Earth's shape may have come up then. In any case, as will be seen in chapter 5, Powell and Porter would not be the only medical doctors to wander into the field of Flat Earth speculation, or to pair the search for the Earth's true shape to the search for immortality via science.[118]

But outside of receiving a letter from Porter, another result of Jasper's visit to Baltimore was a story by the local *Baltimore Sun* newspaper. The *Sun*'s report mainly used Jasper as a way to discuss local Flat Earther William Carpenter, who was quoted as saying that "Jasper is right about the matter, although he has not a complete understanding of it. There are plenty of men who agree with him, but there are few who are willing to acknowledge the belief. It is unpopular and keeps a man back in the world, but I care nothing for that." The reporting finished by stating that according to Carpenter, the Flat Earth idea "is the only astronomical theory that accords with Holy Writ, which is what commends it to the Rev. John Jasper."[119]

No. 4, eds. Dennis P. Quinn and Elena Tchougounova-Paulson (New York: Hippocampus Press, 2022), 149–60.

[117] *Journal of the Seventy-Fourth National Encampment of the Grand Army of the Republic* (Washington, DC: United States Government Printing Office, 1941), 274. It is also worth noting that as of 1884, the only one of Jasper's children or grandchildren who weren't resident in Virginia was his daughter, Irene Johnson, who by then had been resident in New York City for some time. It is entirely speculative, but interesting, to consider whether she ever approached the members of the New York Zetetic Society. Randolph, *The Life of Rev. John Jasper*, 13.

[118] Of note is that Cyrus Teed (1839–1908), the most prominent Hollow Earther of the fin de siècle United States, was also a Union Army veteran who served in its medical corps, before practicing alternative medicine and becoming a socialist like several contemporary and future Flat Earthers. Holly Folk, "Raymond W. Bernard, Hollow Earth, and UFO s," in Ben Zeller, ed., *Handbook of UFO Religions* (Leiden: Brill, 2021), 320; Standish, *Hollow Earth*, 143–86.

[119] Randolph, *The Life of Rev. John Jasper*, 43–47.

This appears to be the only evidence that Carpenter and Jasper intersected along their separate Flat Earth trajectories. Even so, from the reporting it is not certain that Carpenter attended Jasper's sermon, let alone met him. This would be hard to discern from most of the existing histories of the Flat Earth idea. In Christine Garwood's 2007 *Flat Earth*, Jasper is given a total of half a paragraph in the middle of the section of Carpenter, identified as one of the "ardent campaigners across the United States" who "joined" Carpenter in promoting the Flat Earth notion.[120] In Robert Schadewald's 2015 *The Plane Truth*, Carpenter gets a twenty page chapter devoted to him. The book's sole reference to Jasper is in the middle of that chapter, where Schadewald asserts that "Carpenter was in touch with the leading flat-earthers of his day, [including] Reverend John Jasper. A former slave, John Jasper was well known in the mid-Atlantic states for his favorite sermon, "De Sun Do Move," and he and Carpenter apparently hit it off immediately."[121] Jasper is entirely absent from Kelly Weill's 2022 *Off the Edge*, although she gives Carpenter two pages.[122] Ironically, while Jasper is absent from Carpenter's later *One Hundred Proofs*, Carpenter does suggestively state that "he who may be called an intellectual man will know that the demonstration of the fact that the Earth is not a globe is the grandest snapping of the chains of slavery that ever took place in the world of literature or science."[123]

The result of these modern accounts is a narrative by which Flat Earth thought was brought to the United States by Carpenter, who was its chief proponent; Jasper was merely an adjunct to his friend Carpenter in spreading the Zetetic gospel. The reality is that Jasper had become famous for his Flat Earth sermon a year before Carpenter even emigrated to the United States. While Carpenter was largely a marginal figure during his life who spent his time skulking around Johns Hopkins hawking his pamphlets and harassing the university president and lecturers (tactics which, admittedly, bring to mind modern conservative attacks on higher education), Jasper was bringing in thousands of dollars and drawing an international audience to his sermons. Although Flat Earthers existed in

[120] Garwood, *Flat Earth*, 163–64.

[121] Schadewald, *The Plane Truth*, 95.

[122] Weill, *Off the Edge*, 25–26.

[123] William Carpenter, *One Hundred Proofs that the Earth is Not a Globe, Fifth Edition* (Baltimore: William Carpenter, 1886), 26.

the United States before either Jasper or Carpenter, it was Jasper who more mainlined it as a concept, was more popular—and influential—in terms of audience, and who linked it with a political ideology and goals. Indeed, if Flat Earth is viewed as a populist belief in the twenty-first century, it makes more sense for a slave preacher to be its popularizer than the various educated figures among the NYZS. If history has remembered Carpenter as the more influential of the two, that says more about the racial blindness of histories, especially early twentieth-century American histories, than of the two men.

Even so, Jasper's absence is perhaps forgivable on the part of the casual scholar of Flat Earth history. Outside of the reference by Carpenter in the local newspaper article, Jasper does not appear to be mentioned in any surviving correspondence or publication by Flat Earthers of the nineteenth or twentieth centuries, although he is sporadically mentioned in the voluminous social media and forum posts among twenty-first-century online Flat Earthers. Instead, Jasper stands as a Flat Earther who is primarily remembered for his Flat Earth belief, whose memory was sustained largely by white scholars who saw him as a misguided but sincere and effective preacher, and Black figures who saw him as either a source of ridicule for Blacks amongst whites, or a promoter of Black independence.[124]

An example of this would be the reception of Jasper by the aforementioned Carter Woodson. Himself born in Virginia to parents who had been slaves, Woodson went on to become the second Black man (after W. E. B. Du Bois) to earn a PhD from Harvard University.[125] Woodson was a critic of Christianity in failing to truly advocate for or provide a route to liberation for Blacks. However, in his 1921 monograph *The History of the Negro Church*, Woodson describes Jasper as "known to fame as the Baptist preacher of the Sun-Moving theory, established for himself throughout Virginia and adjacent States a reputation for piety and sincerity, which, without political influence, made him a power in the country."[126] Woodson later used Jasper as an example of a leader of the

[124] Randolph, *The Life of Rev. John Jasper*, 36–42; Strange, "A black and white sermon," 278–82; Hobson, "The Lord Is a Man of War," 629; Brown, *My Southern Home*, 204.

[125] Sharps, *Black Folklorists in Pursuit of Equality*, 246–51.

[126] Carter Godwin Woodson, *The History of the Negro Church* (Washington, D.C.: The Associated Publishers, 1921), 238–39.

larger, more conservative faction of postwar Black Baptists whose refusal to compromise drove the smaller, more educated, progressive Baptists to split their congregations.[127]

White authors like Hatcher tended to highlight Jasper's extremely limited education and emphasize his vernacular speech. For example, Hatcher writes how Jasper stood fast against "furlosofurs" (philosophers) on astronomical matters, because "I knows de way uv de sun, as de Wurd of Gord tells me, an' ef I don' pruv' dat de sun moves den yer may pos' me as er lier on ev'ry street in Richmun'."[128] Nor were whites the only ones who made fun of Jasper's ideas and dialect. In 1904, Powell W. Gibson (1875–1959), a Black principal of the segregated Brown School in Manassas, Virginia wrote a poem using "The Sun Do Move" as a way of mocking Black folk beliefs. In 1906, Black activist James Weldon Johnson (1871–1938) and his composer brother J. Rosamond Johnson (1873–1954)—who six years earlier had composed the song "Lift Every Voice and Sing," known as the Black national anthem—wrote a song called "De Bo'd of Education," where Jasper's belief in the account of "Joshway" is emphasized over the lessons of books and schools that the Earth is round and rotating. However, while patronizing, the parodies of Gibson and the Johnsons attempted to build a Black literary tradition that was both separate from white minstrel performances and rooted in authentic Black traditions.[129]

While those caricatures may have had elements of nuance, Jasper also drew Black defenders, including from surprising corners. In 1942, Woodson wrote how he had heard Jasper preach six months before his 30 March 1901 death. Woodson states that

> John Jasper was laughed at by almost everybody in his day because he preached, as they said, "The sun do move." Jasper, however, never used such language. He was a self-educated man of considerable mental development. He used correct English, and his words were most picturesque. Furthermore, he was not given to intonation as were so many preachers of that day. He could paint beautiful pictures with words.... He believed that

[127] Woodson, *The History of the Negro Church*, 255.

[128] Hatcher, *The Unmatched Negro Philosopher*, 123.

[129] Dickson D. Bruce, *Black American Writing from the Nadir: The Evolution of a Literary Tradition, 1877–1915* (Baton Rouge: Louisiana State University Press, 1989), 108–09.

> he was God's representative on earth, and any one who opposed him was an enemy of God. In this he was narrow. But he was a kindhearted man and tried to do good.[130]

Even the Black critic of conservative Black Christian leaders, as Woodson categorizes Jasper, came to his defense decades after the fact, pointing out his erudition and kindness along with his religious fervor—and the political influence, however indirect, that came with it. Black historian Alrutheus Ambush Taylor (1893–1954) made a similar observation in his 1926 *The Negro in the Reconstruction of Virginia*, where in discussing "The Sun Do Move" he argued that "The thinking class of people usually referred to this sermon as Jasper's weakest. In him, however, they had to recognize the power and the force of a great preacher."[131]

Even more effusive is the praise given to Jasper by William Simmons in his 1887 collection of biographies of prominent Blacks:

> This sermon has had a checkered but an interesting history from its birth to the present time. It created a lively enthusiasm among the members of his own church. Then its fame rapidly overspread Richmond. Several ministers who honestly differed from Rev. Jasper vigorously opposed the idea that the Bible inculcates that "The Sun Do Move." This led to an estrangement between Rev. Jasper and several ministers; but the name of the eloquent divine and his sermon were mentioned in scientific journals of the North. He journeyed to Baltimore, where he preached with great success before white and colored. At Washington and Philadelphia, too, he met flattering audiences.... The scientists of London, Berlin and Paris—the philosophers, scholars and students of America and Europe—have discussed orally and in print the opinions of Rev. John Jasper.... That prince dialectician of Greece, the famous Aristotle, was an enthusiastic teacher of the same thing. He conceived the moon, sun and planets set in a hollow crystalline sphere, by which they were borne around the earth. Illustrious Socrates and learned Plato diligently inculcated that the earth was the centre of the solar system, and that the sun constantly revolved around it. The philosophers and scholars of Rome accepted the teachings of Greece and diligently taught it. The great and good men of the early

[130] Carter Godwin Woodson, "John Jasper," *Negro History Bulletin*, Vol. 5, No. 4 (Jan., 1942), 95.

[131] Alrutheus Ambush Taylor, *The Negro in the Reconstruction of Virginia* (New York: Russell and Russell, 1926), 189–90.

> church were likewise convinced. Origen, Athanasius, Augustine, Clement, Ignatius, Polycarp and others simply accepted the belief of the age.[132]

Being compared to Aristotle, Socrates, Plato, and St. Augustine, among others, was certainly no small feat for a self-educated former slave.

Woodson the historian, Simmons the biographer, and others like them kept the memory of Jasper, his Flat Earth views, and their political benefit for the Black population alive long after white Flat Earthers on either side of the Atlantic had forgotten him. But Jasper was not the only figure in the Anglophone African diaspora to become active in fin de siècle Flat Earth beliefs. In December 1883, a decade after the New York Zetetic Society, Rowbotham established a Universal Zetetic Society (UZS) in London. Although Rowbotham would die a year later, the UZS would far outstrip the influence of the NYZS, lasting into the twentieth century under the leadership of Lady Elizabeth Anne Mould de Sodington Blount (1850–1935), who would take the nom de plume Zeteo. The members of the UZS came from not only London, but across the British Empire. Those from South Africa will be discussed more in chapter 5—but several members were also from the Black British community in the Caribbean. These will be the focus of the next chapter.

[132] Simmons, *Men of Mark*, 1070–71.

CHAPTER 4

Pan-African and Flat Earther

Slavery in the Caribbean was far more concentrated—and lucrative—than slavery in what became the United States, based as it was on sugar production. Slave rebellions were common, particularly with slaves who had been brought over from Africa, especially those from the Kingdom of Kongo in modern Angola. The Haitian Revolution of 1791–1804 led to the first successful slave revolution, as well as the first successful independence movement from a European empire after the American Revolution, but it would ultimately fail to lead to the elimination of slavery in the Caribbean from revolution. Instead, reforms from above—albeit pressured by both free and enslaved Blacks and their white allies from below—would ultimately lead to the elimination of slavery in the region (Fig. 4.1).

In 1833, the British Parliament passed an act abolishing slavery in much of the British Empire, including Canada, the Cape Colony (the core of modern South Africa, as discussed next chapter), Mauritius, and the West Indian colonies in the Caribbean. The act went into effect in 1834, turning 800,000 enslaved West Indians into unpaid "apprentices," a status which ended in 1838 with their becoming legally equal subjects of the Crown—in contrast to the status of Blacks in the postwar American South, despite the efforts of the project of Reconstruction. That year, plantation owners in the West Indies began attempting to scout out free

E. Guimont, *The Power of the Flat Earth Idea*,
Palgrave Studies in the History of Science and Technology,
https://doi.org/10.1007/978-3-032-23981-5_4

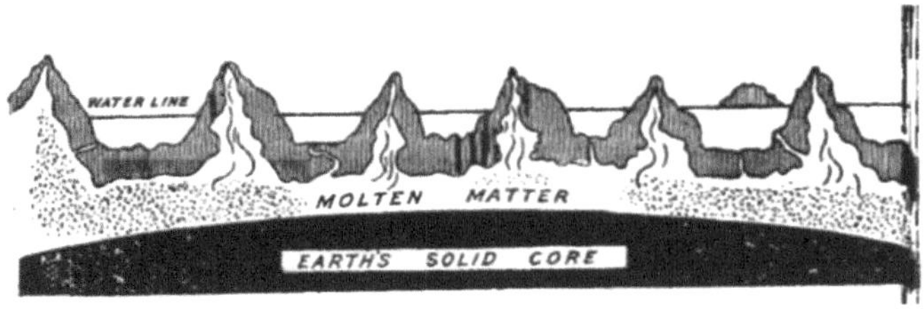

Fig. 4.1 Illustration of the geology of the Mount Pelée eruption, which *The Earth* took offense to[1]

Blacks in the United States as a potential pool of new migrant laborers. For their part, in the early 1840s a number of free Blacks in Baltimore seriously weighed emigration to the British Caribbean colonies of British Guiana (modern Guyana) and Trinidad, seeing the abolition of slavery there as a positive contrast to the expansion of slave power in the United States.[2]

In London, the Universal Zetetic Society (UZS) had an interest in the West Indies dating from the start of the twentieth century. On 8 May 1902, the volcanic Mount Pelée on the island of Martinique erupted, destroying the town of Saint-Pierre and killing all but two of its 30,000 inhabitants, a death toll higher than the infamous 79 CE eruption of Mount Vesuvius. As far back as Rowbotham, it was an accepted tenet of Zetetic geology that the Flat Earth floated over a sea of liquid fire, which occasionally bubbled to the surface as volcanism.[3] The UZS journal, *The Earth*, published an article on the eruption—taking pains to dispute how more mainstream organs depicted the eruption. Objecting to an illustration (Fig. 4.1) of Caribbean geology, the author, John Williams, wrote that

[1] John Williams, "The West Indies," *The Earth*, Vol. III, Nos. 25/26 (Jun., 1902), 18.

[2] Dexter J. Gabriel, "The British Emigration Scheme and the African American Emigration Movement to the Caribbean," in Ronald Angelo Johnson and Ousmane K. Power-Greene, eds., *In Search of Liberty: African American Internationalism in the Nineteenth-Century Atlantic World* (Atlanta: The University of Georgia Press, 2021), "The British Emigration Scheme," 165–96.

[3] Samuel Rowbotham, *Zetetic Astronomy: Earth Not a Globe, Third Edition* (London: Simpkin, Marshall, and Co., 1881), 181–200.

> One would think that the sad and terrible disaster in the West Indies, of last month (May), needed our deepest sympathy, and that none, however "learned," would have been so inhuman as to have made such a calamity an occasion for an ostentatious display of Phantasmal Science; but such has been the case as the following "Geological Section of the West Indies" plainly and irrefutably proves. If it is intended to prove the Earth a globe it is a miserable failure; the so-called coast is curved, but the sea-line above is a straight line! I take it from the *Birmingham Weekly Mercury*, May 17th. In the diagram, the volcano at the left-hand corner is Antigua, the next Guadeloupe, next Dominica, next Martinique, next St. Lucia, next St. Vincent. Barbadoes [*sic*] is between the last two in the distance, on the "water line." The "water line" looks very level, horizontal, straight, flat. Ah! perhaps this is the production of a telephotographic lens and camera. If so, it has "recorded in an *unmistakable manner*" that Mr. H. Yule's *so-called* experiment on the Old Bedford Level was, to say the least, an "optical delusion." What have the members of the Glasgow British Association to say to this absolutely contradictory evidence? But look where the globe has gone to! Absolutely buried beneath the three elements—air, water, and fire! And so the poor globe is buried at last beyond all recovery, and the "flat earth" and horizontal water line are on the top.[4]

Henry Yule Oldham (1862–1951) was a Cambridge geographer who, in 1901, used cameras positioned on the Old Bedford canal to definitively establish that the water along it is curved, not perfectly flat, as claimed by Rowbotham since 1838. As Rowbotham's Bedford Level Experiment was a cornerstone of Flat Earth claims, this set Oldham up to be a primary villain of Zetetics.

But more importantly, Williams' article establishes that only Flat Earthers have a true knowledge of geology, and therefore only a Flat Earth geology can help predict future eruptions and avoid similar mass deaths. As there are no active volcanoes in the British Isles themselves, the abrogation of Flat Earth geology and its predictive abilities by imperial authorities in the metropole was an abdication of their responsibilities to colonial subjects, particularly the mostly Black population in the West Indies. To the UZS, Flat Earth geology was a tool for protecting the

[4] Williams, "The West Indies," 17–18.

lives and livelihood of non-white colonial subjects from further natural disasters.[5]

Interestingly, there may be a grain of truth to this, even though Williams would not have known it at the time. A 2020 study found that electromagnetic geophysical modeling of the Earth's interior could be modeled on a two-dimensional plane without major differences in outcome.[6] Of course, deeper observations of the Earth's gravity and its changes in the interior, vital for geological modeling, rely on the very different nature of a spherical Earth than a flat one.[7] There are also several studies demonstrating how meteorological predictions, also vital for preventing loss of life (particularly in the Caribbean, but elsewhere also), rely on a spherical Earth rather than the alternative weather patterns of a flat model for accuracy.[8] In a demonstration against the Flat Error, even medieval European and Arab meteorologists understood the spherical nature of the atmosphere.[9] Even most modern Young-Earth Creationist meteorologists accept the sphericity of the Earth and its atmosphere, even if they give an active role to God in directing weather.[10]

However, the need to mitigate the effects of a future natural disaster was not the only issue that caused the UZS in London to be interested in the West Indies. After Emancipation, a number of freed Blacks made

[5] Of note is that volcanic eruptions have long been used as a metaphor for the emergence of anticolonial resistance movements. See Geo Maher, *Anticolonial Eruptions: Racial Hubris and the Cunning of Resistance* (Oakland, CA: University of California Press, 2022), 1–7.

[6] Shinya Sato, "Effects of localized source currents on magnetotelluric responses of a flat earth and spherical earth," *arXiv* (Jul., 2020), 1–11.

[7] Olena Kuzii and Andrij Rovenchak, "What the gravitation of a flat Earth would look like and why thus the Earth is not actually flat," *European Journal of Physics*, Vol. 40, No. 3 (Apr., 2019), 1–11.

[8] John P. Boyd, "Meteorology and Oceanography on a Flat Earth," *arXiv* (Mar., 2020), 1–43; Tetu Makino, "On Incompressible Vibrations of the Stratified Atmosphere on the Flat Earth," *arXiv* (Feb., 2020), 1–17; Tetu Makino, "On Adiabatic Oscillations of a Stratified Atmosphere on the Flat Earth," *arXiv* (Sep., 2020), 1–35. See also a defense of Flat Earth meteorology: Edward Hendrie, *The Greatest Lie on Earth: Proof That Our World Is Not a Moving Globe* (Great Mountain Publishing, 2016), 120–29.

[9] Anne Lawrence-Mathers, *Medieval Meteorology: Forecasting the Weather from Aristotle to the Almanac* (Cambridge: Cambridge University Press, 2019), 40–109; Rudolph Simek, trans. Angela Hall, *Heaven and Earth in the Middle Ages: The Physical World before Columbus* (Woodbridge: The Boydell Press, 1996), 98–113.

[10] Michael Oard, *The Weather Book* (Green Forest, AR: Master Books, 1997), 8–9.

their way from the Caribbean to the imperial metropole, most as laborers, but a few as intellectual elites. This included two Black figures from the West Indies who, for a brief period, would play a major role in the UZS: John Edward Quinlan and Emanuel Vaun Mulgrave.

Quinlan (1862–1940) was born in Castries, the capital of the Caribbean island of Saint Lucia, which had been a British colony since its seizure from the French in 1814.[11] Although slavery was abolished in the 1830s, at the end of the century the small population of 800 whites continued to politically and economically dominate the island's 39,000 Blacks. Quinlan claimed that his grandfather had been a slave who purchased his own independence, self-emancipating prior to the British government's abolition of the institution. By 1885, Quinlan was working as a land surveyor on the island, lobbying the British government on behalf of landless Black peasants.

In March 1900, Quinlan was elected to be a delegate of the island's population to the First Pan-African Conference that was to be held in London that July. Other delegates included W. E. B. Du Bois; Benito Sylvain (1868–1915), Haitian ambassador to Ethiopia; Frederick Johnson, former Attorney-General of Liberia; and Dadabhai Naoroji (1825–1917), past and future president of the Indian National Congress. At the conference, Quinlan gave a speech on 24 July titled "The Progress of our People," seemingly the first time a call was made for financial reparations to be paid to the descendants of slaves and which included the claim that "In South Africa the aim of the English capitalist seemed to be to enslave the black people." Quinlan was fluent in both French and English from his upbringing on Saint Lucia, and one imagines that his ability to translate may have been of use at the conference. After the conference, Quinlan remained in London (and on the British government payroll until at least 1901), and became active in the Aborigines' Protection Society and the Pan-African Association.[12] By 1902, Quinlan had joined the socialist movement and published a pamphlet, *The Labour Problem: An Idea of a British Workers' Society*, which supported Labour Party candidates for Parliament and urged working-class British workers

[11] Much of the bibliographical information on Quinlan is courtesy of David Killingray of the University of London, who graciously provided notes for an upcoming biographical article of his on Quinlan.

[12] Marika Sherwood, *Origins of Pan-Africanism: Henry Sylvester Williams, Africa, and the African Diaspora* (London: Routledge, 2010), 45–46, 73, 84–85, 137.

to "return one or two dark Colonials as members of the House of Commons."[13]

Given the fact that most Flat Earthers, both at the time and in the modern day, are evangelical Christians, it was rare for a socialist to adopt Flat Earth views. As will be detailed in the next chapter, Rowbotham's origins were as an Owenite utopian socialist, and his measurements of the supposed flatness of the Earth stemmed from observations of the Old Bedford River canal near his commune, but his negative experience there seemed to have turned him away from not only socialism, but any positive notion of community whatsoever, later comparing communitarianism to slavery, setting him up to be politically closer to modern individualist Flat Earthers.[14] Lady Blount's 1898 novel *Adrian Galilio* opens with a denunciation of "lawlessness, Socialism, Anarchism, dynamitism, and many other "isms"" as leading to "the adoption of democracy, bloodshed, and destruction."[15]

In contrast, the British Trotskyist group International Marxist Tendency has argued that the resurgence of Flat Earth thought is a byproduct of the capitalist distortion of not only education, but entertainment and social relations.[16] The same "Protestant work ethic" which was promoted in the early twentieth century as the cornerstone of industrial capitalism and time management was an outcome of the same Protestant cultural propaganda which established the Flat Error.[17] There has been

[13] John Edward Quinlan, *The Earth a Plane* (London: John Edward Quinlan, 1906), 1; David Killingray, "The Telephone," in David Dabydeen, John Gilmore, and Cecily Jones, eds., *The Oxford Companion to Black British History* (Oxford: Oxford University Press, 2007), 476; Earl Bousquet, "From Emancipation Through Recognition, Reconciliation, Respect," *The Voice*, August 20, 2015, accessed July 21, 2020: https://www.thevoiceslu.com/2015/08/from-emancipation-through-recognition-reconciliation-respect/.

[14] John Langdon, ""A Monument of Union": Social Change and Personal Experience at the Manea Fen Community, 1839–1841," *Utopian Studies*, Vol. 23, No. 2 (2012), 525.

[15] Elizabeth Blount, *Adrian Galilio; or, a Song Writer's Story* (Malvern: The Faith Press, 1898), 6.

[16] Bryce Gordon, "Flat-Earth Revival: A Distortion of Capitalism," *Socialist Revolution*, No. 21 (Mar., 2020), 12–13. See also George Monbiot, "Conspiracist Thinking Is the Result of 40 Years of Social Atomization," *Jacobin*, September 27, 2021, accessed September 27, 2021, https://jacobin.com/2021/09/conspiracy-theories-qanon-covid-left-anarchism-counterculture-extreme-right-restoration-stories-bridging-communities/.

[17] Peter Razzell, "The Protestant Ethic and the Spirit of Capitalism: A Natural Scientific Critique," *The British Journal of Sociology*, Vol. 28, No. 1 (Mar., 1977), 17–37.

research on the role of neoliberal economics and postcoloniality creating environments where conspiracy theories thrive. This would help explain the prevalence of Flat Earth thought not only in the modern United States, but also heavily neoliberalized countries such as Nigeria, Brazil, and South Africa, as will be discussed in chapters six and seven.[18] When Samuel Shenton died in 1971, his records of the International Flat Earth Research Society were saved by the Trotskyist politician and academic Ellis Hillman (1928–1996).

However, it should be noted that probably the most significant Marxist historian of science, John Desmond Bernal (1901–1971), endorsed the Flat Error in his 1954 exploration of the origins of science.[19] Likewise, during his 1981 campaign for deputy leader of the Labour Party, British socialist politician Tony Benn (1925–2014) said of his opponents on the Labour right, "They can call me a Marxist, a Jesuit, a flat-earther, a Trotskyite, a vegetarian. I don't care, because I know why they call us names. It's because they dare not face our arguments."[20] From the opposite end of the political spectrum, the Society of Saint Pius X, a Traditionalist Catholic organization, claims that modern Flat Earthism is a direct result of Rowbotham's earlier adoption of utopian socialism over Christianity.[21] Nor is this unique to Flat Earthers; Young-Earth Creationism in the United States is also closely linked with anti-communism.[22]

The anti-communist connection to Flat Earth belief will be explored more in chapters six and seven, but suffice to say Quinlan's commitment to both socialism and Flat Earth was not his only contradiction. In 1903, Quinlan spoke against the recruitment of West Indians for work in South African mines, and by 1906 he had established the short-lived National Society for the Protection of the Dark Races, headquartered in the Bayswater area of London. The society promoted good relations

[18] Liane Tanguay, "We, the Sheeple: Making Sense of Conspiracy Theory in the Context of Neoliberalism," *symplokē*, Vol. 29, Nos. 1–2 (2021), 665–76.

[19] J. D. Bernal, *Science in History, Volume 1: The Emergence of Science* (Cambridge, MA: The MIT Press, 1971 [1954]), 303–07.

[20] *TV Eye*, ep. "Benn's Bandwagon" (prod. Jack Saltman), ITV, 28 May, 1981.

[21] "Flat Earth? The Hidden Side of a Hoax," Society of Saint Pius X, July 11, 2023, accessed July 11, 2023, https://sspx.org/en/news-events/news/flat-earth-hidden-side-hoax-84054.

[22] Carl R. Weinberg, *Red Dynamite: Creationism, Culture Wars, and Anticommunism in America* (Ithaca: Cornell University Press, 2021).

between the races at home in Britain and took an active stance in imperial racial politics. In its May 1907 single issue, the society's journal, *The Telephone*, called for economic development in Britain's tropical colonies, the creation of multiracial working-class clubs in Britain, for the Empire to ensure the protection of the rights of the "dark races," and to "put a stop to all punitive expeditions in the British dominions beyond the seas."[23] Despite its single issue lifespan, *The Telephone* is notable for being the only known Black-owned newspaper in Britain in the years 1907–1912.[24]

Although Quinlan was the founder and primary driver of the National Society for the Protection of the Dark Races, he was not its president. That position fell to none other than Lady Blount, who was already President of the Universal Zetetic Society.[25] Robert Schadewald referred to Lady Blount as "a right-wing Shirley MacLaine," but this does not seem an accurate reflection of her politics, especially on race relations and imperial-metropole ties.[26] 1906 would prove to be a busy year indeed for Quinlan, as in addition to founding the National Society for the Protection of the Dark Races, he also published a Flat Earth pamphlet of his own, *The Earth a Plane*. The publication of *The Telephone* a year later therefore drew on Quinlan's experience writing his Flat Earth work, and also meant that for a time, the only Black-owned newspaper in Britain was published by a Flat Earther.

Quinlan begins *The Earth a Plane* by noting his position as "Commissioned Land Surveyor of St. Lucia and St. Vincent, British West Indies," before diving into the heart of the matter (Fig. 4.2):

> I am aware that one of the first questions that globites will put to me is: How can ships sail round the world if it is not a globe? In answer to this query, please note that ships cannot sail round the world on a uniform course, whether the world be a globe or a plane. This is impossible except

[23] Victoria J. Collis-Buthelezi, "Caribbean Regionalism, South Africa, and Mapping New World Studies," *Small Axe*, Vol. 19, No. 1 (Mar., 2015), 54; Killingray, "The Telephone," 475.

[24] David Killingray, "Rights, Land, and Labour: Black British Critics of South African Policies before 1948," *Journal of Southern African Studies*, Vol. 35, No. 2 (Jun., 2009), 389.

[25] Elizabeth Blount, "Ancient and Modern Astronomy," *The Earth*, Vol. VI, Nos. 72 & 73 (Jul./Aug., 1906), 97.

[26] Robert J. Schadewald, *The Plane Truth: A History of the Flat Earth Movement* (Wendy S. Schadewald, 2015), 107.

Fig. 4.2 The cover of *The Earth a Plane*, with Quinlan's portrait

in a latitude which is south of Cape Horn. Land would intervene on every uniform course that lies between this extreme southerly latitude and the Arctic regions. But when airships shall have been brought to a high state of perfection, a globe-trotter will be able to sail round the world on a uniform course by getting into one from a sailing ship when land was in the way, and out of it, back into another sailing ship when water was reached.[27]

Quinlan brings up Cape Horn, at the southern tier of Chile and Argentina, at a time when the British "informal empire" of economic domination over both countries was at its height.[28] Quinlan's establishment of the Society for the Protection of the Dark Races may have

[27] Quinlan, *The Earth a Plane*, 1, 3.

[28] Ben Markham, "The Challenge to 'Informal' Empire: Argentina, Chile and British Policy-Makers in the Immediate Aftermath of the First World War," *The Journal of Imperial and Commonwealth History*, Vol. 45, No. 3 (2017), 449–74.

been inspired by the Bambatha Rebellion, a Zulu uprising in the Natal Colony in 1906.[29] If so, it is also tempting to consider whether Thomas Winship, a UZS member living in Natal at the time—discussed more next chapter—forwarded firsthand accounts of the rebellion to his comrades in London. Quinlan also brings up the longstanding Zetetic interest in air travel; by emphasizing airships over balloons, Quinlan perhaps draws on the same African traditions as African Americans, while also updating the technology to the twentieth century. As mentioned above, Lady Blount made her own airship design three years later; one wonders if the seed to do so was planted here by Quinlan.

After rejecting the "globite" argument on global travel, Quinlan turns to the technology used to navigate that route:

> The magnetic compass, that would help the globe-trotter to keep on his course, always points to the magnetic north when there are no local attractions. The magnetic north is near the North Pole. The North Pole is the centre of the extended plane—the earth—and not one of the two flattened ends of a globular world as is so often asserted. It was not by white scientists and astronomers that the magnetic attraction toward the North Pole was discovered, but by learned Chinese who knew then, as their offspring know now, that the earth was flat.[30]

In other words, the basic science of navigation proves the Earth is flat. What is more, the origin of this science is not the "white scientists and astronomers" who claim the world is spherical, but the "learned Chinese" whose Confucian cosmology was based on a Flat Earth. Recall also the presence of the former US consul to China, George Henry Colton Salter, on the board of the New York Zetetic Society; and the use of the Great Wall and Grand Canal of China in Zetetic arguments will be covered more next chapter. For Quinlan, this statement seems to develop from extending his Pan-Africanism into a wider Afro-Asian solidarity bloc, decades before the 1955 Bandung Conference, and at the same time that his erstwhile fellow conference delegate Du Bois was beginning to articulate his own formation of Afro-Asian solidarity around the "world color

[29] Killingray, "The Telephone," 475; Killingray, "Rights, Land, and Labour," 388–89.

[30] Quinlan, *The Earth a Plane*, 3.

line."[31] One wonders if Quinlan's own Afro-Asian approach to Flat Earth science emerged from the participation of Indians at the Pan-African Congress.

If Quinlan positioned himself as a radical in the above passages with his Flat Earth sentiment seemingly a natural expansion of his views since the Pan-African Conference, he ends his short pamphlet with a surprising twist: that the Empire as an institution is vital to providing the measurements that would prove the Earth flat. Having

> mentioned the meridian of Greenwich, I must point out, in connection with the alleged interference of the London County Council electric generating station at Greenwich with the delicate instruments of the Royal Observatory, that it is the Royal Observatory that should be removed and not the generating station. An observatory ought to be erected on a spot at the level of a large sheet of water, which would serve as a natural horizon. Great Britain is surrounded by such water; but Great Britain is otherwise unsuited for a proper observatory for reasons which I give in the next paragraph, and also because every meridian that passes through Great Britain crosses the Equator in mid ocean, where a sister observatory cannot be erected.... Two observatories are therefore indispensable, one at the equator and the other 45° north or south of the Equator. There are no such observatories in existence, and I venture to say, without fear of contradiction, that there are only two spots in the world on the same meridian—one at the Equator, and the other 45° north or south of the Equator—where the angle of elevation to the sun could be measured from a natural horizon.[32]

It is not just Britain whose technical and industrial capabilities are needed to build observatories; it is the territory of the British Empire itself that is needed to be acquired (and defended) in order to build those observatories. This is somewhat surprising, given the previous allusions to Afro-Asian solidarity and his participation in the Pan-African Conference. A possible explanation can be found in the tendency of West Indian middle classes, from the Victorian era through the end of World War II,

[31] W. E. B. Du Bois, *The World and Africa: An Inquiry into the Part which Africa has Played in World History, New Enlarged Edition* (New York: International Publishers, 1965), 311–16; Bill V. Mullen, *Afro-Orientalism* (Minneapolis: University of Minneapolis Press, 2004), 1–42.

[32] Quinlan, *The Earth a Plane*, 5–7.

to attempt to refashion themselves to be British, while also refashioning Britishness into an ideal that suited them. This included a heavy focus on respectability and Victorian emphasis on middle-class values and propriety, basing their concepts of citizenship and imperial identity on replicating the class system of Britain in the Caribbean.[33] For the first few years after independence in the 1960s, most West Indian governments tended to pursue foreign policies more aligned with Britain than with African or Latin American states.[34] As mentioned above, in *The Telephone* Quinlan also seemed to view the Empire as an institution that could extend legal protections to the natives of southern Africa (even if it was ultimately responsible for many of those problems in the first place). These factors may help explain the seeming contradictions in Quinlan's view of the Empire as an institution.

For all of his importance as a pioneering Pan-Africanist, Quinlan is not well remembered. The widest discussion of him in contemporary sources seems to be Marika Sherwood's (1937-2025) history of Pan-Africanism, where Quinlan is quoted and discussed several times; however, his Flat Earth advocacy is entirely absent.[35] He is arguably better remembered by Flat Earthers, and even then he is not prominent. Notably, as of 2025 the only YouTube video extant about him is by an African American Flat Earther, "Flat Earth Brooklyn Live."[36] However, even so, Quinlan seems to be more remembered by modern Flat Earthers than his contemporary Black British Zetetic of the time, the Jamaican immigrant Emanuel Vaun Mulgrave (1859–1916). This is ironic given that, despite his connection with Lady Blount via the leadership of the Society for the Protection of the Dark Races, Quinlan does not seem to have been very active in the UZS. Mulgrave, however, was at least for a time one of its leaders.

[33] Anne Spry Rush, *Bonds of Empire: West Indians and Britishness from Victoria to Decolonization* (Oxford: Oxford University Press, 2011), 21–68.

[34] Martin Ira Glassner, "The Foreign Relations of Jamaica and Trinidad and Tobago, 1960–1965," *Caribbean Studies*, Vol. 10, No. 3 (Oct., 1970), 127–31.

[35] Sherwood, *Origins of Pan-Africanism*, 260.

[36] Flat Earth Brooklyn Live, "How Could The Earth be Flat | The Earth Is A Plane John Edward Quinlan," YouTube, November 22, 2017, accessed July 22, 2020, https://www.youtube.com/watch?v=cZ4VUJ94J1s. At least one other YouTube video about Quinlan, also by an African American Flat Earther, existed as of 2020 before being subsequently deleted in YouTube's suppression of conspiratorial videos after the covid pandemic and QAnon-inspired 6 January 2021 attack on the US Capitol.

Writing in a 1905 publication they coauthored, Lady Blount noted of Mulgrave that "Until somewhat recently this gentleman was known as the noted "Rev." E. V. Mulgrave. He was not only "noted" as a minister of the Church of England, but also he is notable by birth and lineage. His father's mother having been a Princess, and his great-grandfather on his mother's side was Lord Colin Campbell, of the Campbells' Clan of Scotland."[37] Presumably this would indicate that Mulgrave's paternal grandmother had been a princess in Africa; which Colin Campbell Lady Blount is referring to is unclear. However, Lady Blount's claims here are the only source for Mulgrave's early biography, and Black claims of aristocratic descent were useful in deflecting racism (see also the dispute over Benjamin Banneker's background).

Whatever his grandparents' origins, Mulgrave grew up in Jamaica at a time when there was a backlash against the initial attempts to provide extensive education to the former slaves. Education consisted mainly of Christian and trades education.[38] The role that Christian education could play in the development of Flat Earth ideas has already been covered, but the role that technical education plays will be explored in chapter six. Mulgrave was originally a seaman, and it was in that capacity that he came to Britain in 1883. Becoming a Protestant, he spent time as a missionary in the Transvaal Republic in southern Africa, an experience which will be explored more next chapter. By 1896 he was back in Britain, where he married a white woman. By 1898, he was a fixture in Sussex, working both as a preacher and as a speaker on the "Customs and life in Darkest Africa" and urging an improvement in "the treatment of native races in South Africa."[39] Interestingly, one of the few biographical publications on Mulgrave does not mention his Flat Earth advocacy, while his two appearances in Flat Earth studies do not mention his race.[40]

[37] Elizabeth Blount and E. V. Mulgrave, *Mr. Blatchford's Great Mistake, or, Clarion's "Science" Versus God's Truth* (Kingston Hill: Lady Blount, 1905), 33.

[38] Dexter J. Gabriel, *Jubilee's Experiment: The British West Indies and American Abolitionism* (Cambridge: Cambridge University Press, 2023), 203–07.

[39] David Killingray, "Black Baptists and Pan-Africanism in Britain, 1890–1913," *Immigrants & Minorities*, Vol. 35, No. 1–2 (Sep., 2020), 115. As with Quinlan, thanks to David Killingray for graciously sharing some of his biographical research on Mulgrave.

[40] Killingray, "Black Baptists," 115; Michell, *Eccentric Lives*, 28; Schadewald, *The Plane Truth*, 110.

Flat Earth appears to have played a major role in Mulgrave's personal and religious development. In a 1903 letter to the UZS journal *The Earth*, Mulgrave related how

> My experience is contrary to what I was taught to believe, viz: that the earth is a globe. As a boy, in Jamaica, I used to ask my parents many questions about the earth; and my mother told me there were several points in Jamaica from which Cuba could be seen. Later I proved this to be true. In March, 1881, I went from Montigo [*sic*] Bay to St. Ann's Bay; some companions went with me up the hills, and the morning being bright and very clear, I saw two small sails standing out from shore quite distinctly, about five miles off. Another sail was also visible, about 80 miles away, which I and my friends made out to be a barque leaving Cuba. This we saw with the naked eye. One of my companions, a pilot boy, had a small telescope with a range of 40 miles, but on looking through the glass we could not see these ships; so the owner of the telescope, whose turn it was to watch for ships, stayed until he was able to ascertain that the vessels were going to Kingston, Jamaica. I would suggest a few causes which prevent anyone seeing a ship a great distance away, although water is level:
> 1. The nature of the atmosphere. 2. The capacity of our visual organs.[41]

The origins of his Flat Earth belief seem therefore to have stemmed from his mother, and also related to his early career as a seaman. One presumes Mulgrave would be very interested in the issue of how to navigate across the seas of a Flat Earth, a topic that will be discussed in relation to a more famous nineteenth-century sailor in the next chapter.

By 1904, publications of the Zetetic Society listed Mulgrave as being a member of the society's Committee.[42] The next year, Mulgrave and Lady Blount jointly published a pamphlet, *Mr. Blatchford's Great Mistake, or, Clarion's "Science" Versus God's Truth*, which attacked the globularist position of the titular individual, seemingly an atheist gadfly who Mulgrave had tangled with in his public preaching for a decade by that point. As mentioned above, Lady Blount used the pamphlet to give some

[41] E. V. Mulgrave, "Letters," *The Earth*, Vol. III, Nos. 39 & 40 (Oct. and Nov., 1903), 292.

[42] Elizabeth Blount and Albert Smith, *Zetetic Astronomy: or The Sun's Motions North and South; with the Moon's Motions; Fancied and Real. Showing the Uselessness of the Gravitation Theory*, etc. (Kingston Hill, Surrey: Lady Blount, 1904), 74.

genealogical information on Mulgrave. Mulgrave himself states that he had used to be a reverend of the Church of England, but was compelled to leave and become a Baptist pastor (the same faith as Jasper, on the other side of the Atlantic at the same time) due to his Flat Earth beliefs being at odds with the Church's acceptance of the spherical Earth, which he viewed as blasphemy.[43]

For such a devout Baptist, Mulgrave draws from some interesting sources in *Mr. Blatchford's Great Mistake*. He and Lady Blount note that astrology and astronomy were the same in the ancient world, so ancient astrologers were the same as modern astronomers and implicitly can be used to challenge the moderns. As such, Rowbotham

> informs us that "tables of the phases of the sun and moon, of eclipses, and kindred phenomena, have existed for thousands of years, and were formed independently of each other by the Chaldean, Babylonian, Egyptian, Hindoo, Chinese, and other ancient astronomers; and this was long before the globular theory was accepted."[44]

This is a fairly even-handed acceptance of pre- and non-Christian belief from across the world for a strict Baptist. However, while Mulgrave did not have quite the history of activism as Quinlan, he was still a Black Jamaican who had been in southern Africa and had promoted Black causes in Britain. Perhaps some element of Afro-Asian solidarity drifted into Mulgrave's worldview from the wider membership of the Zetetic Society. While there is no indication that Mulgrave and Quinlan interacted, it cannot be dismissed, especially given their joint connection to Lady Blount.

Perhaps more on brand, and also lining up more with where modern Flat Earthers would go, was the invective that

> Those who are acquainted with the books which such a man has read, know that they present a *resumé* of what he repeats, without even the shadow of an original thought to garnish the reproduction. Hence it follows that if these men read scientific and evolutionary fiction, they will repeat the self-same fiction. Consequently, when Dr. Chalmers was "bitten"

[43] Blount and Mulgrave, *Mr. Blatchford's Great Mistake*, 32; Killingray, "Black Baptists," 115, 128n60-62.

[44] Blount and Mulgrave, *Mr. Blatchford's Great Mistake*, 23n2.

> with pseudo-science, and published his grandiloquent essays and "high falutin" anent myriads of glorious inhabited worlds wandering through space, a "rush" was made for his essays, which were eagerly absorbed and repeated throughout the length and breadth of the land, as if his high-flown fiction were veritable Gospel Truth. The inevitable result of this sort of preaching is that most of the frequenters of the various Churches and Chapels are in the condition of voracious devourers of sensational serials and novels of the Marie Corelli type.[45]

Thomas Chalmers (1780–1847) was a major figure in the Church of Scotland who attempted to integrate gap creationism with the evolving notion of the geological age of the Earth, and also promoted the notion of the "plurality of worlds," an early view of extraterrestrial life. Marie Corelli (1855–1924) was a British novelist most notable for her 1886 debut, *A Romance of Two Worlds*, a proto-New Age novel touching on Spiritualism and evolution. In attacking their views, Mulgrave and Lady Blount reject the idea of life on other worlds, as in the Zetetic cosmogony, there are no other worlds, from both mechanical and theological standpoints, God having placed his greatest creation at the center of the universe.

From the center of their empire, therefore, Lady Blount and Mulgrave attack a theologian's acceptance of the displacement of their world from the center of creation. But a century prior, the dissolution of another empire had embraced a similar dislocation. In 1809, the population of the South American city Quito rose up against the rule of Spain, which itself had been invaded by Napoleonic France the year earlier. Quito's political leaders declared an independent state, the existence of which was crushed less than a year later under a royalist counterattack. Nevertheless, the First Autonomous Government Junta of Quito was the first splinter in the political unity of the Spanish Empire, and in 1830 the territory of Quito would become the independent Republic of Ecuador, named after its position on the global equator.[46] The dissolution of the Spanish Empire, whose establishment was connected to Columbus and the Flat Error stemming from him, itself drew on the fact that the world is a globe.

[45] Blount and E. V. Mulgrave, *Mr. Blatchford's Great Mistake*, 27.

[46] Carlos de la Torre and Steve Striffler, "A New Nation," in Carlos de la Torre and Steve Striffler, eds., *The Ecuador Reader: History, Culture, Politics* (Durham, NC: Duke University Press, 2008), 99.

The connection between Flat Earth belief and criticism of imperialism and its racial order was not limited to the UK, US, West Indies, and Latin America, however. Just as colonization of the Americas and Caribbean was connected to the colonization and exploitation of Africa, Flat Earthers of the late nineteenth and early twentieth centuries had much to say in criticism of that project also. This not only included criticism from the Zetetics of London, but from amongst the settler project as well—and not only from those of British descent.

CHAPTER 5

"The World According to Kruger"

It is ironic, but apt, that perhaps the greatest novelists of the prior two centuries were both born in colonial India: the defender of empire Rudyard Kipling (1865–1936) and critic of empire Eric Blair (1903–1950), better known by his pen name George Orwell. Indeed, Orwell, the radical socialist, was an admirer of the conservative Kipling's mastery of storytelling.[1] Notably—given the movement was at a low point at that time—both authors touched upon the Flat Earth in their writings. For Kipling, this was his 1913 story "The Village that Voted the Earth was Flat." Set in the fictional English village of Huckley, which after hosting the annual meeting of "the Geoplanarian Society—a society devoted to the proposition that the earth is flat," is convinced to vote in favor of the Earth being flat. This is the result of Huckley's press, judicial, and political institutions falling under the control of a Radical MP.[2]

Orwell's initial dip into the topic came in the 27 December 1946 edition of his regular *Tribune* column "As I Please." In it, Orwell reflected on Irish playwright George Bernard Shaw's (1856–1950) comment that "The average man... can advance not a single reason

[1] J. Carlos Viana Ferreira, "Orwell on Kipling: an imperialist, a gentleman and a great artist," *Anglo Saxonica*, Ser. III, No. 13 (2017), 61–71.

[2] Rudyard Kipling, "The Village that Voted the Earth was Flat," in Stephen Jones, ed., *The Mark of the Beast and other Fantastical Tales* (London: Gollancz, 2007), 637–74.

E. Guimont, *The Power of the Flat Earth Idea*,
Palgrave Studies in the History of Science and Technology,
https://doi.org/10.1007/978-3-032-23981-5_5

for thinking that the earth is round. He merely swallows this theory because there is something about it that appeals to the twentieth-century mentality." Orwell spends the rest of the essay musing what evidence a non-scientist like himself might be able to deploy against the claims of Flat Earthers.[3]

This topic clearly resonated with Orwell, as he began work the next year on his most famous work, the science fiction dystopia that would become titled *Nineteen Eighty-Four*, continuing the connection to the Flat Earth and science fiction touched upon by previous chapters. Early in the novel, the main character, regime dissident and "thoughtcriminal" Winston Smith, muses

> as he had many times wondered before, whether he himself was a lunatic. Perhaps a lunatic was simply a minority of one. At one time it had been a sign of madness to believe that the earth goes round the sun; today, to believe that the past is unalterable.... The obvious, the silly, and the true had got to be defended. Truisms are true, hold on to that! The solid world exists, its laws do not change. Stones are hard, water is wet, objects unsupported fall towards the earth's centre.[4]

When Winston is eventually captured by the ruling Party, his interrogator, O'Brien, after endorsing Young-Earth Creationism and rejecting fossils as the creation of nineteenth century biologists, tells Winston that stars are merely

> bits of fire a few kilometres away. We could reach them if we wanted to. Or we could blot them out. The earth is the centre of the universe. The sun and the stars go round it.... For certain purposes, of course, that is not true. When we navigate the ocean, or when we predict an eclipse, we often find it convenient to assume that the earth goes round the sun and that the stars are millions upon millions of kilometres away. But what of it? Do you suppose it is beyond us to produce a dual system of astronomy? The stars can be near or distant, according as we need them. Do you suppose our mathematicians are unequal to that? Have you forgotten doublethink?[5]

[3] George Orwell, "As I Please, *Tribune*, 21 December 1946," in Sonia Orwell and Ian Angus, eds., *The Collected Essays, Journalism and Letters of George Orwell, Volume IV: In Front of Your Nose, 1945–1950* (London: Secker & Warburg, 1968), 259–62.

[4] George Orwell, *1984* (New York: Plume, 1983 [1949]), 70–72.

[5] Orwell, *1984*, 236–37.

After ultimately being broken by O'Brien, Winston ultimately "set to work to exercise himself in crimestop. He presented himself with Propositions—'the Party says the earth is flat', 'the party says that ice is heavier than water'—and trained himself in not seeing or not understanding the arguments that contradicted them."[6]

Of these three works, the most influential—not merely in a Flat Earth sense—is obviously *Nineteen Eighty-Four*. But with its depiction of a government suppressing the truth to support its totalitarian rule, Flat Earthers (as with many conspiracists and conservatives) have found the novel useful. In at least one case, a prominent Flat Earther seems to have believed that Orwell's novel was an admission of the coverup by the elites. In a 3 May 1973 letter to Charles Barren (1913–1999), a professor at North East London Polytechnic (now University of East London) and chairman of the Science Fiction Foundation, Charles K. Johnson, then-president of the International Flat Earth Research Society (IFERS), wrote how "I know I don't have to mention to you the Society of Jesus, Jesuits are the Max Factor in all. As far as I can tell it is they who have been the main purvayers [*sic*] of the completely insane idea of a round spinning world., [*sic*] of couse [*sic*] in '1984' they clean it up and confess they lied all the while."[7] Johnson would return to this in his 1978 article "Backward is forward or this Flat Earth is a spinning ball," where he claimed that O'Brien represented the US Democratic Party, that Winston could have held out against O'Brien if he had been buttressed with the knowledge of the Flat Earth, and the British public school system caused the rise of the Party by destroying truth with their promotion of the globular Earth.[8] Needless to say, this is what might be considered fanfiction, as it does not follow the backstory provided in the novel. Two months later, Johnson would write the article "Australia Not Down Under," which opened by quoting Kipling.[9]

[6] Orwell, *1984*, 248–49.

[7] Charles K. Johnson, "Outgoing Correspondence 2348," The Flat Earth Society Library, March 12, 2014, accessed November 30, 2019, https://www.theflatearthsociety.org/library/correspondence/Charles%20K.%20Johnson%20-%20Outgoing%20Correspondence%202348.pdf.

[8] Charles K. Johnson, "Backward is forward or this Flat Earth is a spinning ball," *Flat Earth News* (Mar., 1978), 4.

[9] Charles K. Johnson, "Australia Not Down Under," *Flat Earth News* (May, 1978), 1.

While Johnson was wrong about O'Brien representing US Democrats, he was not wrong in that the Party of *Nineteenth Eighty-Four* was meant as an unsubtle critique of the Soviet communist party, which at the time was at the height of the promotion of scientifically erroneous Lysenkoist genetics. Orwell named Winston Smith after British Conservative Party leader Winston Churchill. Similarly, in the Kipling story, the protagonists are a Conservative MP and his friends, and the antagonists who lead to the spread of the Flat Earth belief in town are the Radicals. The conservative Kipling and the socialist Orwell both associated the Flat Earth movement with the political left—not the right, as it generally aligned with by the time Johnson took over IFERS. This is perhaps not too surprising in the British context. Both Edwin Abbott Abbott's (1838–1926) *Flatland* (1884) and H. G. Wells' (1886–1946) *The Time Machine* (1894) were novels which used alternative geometries of the world to critique Victorian British society and its hierarchies; and what is the Flat Earth if not a very alternative geometry? But also of note is that both Kipling and Orwell were products of the British Empire, born in India and working there before returning to the metropole. Although Johnson was American, his wife, Marjory (d. 1996), who helped him run the Society, was Australian. The Flat Earth idea emerged in Britain in the mid-1800s, at the time the British Empire was rapidly expanding across the globe, and its proponents included clergy searching for a new cause after having succeeded in the slavery abolition movement, as in the Black Flat Earthers of the previous chapters. A distrust of the scientific and technical knowledge used to prop up the Empire whose racial and military stances they disagreed with—from this point of view, there is a valid reason that those who might wish for the Sun to set on the British Empire might have a very different notion of what exactly was meant by a sunset (Fig. 5.1).

Modern Flat Earth sentiment emerged through the figure of Samuel Birley Rowbotham (1816–1884), who had originally been an Owenite utopian socialist. Rowbotham helped organize an Owenite commune at Manea Fen in Cambridgeshire, along the Old Bedford Canal.[10] In

[10] Robert J. Schadewald, *The Plane Truth: A History of the Flat Earth Movement* (Wendy S. Schadewald, 2015), 8–11; Kelly Weill, *Off the Edge: Flat Earthers, Conspiracy Culture, and Why People Will Believe Anything* (Chapel Hill: Algonquin Books, 2022), 6–12; John Langdon, ""A Monument of Union": Social Change and Personal Experience at the Manea Fen Community, 1839–1841," *Utopian Studies*, Vol. 23, No. 2 (2012), 511.

Flat Earth News

International Flat Earth Research Society
$7.00 yearly to Associate Members only.
Box 2533, Lancaster, California 93534
Phone: (805) 946-1595

Quarterly
RESTORING THE WORLD TO SANITY
May 1978
Charles K. Johnson, President
Marjory Waugh Johnson, Secretary

AUSTRALIA NOT DOWN UNDER

A Compass Always Points North

Australia...

AUSTRALIA could very well be the Key or the catylist to the DESTINY of the human race and the Future of the World! The answer could be my friend in this IF, so much hinges on this little word as Kipling has so well said.... "If you can keep you sanity, when all about you are losing theirs..." If Aussies could as a people overcome their natural reticence and their deference to Crazy English Scientists... who have kept them "down under" you, for so very long, come out of the closet and Lift up their voice and speak up with one voice and state what they all know to be true, they are not hanging head down by their feet... under the world!

Well anyway, the doors have opened over there. How it will all end, I don't know. Early April I was interviewed by a man in Washington D.C.. April 14, the interview verboten was on front page of Washington Star, D.C.! Next, the news went all over. Then Australia got it, and had UPI news service come over from L.A. to get pictures of us both. Australia wanted them! April 24, Aussie time I have a live interview on ABC Radio, Sydney, Australia, via telephone with Carylin Jones (Australian Broadcasting Co.), 15 minutes I told ALL! A couple of hours ago I did again broadcast in Australia, Melbourne. This time an interview with Bert Newton, Radio Australia, Melbourne. It will be MAY DAY there, in more way than one, May 1. Also newspaper articles are running ALL OVER AUS-

Fig. 5.1 May, 1978 issue of *Flat Earth News*, only one of many in which the Johnsons would proclaim the idea of a globe Earth to be inherently derogatory to Australians.[11] Ironically, by the early twenty-first century, later Flat Earthers would claim that Australia did not even exist and was invented to cover up the British executing the prisoners supposedly exiled there[12]

[11] Johnson, "Australia Not Down Under," 1.

[12] Dan Ariely, *Misbelief: What Makes Rational People Believe Irrational Things* (New York: HarperCollins, 2023), 89–90.

Chapter 3, the fact that African American Flat Earth belief developed in the context of the fringe basis for Antebellum science among the white Southern elite was discussed. In contrast, the materialism of the Owenites fueled Spiritualism among the religious English middle class as a conduit to oppose them, an account of which was given by no less than Friedrich Engels (1820–1895) in his 1878 essay "Natural Science and the Spirit World."[13]

By 1841, Manea Fen collapsed, and Rowbotham ultimately transitioned to Flat Earth advocacy with his adoption of the pseudonym "Parallax" and his 1849 publication of the pamphlet *Zetetic Astronomy: Earth Not a Globe*. The term "Zetetic" means "to question," and was originally used by proponents of atheism in 1820s Scotland. In the 1880s, a Zetetic Society of Spiritualists would be established in England; the term would later be incorporated into the title of *Zetetic Scholar*, a 1978–1987 journal in the United States that focused on topics such as UFOs, cryptozoology, and psychic powers. As will be seen in Chapter 7, this is not the only time that Biblical Flat Earthers would adopt terminology of social groups they oppose. Rowbotham's usage of the term was unusual in that he had turned to Biblical literalism to shore up his Flat Earth views.[14] However, Rowbotham was not entirely done with Manea Fen. In 1838, Rowbotham had gone into the Old Bedford Canal where he had chosen to place the colony, and using the thirty-four-kilometer straight waterway, claimed that he could observe a floating flag for ten kilometers, meaning that the curvature of the Earth was false. This would become known as the Bedford Level Experiment and would be repeated for decades by both Flat Earth proponents and detractors, making them some of the earliest examples of what has been termed ARIGs—Amateur Research Investigation Groups—untrained amateurs who use flawed scientific methods to try to prove pseudoscience. The most famous case would be by evolutionary advocate Alfred Russel Wallace (1823–1913) in 1870, leading to

[13] Friedrich Engels, trans. Clemens Palme Dutt, *Dialectics of Nature* (Moscow: Foreign Languages Publishing House, 1954 [1883]), 70.

[14] Christine Garwood, *Flat Earth: The History of an Infamous Idea* (New York: Thomas Dunne Books, 2007), 36–78; Schadewald, *The Plane Truth*, 11–22; Weill, *Off the Edge*, 15–16; Gordon Pentland, "The freethinkers' zetetic society: an Edinburgh radical underworld in the eighteen-twenties," *Historical Research*, Vol. 91, No. 252 (May, 2018), 314–32.

a major lawsuit and a refusal to admit defeat by the Zetetics.[15] The irony of the Zetetic use of canals as a means of confirming the Biblical flatness of the Earth is that the era of canal construction led to the development of notions of deep time. Surveyor William Smith (1769–1839) used the surveys for the creation of the Somersetshire Coal Canal, the Thames & Severn Canal, the North Wilts Canal, the Kennet & Avon Canal, and the Wilts & Berks Canal to collect remains of mastodon and dinosaur fossils in support of his theory on the "Order of Strata," which argued "that the earth had been an infinite time in arriving at perfection" at odds with the Young-Earth Creationism that later Zetetics would attach to the Flat Earth measurements of the canals.[16]

Nor were canals the only straight lines that could be used to confirm the Earth's flatness. Writing in 1896, a Zetetic named Ebenezer Breach began his pamphlet by offering several constructs to be used for the purpose: "Level Seas, Level Walls, Level Railways, prove a Level Earth, as the great wall of China, 22,000 kilometers long, proves it."[17] This seems to have been written from the erroneous assumption that the Great Wall was built in a straight line, but given that this was an Englishman—white, in contrast to John Quinlan's use of China—using ancient Chinese architecture in an approving way during the era of the Unequal Treaties, Breach's lack of knowledge of the Great Wall is second to his willingness to use it at all.

To the Zetetics of the late nineteenth century, therefore, canals, railways, and steamships were key to proving the Earth was flat—and not coincidentally, those were also the infrastructure needed to keep the

[15] Samuel Rowbotham, *Zetetic Astronomy: Earth Not a Globe, Third Edition* (London: Simpkin, Marshall, and Co., 1881), 13–17; John Michell, *Eccentric Lives and Peculiar Notions* (London: Thames and Hudson, 1984), 22–26; Garwood, *Flat Earth*, 79–118; Weill, *Off the Edge*, 26–29; Schadewald, *The Plane Truth*, 42–66.

[16] Jill Darrell and Diana Clements, "Fossil Collector," in Oxford University Museum of Natural History, eds., *Strata: William Smith's Geological Maps* (Chicago: The University of Chicago Press, 2020), 184–85, 188; Mick Ashworth, *Why North is Up: Map Conventions and Where They Came From* (Oxford: Bodleian Library Publishing, 2019), 167–73.

[17] Ebenezer Breach, *Fifty Scientific Facts For Surrender to Nature's Fixed Truths* (London: John Williams, 1896), i.

Empire running.[18] The promotion of Zetetic science to protect the colonial subjects of the West Indies from volcanoes was addressed last chapter, but the connection of Zetetic navigation with maintaining the Empire is explicit in an undated, unsigned UZS pamphlet addressed to the admiralty and merchant marines of both the United States and United Kingdom:

> If those whom it may concern are really in earnest when they propose to take into consideration any means that may be suggested for affording greater security in ocean travelling, and leave them less dependent on systems of navigation which may have been considered efficient for coasting purposes, or before our colonial empire was so extensive as it is now, they will, I trust, be willing to admit that not merely the substantial construction of the vessels themselves should be practically considered, but shipowners and naval men generally should honestly and intelligently ascertain how far their existing charts and projections may be regarded as incapable of improvement, or be practically relied on as affording the most accurate information as to distances and relative position of every principal spot on the navigable ocean.... If other countries are no wiser, is that any reason why we should persist in setting such an example of barbarous ignorance? Our Admiralty and Boards of Trade are, unfortunately, regarded as oracular on all subjects relating to seamanship and naval science. We are, therefore, the more inexcusable for encouraging others to retain what we dare not defend in our own principles.[19]

Without knowledge of the Flat Earth, in other words, how can the Empire be maintained, and how can British science (once Zetetically corrected, of course) continue to inspire true development? The shape of the Earth and its relation to celestial bodies also has a particularly imperial context among the sciences. Astronomy, as with other sciences, both helped to support the Empire through the development of longitude, and provided justification for imperial expansion through the establishment of

[18] Rachel Mairs and Maya Muratov, *Archaeologists, Tourists, Interpreters: Exploring Egypt and the Near East in the Late 19th–Early 20th Centuries* (London: Bloomsbury Academic, 2015); Holger Hoock, *Empires of the Imagination: Politics, War, and the Arts in the British World, 1750–1850* (London: Profile Books, 2010), 206–18; Philippa Levine, *The Amateur and the Professional: Historians, Antiquarians, and Archaeologists in Victorian England, 1838–1886* (Cambridge: Cambridge University Press, 1986), 9–10.

[19] Zetetic Society, *Chart & Compass, Sextant & Sundial, Latitudes & Longitudes, Plumbline & Pendulum, Globe or Plane?* (London: G. W. Cook, n. d.), 2, 4.

observatories in the southern hemisphere to observe stars not visible from Britain, more on which next chapter.[20]

Of course, Britain was not the only empire which drew support from astronomy. The first two chapters discussed the role of state-supported astronomy in various classical and Muslim empires of North Africa, West Africa, the Near East, and India. The Jesuit introduction of Western astronomy, including heliocentrism, into China during the period of transition between the Ming (1368–1644) and Qing (1644–1911) dynasties allowed for Emperor Kangxi (r. 1667–1722) to introduce a new calendar based on Western astronomy, helping to stabilize the nascent Qing rule.[21] Traditional Chinese Confucian cosmology at the time of the Jesuit arrival was based on a geocentrist Flat Earth with a celestial dome overheard, and there was considerable resistance in the Ming era to the adoption of the Western worldview; the same among Japanese Buddhists who were exposed to Jesuit notions of the spherical Earth starting in 1552.[22] Taking advantage of this local resistance, some European Jesuits opposed to

[20] John McAleer, "'Stargazers at the world's end': telescopes, observatories and 'views' of empire in the nineteenth-century British Empire," *The British Journal for the History of Science*, Vol. 46, No. 3 (Sep., 2013), 389–413; Sadiah Qureshi, "Science, Empire and Globalization in the Nineteenth Century," in John Holmes and Sharon Ruston, eds., *The Routledge Research Companion to Nineteenth-Century British Literature and Science* (London: Routledge, 2017), 19–29; Helen Tilley, *Africa as a Living Laboratory: Empire, Development, and the Problem of Scientific Knowledge, 1870–1950* (Chicago: The University of Chicago Press, 2011), 31–114; Michael F. Robinson, "Scientific Travel," in Nandini Das and Tim Youngs, eds., *The Cambridge History of Travel Writing* (Cambridge: Cambridge University Press, 2019), 493–95; Michael T. Bravo, "Precision and Curiosity in Scientific Travel: James Rennell and the Orientalist Geography of the New Imperial Age (1760–1830)," in Jás Elsner and Joan-Pau Rubiés, eds., *Voyages and Visions: Towards a Cultural History of Travel* (London: Reaktion Books, 1999), 162–83.

[21] Xiaoqiao Xu, "Stabilizing the Empire: Western Calendric Astronomy as Statecraft during the Kangxi Reign, 1667–1720," *The Quiet Corner Interdisciplinary Journal*, Vol. 3, No. 1 (Aug., 2022), 22–41.

[22] George H. C. Wong, "China's Opposition to Western Science during Late Ming and Early Ch'ing", *Isis*, Vol. 54, No. 1 (Mar. 1963), 44–47; Christopher Cullen, "A Chinese Eratosthenes of the Flat Earth: A Study of a Fragment of Cosmology in Huai Nan tzu," *Bulletin of the School of Oriental and African Studies, University of London*, Vol. 39, No. 1 (1976), 106–27; Christopher Cullen and Hiraoka Ryuji, "The Geneva Sphere: An Astronomical Model from Seventeenth-Century Japan," *Technology and Culture*, Vol. 60, No. 1 (Jan., 2019), 224, 233; John North, *Cosmos: An Illustrated History of Astronomy and Cosmology, Revised Edition* (Chicago: The University of Chicago Press, 2008), 151–52; Karen C. Pinto, *Medieval Islamic Maps: An Exploration* (Chicago: University of Chicago Press, 2016), 109–11.

heliocentrism were able to continue to teach geocentrism in China as late as 1738.[23] Because of this, Confucian Chinese cosmology, along with the aforementioned Great Wall and 1800 kilometer long Grand Canal built during the Sui dynasty (581–618), were long staples of Flat Earth evidence for British Zetetics.[24]

Before exploring more of the role of colonial astronomy in Zetetic thought, it is worth turning to another field of science whose development was related to colonialism and which blended with the Flat Earth movement: medicine. After Manea Fen, but before becoming Parallax, Rowbotham reinvented himself for the first time, as a medical huckster promoting miracle cures, particularly soda water as life extension.[25] In this, as in so many other things, Rowbotham was a trendsetter, pioneering not only the Flat Earth idea but its connection with medical fraud, a trend whose most recent emergence was the intermixing of Flat Earth belief with covid conspiracies.[26] As mentioned in Chapter 3, several major medical figures in the nineteenth-century US—Hans Powell, George Henry Colton Salter, and George Loring Porter—expressed support of Flat Earth views. Kelly Weill believes that they satirically joined the Flat Earth movement, and the gullible Rowbotham took it at face value.[27] However, this is made less likely by the number of well-documented British medical professionals who followed Rowbotham's views.

The governing council of the UZS included a medical doctor named Edward Haughton, who was Senior Moderator of the Natural History Society at Trinity College, Dublin. As with Lady Blount, he opposed vivisection; like Rowbotham and Porter, he believed that life could be

[23] Stephen P. Blake, *Astronomy and Astrology in the Islamic World* (Edinburgh: Edinburgh University Press, 2016), 124.

[24] Breach, *Fifty Scientific Facts*, i, 12; William Carpenter, *One Hundred Proofs that the Earth is Not a Globe, Fifth Edition* (Baltimore: William Carpenter, 1886), 19; Albert Smith, "The Shape of the Earth," *The Earth not a Globe Review*, No. 1 (Jan., 1893), 20; Elizabeth Blount, *Adrian Galilio; or, a Song Writer's Story* (Malvern: The Faith Press, 1898), 134; Elizabeth Blount and E. V. Mulgrave, *Mr. Blatchford's Great Mistake, or, Clarion's "Science" Versus God's Truth* (Kingston Hill: Lady Blount, 1905), 23n2; John Edward Quinlan, *The Earth a Plane* (London: John Edward Quinlan, 1906), 3.

[25] Weill, *Off the Edge*, 13–15.

[26] Meredith Mattlin, "Examining the Social Construction of Health, Illness, and Wellness in Anti-Science Communities," MA diss., Vanderbilt University, 2018; Weill, *Off the Edge*, 32–33.

[27] Weill, *Off the Edge*, 25.

extended indefinitely, with science allowing the fulfillment of the eternal life promised by Biblical salvation.[28] But more prominent in the Society was its vice president, Charles Watkyns de Lacy Evans.[29] Like Porter, Evans was a former army surgeon who had developed an interest in the extension of life, having authored the 1897 book *How to Prolong Life*. Evans served in the British colony of the Gold Coast in West Africa, which in 1957 would inaugurate the disintegration of European colonization of the continent when it became the state of Ghana. In the nearby British West African colony of Sierra Leone, the Yoruba missionary Orishatukeh Fadama (1855–1946) would publish a 1915 article stating that the Book of Joshua more than the Book of Exodus served as a spiritual guide for African independence from British rule, mirroring the combined Flat Earth and liberation theology that inspired John Jasper, as mentioned in Chapter 3.[30] In his book, Lacy Evans wrote that, "By experiments on ourselves, on friends, and on natives of tropical regions, including Western Africa, we find a comparatively small quantity of nitrogen necessary to sustain life in good bodily health… we have ourselves observed on the West Coast of Africa many tribes who subsist upon foods which contain comparatively little nitrogen."[31]

Lacy Evans' description of experiments on West African natives brings to mind the Night Doctors and American medical apartheid from Chapter 3. Indeed, there was a long history of Europeans connecting medical experimentation with African colonial subjects, while also using medical aid and related botanical collection as a justification for both expanding colonial powers and promoting Christian missionary efforts.[32]

[28] Elizabeth Blount and Albert Smith, *Flat or Spherical?* (Kingston Hill: Lady Blount, 1905), vi; *Report of the "Conditional Immortality" Conference, held at the Cannon Street Hotel, London, Monday, May 15th, 1876* (London: Kellaway & Co., 1876), 3–4; Schadewald, *The Plane Truth*, 81.

[29] Blount and Smith, vi.

[30] Allen Dwight Callahan, *The Talking Book: African Americans and the Bible* (New Haven: Yale University Press, 2006), 137.

[31] Charles Watkyns de Lacy Evans, *How to Prolong Life: An Inquiry into the Cause of "Old Age" and "Natural Death"* (London: Central Publishing Company, 1897), 69–70.

[32] Sujit Sivasundaram, "A global history of science and religion," in Thomas Dixon, Geoffrey Cantor, and Stephen Pumfrey, eds., *Science and Religion: New Historical Perspectives* (Cambridge: Cambridge University Press, 2010), 183–84; Richard Drayton, *Nature's Government: Science, Imperial Britain and the'Improvement' of the World* (New Haven: Yale University Press, 2000), 221–69; Jim Endersby, *Imperial Nature: Joseph Hooker and*

The separation of indigenous religion and medicine in Africa was largely a product of the colonial period.[33] The nature of Lacy Evans' experiments on Africans is not known, but it can be presumed they were not particularly invasive, given that he carried out the same experiments on "ourselves" and "friends," but also given the fact that, as discussed last chapter, Emanuel Vaun Mulgrave also sat on the Society's governing council, and John Quinlan was an active member at the time. However, Lacy Evans was not the only army surgeon in colonial service who was connected to the Zetetic Society.

In 1918, Lady Blount's collaborator Albert "Zetetes" Smith published the pamphlet *The Sea-Earth Globe and its Monstrous Hypothetical Motions*, which in its preface equated the Zetetic search for "justice and truth" with the righteousness of the Entente cause in the ongoing Great War.[34] However, this was not the only war discussed by Smith, who recounted how

> During the Boer war Dr. Robertson, a medical gentleman, sailed with our troops from England to South Africa, and in 14 degrees N. latitude the vessel at noon came under the vertical rays of the sun. He discovered the fact above mentioned, and published it in a book he wrote. He was a globularist at the time; but as I lost touch with him soon after reading his book, I cannot say how his discovery affected his subsequent belief. It ought to have brought him into the ranks of the planists; and I posted him some of our literature. His book was entitled *The Mutual Relations of the Sun and the Earth*. I do not now possess a copy, so I cannot quote directly from it, and our space is very limited; but Dr. R., by diagrams and arguments, demonstrated that the diameter of the sun is only 32 miles across. Thus the sun is a small body as compared with the size of the earth;

the Practices of Victorian Science (Chicago: The University of Chicago Press, 2008), 276–310; Kathleen S. Murphy, *Captivity's Collections: Science, Natural History, and the British Transatlantic Slave Trade* (Chapel Hill, NC: University of North Carolina Press, 2023), 101–23; Tilley, *Africa as a Living* Laboratory, 169–216; Downs, *Maladies of Empire*, 50–87.

[33] Steven Feierman and John M. Janzen, "African Religions," in John Hedley Brooke and Ronald L. Numbers, eds., *Science and Religion Around the World* (Oxford: Oxford University Press, 2011), 240–44.

[34] Albert Smith, *The Sea-Earth Globe and its Monstrous Hypothetical Motions; or Modern Theoretical Astronomy* (Abington: Albert Smith, 1918), 2.

yet as compared with the planets it is a giant, and, as the Psalmist says, "a giant rejoicing to run his race!" (Psa. xix. 5)[35]

Although not specified by Smith, the conflict in question appeared to be the First Boer War of 1880–1881, as Robertson's undated manuscript was reviewed in an 1893 volume of the Royal Scottish Geographical Society's *Scottish Geographical Magazine*.[36] At the time in Britain, there was a fear that spending too long among Africans would cause a degeneration among Europeans, exemplified by David Livingstone's (1813–1873) 1856 report on the Portuguese settler prazo system along the Zambezi River in what is now Mozambique.[37] In Smith's account, service in southern Africa during the First Boer War caused otherwise mainstream British thinkers to be converted to Flat Earth belief. However, the vector for this conversion were not native Africans, but Britain's rival Europeans (Fig. 5.2).

The Dutch established their settler colony at the Cape of Good Hope in southern Africa in 1652, permanently losing it to the British in 1806. This ultimately resulted in the start of the Great Trek in September 1835, when large numbers of the Dutch-language Boer[38] (meaning "Farmer") settlers departed from the Cape Colony and established the so-called Boer republics of the Transvaal (formally, the South African Republic) and the Orange Free State in the African interior. Cape Colony and the Boer states would exist in a cold war broken by the aforementioned First Boer War, resulting in a largely status quo ante bellum, and the 1899–1902 Second Boer War, resulting in the conquest of the Boer republics by the British. In 1910, the former Boer states, along with the British Cape and Natal colonies, would become united into the Union of South Africa.

Almost from the start of British rule over Cape Colony, southern Africa had been the center of strange developments, astronomical and otherwise.

[35] Smith, *The Sea-Earth Globe*, 21–22.

[36] C. Robertson, *The Mutual Relations of the Sun and the Earth: A Study in Physical Geography* (Edinburgh: St. Giles Printing Company, n.d.); *Scottish Geographical Magazine*, Vol. 9 (1893), p. 387 in "New Books" section.

[37] Edward Guimont, "From King Solomon to Ian Smith: Rhodesian Alternate Histories of Zimbabwe" (PhD diss., University of Connecticut, 2019), 53–54, 70.

[38] For ease and standardization, I will refer to the Dutch- (and later Afrikaans-) speaking settlers of southern Africa as Boers, rather than as Afrikaners, a term more often applied later than the major setting of this chapter.

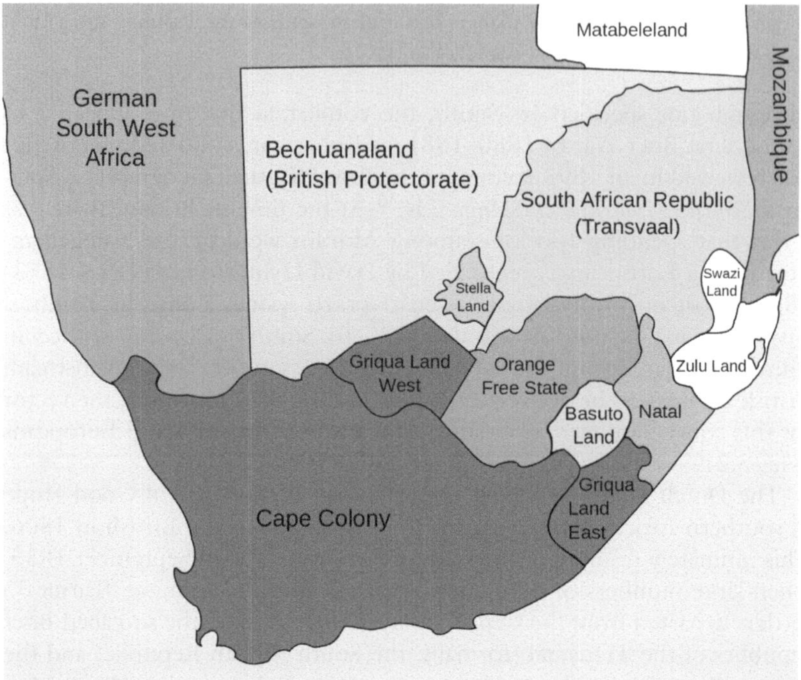

Fig. 5.2 Map of southern Africa in 1885, between the two Boer Wars (Wikimedia Commons)

Africans looked to skies of what would become Cape Town since humans arrived in the area around 10,000 BCE. In the 1870s, scholars at what is now the University of Cape Town transcribed 12,000 pages of mythology of the local |Xam speakers. Of those 12,000 pages, approximately 17% involved celestial bodies.[39] Some of the oldest extant art in southern Africa are petroglyphs of comets, meteors, and supernovae made by the

[39] J. C. Hollmann, "'The Sky's Things': |xam Bushman 'Astrological Mythology' as recorded in the Bleek and Lloyd Manuscripts," *African Skies/Cieux Africains*, No. 11 (Jul., 2007), 8–12; W. P. Koorts and A. Slotegraaf, "/Xam astronomical references in G R von Wielligh's "*Boesman-Stories*"," *African Skies/Cieux Africains*, No. 11 (Jul., 2007), 13–15.

San people.[40] Nor is this scientific interest limited to astronomical observations; in the early nineteenth century, San people in modern Lesotho left illustrated reconstructions of dinosaurs based on local fossilized footprints in the Mokhali cave. These depicted ornithopods as upright bird-like creatures, a depiction which anticipates by centuries the acceptance by Western science of ornithopods' appearance.[41]

But in terms of the late-coming European settlers, the first to look to the skies in the region of what is now Cape Town were the navigators of the first Portuguese ship which landed there in 1488. Professional astronomy had been conducted at Cape Town since at least 1751, when an Abbé Nicholas de la Caille of the Paris Observatory visited for a year; his observations included helping to determine the precise shape of the Earth.[42] In 1820, a Royal Observatory was established in Cape Town for the purpose of mapping star charts of the southern hemisphere for navigation. In 1833, the eminent British astronomer Sir John Herschel (1792–1871) departed from England to take up residence at the Cape Town observatory, remaining there until 1838. In the 25 August 1835 issue of the *New York Sun* newspaper, an announcement was made that Herschel had made amazing discoveries on the Moon's surface with the use of a new telescope. The reports in the *Sun* continued for a total of six issues, appearing under the headline "Great Astronomical Discoveries Lately Made by Sir John Herschel at the Cape of Good Hope." The fantastic discoveries—which included enormous quartz obelisks near forests inhabited by giant single-horned goats and intelligent bat-winged orangutans—became too incredible to maintain, and the story was eventually revealed to be a hoax written by a reporter named Richard Adams Locke (1800–1871), descendant of the uncle of famed philosopher John Locke (1632–1704).[43] One of the first to uncover the hoax's nature was

[40] B. Fraser, "Comets in Bushman Paintings," *African Skies/Cieux Africains*, No. 11 (Jul., 2007), 16.

[41] Paul Ellenberger, David J. Mossman, Alexander D. Mossman and Martin G. Lockley, "Bushmen Cave Paintings of Ornithopod Dinosaurs: Paleolithic Trackers Interpret Early Jurassic Footprints," *Ichnos: An International Journal for Plant and Animal Traces*, Vol. 12, No. 3 (2005), 223–26.

[42] H. E. Houghton, "Some South African Amateur Astronomers," *Monthly Notes of the Astronomical Society of South Africa*, Vol. 6, No. 5 (May, 1947), 41.

[43] Richard Adams Locke, "The Moon Hoax," *Amazing Stories*, Vol. 1, No. 6 (Sep., 1926 [1835]), 556–75.

none other than Edgar Allan Poe, who believed Locke plagiarized from his own "Hans Pfaall" story, mentioned in Chapter 2.[44]

This so-called Great Moon Hoax became possible in the United States not only due to the distance of the colonial observatory in southern Africa, but also because the lunar inhabitants being supposedly reported on reinforced contemporary imperial stereotypes of colonized peoples being less developed and more animalistic.[45] In this case, the *New York Sun*'s reporting had another New York predecessor. In 1809, a decade prior to his authoring "Rip Van Winkle" and "The Legend of Sleepy Hollow," Washington Irving published the satirical *Knickerbocker's History of New York*, portraying itself as a history of New York State from the perspective of one of the Dutch-speaking descendants of the New Netherlands colony. The history contains a section satirically defending the Dutch conquest of Native Americans in New York by portraying a hypothetical invasion of "Lunatics"—inhabitants of the Moon—against the Dutch settlers as justified because the Lunatics would be bringing civilization to the backwards Dutch farmers.[46]

In the case of the Great Moon Hoax, the credulity of reporting from the Cape also drew on what was by then a longstanding connection of southern Africa with strange monuments.[47] While Herschel was not amused when learning about his association with the Great Moon Hoax, his time at the Cape Observatory led to him becoming associated with another monument-based speculative theory. From 1835 to 1845, a young man named Charles Piazzi Smyth (1819–1900) served as assistant to the observatory. Smyth would later return to Britain and

[44] John Tresch, *The Reason for the Darkness of the Night: Edgar Allan Poe and the Forging of American Science* (New York: Farrar, Straus and Giroux, 2021), 85–87.

[45] Matthew Goodman, *The Sun and the Moon: The Remarkable True Account of Hoaxers, Showmen, Dueling Journalists, and Lunar Man-Bats in Nineteenth-Century New York* (New York: Basic Books, 2008); Michael J. Crowe, *The Extraterrestrial Life Debate 1750–1900: The Idea of a Plurality of Worlds from Kant to Lowell* (Cambridge: Cambridge University Press, 1986), 202–15; Roland Wenzlhuemer, *Doing Global History: An Introduction in 6 Concepts* (London: Bloomsbury Academic, 2019), 19–45; Bernd Brunner, *Moon: A Brief History* (New Haven: Yale University Press, 2010), 18–22.

[46] Washington Irving, *A History of New-York from the Beginning of the World to the End of the Dutch Dynasty, by Diedrich Knickerbocker* (Philadelphia: J. B. Lippincott & Co., 1871 [1809]), 90–97. With thanks to Max Chapnick.

[47] Guimont, "From King Solomon," 151–53.

become Astronomer Royal of Scotland, but is best known for his speculative theories about the Egyptian pyramids, including that by analyzing its dimensions, major religious truths from the Bible could be proven, and which in turn invalidated the metric system, which he viewed as a communist plot. Smyth developed these views with the help of Herschel, and published them in an 1864 book, despite not actually having visited the pyramids in person at that time.[48]

Herschel and Smyth were not the only Europeans in southern Africa to take an interest in fringe religious and archaeological notions. The Dutch who settled southern Africa were Calvinists, much like the Maryland followers of Jasper Danckaerts mentioned in Chapter 2. In the mid-nineteenth century, after the creation of the Boer republics, and partially due to the radicalizing effect of the Great Trek, the Boers' religion developed into Afrikaner Calvinism, with a prominent sect called the Doppers emerging in the Transvaal Republic in 1859. Dopper means "extinguisher," and was intended to signify the extinguishing of the light of the Enlightenment. The Dopper religion was closely tied to Boer national identity, de facto (though not de jure) the state church for the Transvaal, and integrated a number of folk beliefs. The Calvinist view of the elect and the condemned was given a racial schema, identifying the Boers as the former and the African natives as the latter. The Boers were further seen as modern Israelites, with the Great Trek as the new Exodus.[49] However, the association did not end there.

Since the sixteenth century, Europeans in southern Africa had heard legends of a ruined city in the interior. Now called Great Zimbabwe

[48] Charles Piazzi Smyth, *Our Inheritance in the Great Pyramid* (London: Alexander Strahan, 1864); Larry Schaaf, "Charles Piazzi Smyth's 1865 Conquest of the Great Pyramid," *History of Photography*, Vol. 3, No. 4 (Oct., 1979), 332–34.

[49] Paul Kruger, ed. August Schowalter, trans. Alexander Teixeira de Mattos, *The Memoirs of Paul Kruger, Four Times President of the South African Republic, Told by Himself* (New York: The Century Co., 1902), 74–75; Stuart Cloete, *Against These Three: A Biography of Paul Kruger, Cecil Rhodes, and Lobengula, Last King of the Matabele* (Cambridge, MA: Riverside Press, 1945), 30–33; Irving Hexham, *The Irony of Apartheid: The Struggle for National Independence of Afrikaner Calvinism Against British Imperialism* (Lewiston, NY: Edwin Mellen Press, 1981), 46–52; T. Dunbar Moodie, *The Rise of Afrikanerdom: Power, Apartheid, and the Afrikaner Civil Religion* (Berkeley: University of California Press, 1975), 60–62, 69–72; Pieter Duvenage, "Afrikaner intellectual history: An interpretation," in Joel Netshitenzhe, ed., *Whiteness Afrikaans Afrikaners: Addressing Post-Apartheid Legacies, Privileges and Burdens* (Johannesburg: Mapungubwe Institute for Strategic Reflection, 2018), 96, 103–04.

and firmly established as having been built by native Africans in the ninth century, Europeans adopted the myths of local Arab traders and associated it with King Solomon's Biblical city of Ophir. The city was eventually rediscovered by Europeans in 1871, leading to a new wave of speculation over ancient Israelites having settled in southern Africa.[50] In Britain, this view became tied in with the notions of British Israelism, the idea that Jesus had visited Britain early in his life and the British were the new Israelites, as exemplified by the poet William Blake (1757–1827)'s 1804 poem "And did those feet in ancient time" (better known as "Jerusalem").[51] Edward Hine (1825–1881), the leading proponent of British Israelism, was also a Flat Earther.[52] Paul Kruger (1825–1904), who served as both the State President of the Transvaal Republic and founder of the Dopper Church, further asserted that the Lemba people of southern Africa were descended from the Lost Tribes of Israel of medieval myth. As such, Boer society, politics, and religion were all steeped in a pseudohistorical narrative where God had delivered them, the new Israelites, to the region where the ancient Israelites had once lived, with local cultures deemed the product of their white forebears.[53]

While the myth of Great Zimbabwe as Ophir was also present among the British, their interest was less in the religious connotations as it was for the fact that Ophir would presumably be the location of rich natural resources to exploit. Especially after the First Boer War, tensions between the British colonies and Boer republics remained high, escalating with the rising career of Cecil Rhodes (1853–1902), who formed the De Beers company in 1888 and the British South Africa Company (BSAC) in 1889. In June 1890, the BSAC dispatched the Pioneer Column, a paramilitary force, to seize the territory of the Mashona, in the north of modern-day Zimbabwe, establishing the settler state of Rhodesia.

[50] Michael F. Robinson, *The Lost White Tribe: Explorers, Scientists, and the Theory that Changed a Continent* (Oxford: Oxford University Press, 2016), 108–19; Guimont, "From King Solomon," 17–58.

[51] Michael Barkun, *Religion and the Racist Right: The Origins of the Christian Identity Movement* (Chapel Hill: University of North Carolina Press, 1997), 3–46; James E. Landing, *Black Judaism: Story of an American Movement* (Durham, NC: Carolina Academic Press, 2002), 24–28; Michell, *Eccentric Lives*, 163–77.

[52] Michell, *Eccentric Lives*, 22.

[53] Kruger, *Memoirs*, 352, 379–80, 399–400; Robinson, *Lost White Tribe*, 111; Tudor Parfitt, *Journey to the Vanished City: The Search for a Lost Tribe of Israel* (New York: St. Martin's Press, 1992), 22; Guimont, "From King Solomon," 54–56.

The next month, Rhodes became Prime Minister of the Cape Colony. Following the First Matabele War of 1893–1894, the BSAC seized the territory of the Ndebele, in the west of modern Zimbabwe.

By 1894, British territory under Rhodes' political and economic control surrounded the Boer republics. Kruger's government was not willing to allow the British to build railways through his territory, or mine the region's gold, preferring a more pastoral pre-industrial lifestyle in line with his view of Calvinist austerity.[54] The conflict between the British and the Boers was therefore religious in basis—but only to the extent that from the British perspective, the Boers' religious fundamentalism was preventing the march of progress, signified in this case by industrialized extraction capitalism. Rhodes' desire to seize the Boer territories culminated in the 1895–1896 Jameson Raid, named for its leader, Rhodes' close friend and administrator of Rhodesia, Leander Starr Jameson (1853–1917). The intention of the Jameson Raid was to launch an invasion of the Transvaal with several hundred BSAC soldiers, pairing it with an uprising of British workers in the mining center of Johannesburg, hoping to overthrow the Boer rule of the Transvaal. This was illegal under British law, and due to a combination of bad luck and Jameson's ineptitude, it was also a complete failure. Kruger's government captured Jameson and his paramilitary soldiers, repatriating them to Britain to face trial. Jameson and Rhodes were both forced to resign from their positions in the BSAC and Cape governments.

The immediate impact of the Jameson Raid in Cape Town was panic over an imminent Boer counter-invasion, which ultimately did not materialize. One observer of this panic was a young American on vacation, Charles Fort (1874–1932). Writing after the fact in the 11 April 1909 issue of the Albany, New York *Argus* newspaper, Fort recounted the sensation in Cape Town during the raid, and personally seeing Jameson's bullet-riddled belt when it was brought to the city.[55] In the early 1910s, Fort was inspired by reading such works as Americus Symmes' book on the Hollow Earth and Smyth's book about on the pyramids to begin compiling and publishing musings on strange news and bizarre

[54] Kruger, *Memoirs*, 181–92.

[55] Damon Knight, *Charles Fort: Prophet of the Unexplained* (New York: Doubleday & Company, 1970), 28–29.

occurrences, a topic now referred to as Forteana.[56] Among those influenced by Fort's writings was H. P. Lovecraft, discussed in Chapter 3. In Fort's 1923 second book, *New Lands*, Fort discusses Flat Earth views in Chapter 6. While he does not explicitly accept the Flat Earth, he does critique heliocentrism and believes that the Bedford Level Experiment is just as valid as the claims of astronomers as to the Earth's shape and motions, and that telescopes are "the polished and mysterious-looking instruments that, in the new ikonology, have replaced the images of the saints," in a perhaps unknowing reflection of the religious claims of Flat Earthers. Ultimately, Fort decides that there is more evidence for the existence of Noah's Ark and its petrified remains on Mount Ararat than for the precise shape of the Earth.[57] That being said, in a later chapter Fort does seem to accept that the Earth has a spherical shape and "rip[s] space at a rate of about 19 miles a second," a reflection of Fort's broad critique of whatever authority has caught his ire at the time more than any specific ideology or belief.[58] In 1931, New York-based writer Tiffany Thayer (1902–1959) organized a Fortean Society based on spreading Fort's ideas and methodology; Thayer not only specifically attempted to bring Flat Earthers into the Society, but himself believed that Flat Earth ideas—along with concepts as varied as Atlantis, the Noachic flood, and continental drift—came from the fact that the Earth was continually growing and shifting form from sphere to cube and back.[59]

South Africa would have no shortage of Fortean phenomena well after the Jameson Raid. In the previous chapter, the connection between African American and Flat Earther airship folklore was discussed, and the frequent usage of the observations of James Glaisher from his balloon to shore up Zetetic arguments. In 1863, Jules Verne published his debut novel, *Five Weeks in a Balloon*, which linked balloon expeditions to the exploration of Africa in European public thought, just as Verne's immediate follow-up, *Journey to the Center of the Earth*, helped popularize the Hollow Earth notion. As such, it should not be a surprise that sightings of mysterious balloons, and later those of "phantom aeroplanes,"

[56] Knight, *Fort*, 55–56.

[57] Charles Fort, *New Lands* (New York: Boni & Liveright, 1923), 42–46.

[58] Fort, *New Lands*, 106.

[59] Joshua Blu Buhs, *Think to New Worlds: The Cultural History of Charles Fort and His Followers* (Chicago: University of Chicago Press, 2024), 53–54, 254.

were common in southern Africa from the 1880s through the 1910s, particularly by Boers during the Second Boer War who feared British aerial attacks, and then by the English-speaking settlers during World War I, who feared attacks from the neighboring colony of German South-West Africa (modern Namibia). These reports have much in common with those of the flying saucer phenomenon, which also began in the context of the tension of the start of the Cold War in the late 1940s.[60] Indeed, the flying saucer emerged in the United States in the Southwest, a place located along both national and settler colonial frontiers, at the height of both international and racial tensions, and where the new technology of war and exploration (radar, rockets, atomic weapons) was being developed. Many of these similar themes were also true of southern Africa from the 1880s–1910s, making it an even more fertile ground for the development and propagation of alternative views than the British metropole.

Of wider popular knowledge than South African balloon UFOs is the coelacanth. Westerners first discovered the coelacanth in 1839, in the form of fossil remains. For almost a century, it was assumed that the fish, hundreds of millions of years old, had gone extinct in the same event that had led to the demise of the dinosaurs. However, in 1938, museum official Marjorie Courtenay-Latimer (1907–2004) discovered one in a local fish market in the South African town of East London, in the Eastern Cape province. What was excitedly hailed by the European press as the discovery of a "living fossil" was a fish which drew little attention from African fishers who had known about it for centuries, largely in the context of avoiding it due to the gastric disruptions its oily flesh causes. The coelacanth's discovery led credence to the idea that Africa was a continent that existed outside of time and evolutionary pressures, an idea which mixed and reinforced racist notions of African backwardness. It also gave justification to the existence of the mokele-mbembe, a supposed living dinosaur encountered by hunters in Rhodesia, and which itself drew from euhemeristic interpretations of Babylonian legendary dragons connected with alternative histories of Great Zimbabwe's construction. More generally, the coelacanth's European rediscovery of an animal

[60] Nigel Watson, *UFOs of the First World War: Phantom Airships, Balloons, Aircraft and Other Mysterious Aerial Phenomena* (Stroud: The History Press, 2015), 97–117.

long thought dead gave a lasting boost to the field of cryptozoology.[61] However, the perception of coelacanth as cryptid has a major flaw. To the African fishermen, coelacanths had never been extinct; to Africans (and others), there were no mythologies or legendary beasts inspired by the coelacanth; among Europeans, between 1839 and 1938 there had been no suspicion that the coelacanth still survived. This is a stark separation from similar cases—for example, alleged sightings of the thylacine by both settlers and Aboriginals in Tasmania after its presumed 1936 extinction, and with many supposed facts about the final thylacine in captivity actually invented decades after her death.[62]

One final example of South African Forteana is somewhat less direct, but still illustrative for the connection between the colonial and alternative science: the case of the Cottingley Fairies. In 1917, two girls in the English town of Cottingley claimed to have taken photographs of fairies, leading to a debate which drew in leading Spiritualists and skeptics. The younger of the two girls, Frances Griffiths (1907–1986), had just relocated with her mother from South Africa; the emotional pressures of the relocation, as well as a condescending belief in the gullibility of colonial settlers so far from metropolitan England, were both held up as explanations in the British press. However, one prominent supporter of Griffiths was Sherlock Holmes creator and Spiritualist Arthur Conan Doyle (1859–1930), who wrote in the girls' defense, "When Columbus knelt in prayer upon the edge of America, what prophetic eye saw all that a new continent might do to affect the destinies of the world? We also seem to be on the edge of a new continent, separated not by oceans but by subtle and surmountable psychic conditions."[63] It is perhaps worth noting that in the original 1887 Holmes story, "A Study in Scarlet," Dr. Watson is shocked to learn that Holmes is a geocentrist, rejecting Copernicanism because it is irrelevant to his detective work, which does not depend on

[61] Edward Guimont, "Hunting Dinosaurs in Central Africa," *Contingent Magazine*, March 18, 2019, accessed August 3, 2022, https://contingentmagazine.org/2019/03/18/hunting-dinosaurs-africa/.

[62] Lydia Pyne, *Endlings: Fables for the Anthropocene* (Minneapolis: University of Minnesota Press, 2022), 19–20.

[63] Arthur Conan Doyle, *The Coming of the Fairies* (New York: George H. Doran Company, 1922), 58.

the shape or centrality of Earth.[64] It is also curious that the archenemy of the geocentrist Holmes, Professor Moriarty, was famous for his book on the dynamics of asteroids.[65]

By the time of the Cottingley Fairies, Kruger had died in Swiss exile, leaving us only to speculate what he might make of the claims of his fellow South African expatriate. However, Doyle's related comments of the sailor on the edge of a new continent—reflective of Washington Irving's Columbus biography—bring to mind one pioneering sailor with whom Kruger did interact with. From 1895 to 1898, Joshua Slocum (1844–1909) became the first person to sail around the world solo, in his sloop *Spray*. Slocum, born in what was then the Colony of Nova Scotia, emigrated to the United States in 1865, at the end of the Civil War, and became a citizen four years later. In the last leg of his journey, on 17 November 1897, Slocum made landfall in Durban, the port city of Natal. Initially founded by a group of Boers (including a young Kruger) in 1839 during the Great Trek, Natal had been annexed by Britain in 1843 and governed as a colony since then. While docked in Durban, Slocum received some unexpected visitors (Fig. 5.3):

> It sounds odd to hear scholars and statesmen say the world is flat; but it is a fact that three Boers favored by the opinion of President Krüger [*sic*] prepared a work to support that contention. While I was at Durban they came from Pretoria [capitol of the Transvaal] to obtain data from me, and they seemed annoyed when I told them that they could not prove it by my experience. With the advice to call up some ghost of the dark ages for research, I went ashore, and left these three wise men poring over the *Spray*'s track on a chart of the world, which, however, proved nothing to them, for it was on Mercator's projection, and behold, it was "flat." The next morning I met one of the party in a clergyman's garb, carrying a large Bible, not different from the one I had read. He tackled me, saying, "If you respect the Word of God, you must admit that the world is flat." "If the Word of God stands on a flat world—" I began. "What!" cried he, losing himself in a passion, and making as if he would run me through with an assagai [Zulu spear]. "What!" he shouted in astonishment and rage, while I jumped aside to dodge the imaginary weapon.

[64] Arthur Conan Doyle, *A Study in Scarlet: A Detective Story* (London: Ward, Lock, Bowden, and Co., 1892), 19–20.

[65] Arthur Conan Doyle, *The Valley of Fear: A Sherlock Holmes Novel* (New York: George H. Doran, 1914), 13.

> Had this good but misguided fanatic been armed with a real weapon, the crew of the *Spray* would have died a martyr there and then. The next day, seeing him across the street, I bowed and made curves with my hands. He responded with a level, swimming movement of his hands, meaning "the world is flat." A pamphlet by these Transvaal geographers, made up of arguments from sources high and low to prove their theory, was mailed to me before I sailed from Africa on my last stretch around the globe. While I feebly portray the ignorance of these learned men, I have great admiration for their physical manhood.... Education in the Transvaal is by no means neglected, English as well as Dutch being taught to all that can afford both; but the tariff duty on English school-books is heavy, and from necessity the poorer people stick to the Transvaal Dutch and their flat world, just as in Samoa and other islands a mistaken policy has kept the natives down to Kanaka.[66]

The association of sailing with Flat Earth thought should not have been alien to Slocum, as during his own voyage, he occupied himself by reading Irving's Columbus biography.[67] While Slocum sadly does not provide what sources the three Boers sent to him in support of their Flat Earth views, this would not be the last time he encountered this sentiment in his journey. Slocum sailed on from Durban on 14 December, reaching Cape Town on 27 December. There he met the astronomer royal, David Gill (1843–1914), who took him on a tour of the Royal Observatory. The *Spray* remained docked at Cape Town until 26 March 1898. Slocum took advantage of his three-month rest to visit by train the cities of Kimberley, Johannesburg, and Pretoria. There,

> At the last-named place I met Mr. Krüger, the Transvaal president. His Excellency received me cordially enough; but my friend Judge Beyers, the gentleman who presented me, by mentioning that I was on a voyage around the world, unwittingly gave great offense to the venerable statesman, which we both regretted deeply. Mr. Krüger corrected the

66 Joshua Slocum, *Sailing Alone Around the World* (New York: Dover Publications, 1956 [1901]), 237–39.

67 Slocum, *Sailing*, 44; Beth Day, *Joshua Slocum, Sailor: A Biography* (Cambridge, MA: Riverside Press, 1953), 216. Other books of interest in his library on the *Spray* included Charles Darwin's *The Descent of Man* (1871) and *The Expression of the Emotions in Man and Animals* (1872), in which his theory of evolution is first adapted to humans; the Nova Scotian astronomer Simon Newcomb's *Popular Astronomy* (1878); and novelist Mabel Loomis Todd's *Total Eclipses of the Sun* (1894): Walter Magnes Teller, *The Voyages of Joshua Slocum* (New Brunswick, NJ: Rutgers University Press, 1958), 200.

Fig. 5.3 Joshua Slocum memorial in Fairhaven, Massachusetts, "birthplace" of the *Spray*. Photograph by the author

> judge rather sharply, reminding him that the world is flat. "You don't mean *round* the world," said the president; "it is impossible! You mean *in* the world. Impossible!" he said, "impossible!" and not another word did he utter either to the judge or to me. The judge looked at me and I looked at the judge, who should have known his ground, so to speak, and Mr. Krüger glowered at us both. My friend the judge seemed embarrassed, but I was delighted; the incident pleased me more than anything else that could have happened. It was a nugget of information quarried out of Oom Paul ["Uncle Paul," Kruger's nickname], some of whose sayings are famous.[68]

The same strict Biblical interpretation which convinced Kruger that it was divine predestination that the Dutch had achieved dominance over the native Africans, and that the Boers were the new Chosen People led on a new Exodus replacing those whom God had led to Ophir millennia prior, also convinced Kruger that the Earth was flat and at the center of the universe. There is of course an irony in the head of state of a country named for its location in the southern hemisphere being a Flat Earther who rejected the premise of hemispheres. What, if anything, Kruger did to try to contextualize that has not been recorded, although the reaction of later Flat Earthers to the modern view of the Antipodes, and South Africa's place within them, will be discussed later (Fig. 5.4).

Regardless of Kruger's thoughts on the name of the South African Republic, his remarks to Slocum were widely circulated. Slocum wrote a travelogue, *Sailing Alone Around the World*, which was first serialized in *The Century Magazine* from September 1899 through March 1900, and then published in book form by the Century Company immediately on 24 March 1900. *The Century Magazine* was widely read, as was the book version; it was kept in print through the seventeenth edition, released on 22 June 1948.[69] Kruger's statements to Slocum were spread in the English press of the British Cape Colony as part of the propaganda wave ahead of and after the outbreak of the Second Boer War in October 1899, and remain cited in biographies of both Kruger and Slocum into the twenty-first century.[70]

[68] Slocum, *Sailing*, 243–45.

[69] Walter Magnes Teller, *The Search for Captain Slocum: A Biography* (New York: Charles Scribner's Sons, 1956), 172, 258.

[70] Johannes Meintjes, *President Paul Kruger: A Biography* (London: Cassell, 1974), 85–87; Derek Wilson, *The Circumnavigators: A History* (New York: Carroll & Graf, 2003),

Fig. 5.4 Cartoon of Kruger by the English-born Constance Penstone (1859–1828), based on the remarks Kruger made to Slocum, printed in the English-language weekly paper *The Owl* (1895–1907) on March 5, 1898. *The Owl* was a comedic paper, and Penstone was its owner and editor. She and her Australian-born husband founded the paper in Johannesburg, but relocated to Cape Town following the Jameson Raid due to fears over Kruger's reprisals at Anglophone settlers. While publishing *The Owl*, Penstone simultaneously worked as cartoonist for the English-language *Cape Times* (1876–present) newspaper[71]

The Second Boer War itself was one of several armed conflicts between the British Empire and locals which embodied the cultural narrative in Britain of a war between science and religion.[72] Of course, very similar

296; Martin Meredith, *Diamonds, Gold, and War: The British, the Boers, and the Making of South Africa* (New York: PublicAffairs, 2007), 9, 76–78; Stan Grayson, *A Man for All Oceans: Captain Joshua Slocum and the First Solo Voyage Around the World* (Thomaston, ME: Tilbury House Publishers, 2017), 250–51; Schadewald, *The Plane Truth*, 101–02; Weill, *Off the Edge*, 37.

[71] Slocum, *Sailing*, 246; Joan Kerr, "A. Constance Roth," Design & Art Australia Online, September 30, 2014, accessed August 27, 2022, https://www.daao.org.au/bio/a-constance-roth/biography/.

[72] Russell, *Inventing the Flat Earth*, 36.

statements had been made by the British Samuel Rowbotham in his foundational pamphlet, but Rowbotham was not a national leader.[73] Perhaps tellingly, references to Kruger's Flat Earth belief are absent from his official memoirs published in 1902; curiously, they are also absent from Garwood's *Flat Earth*.[74] In a centennial tribute to Slocum's voyage endorsed by the Joshua Slocum Society, Kruger and his meeting with Slocum is not mentioned at all; instead, the author ahistorically states—buying into the Flat Error—that upon leaving Durban, Slocum "thought about those who had come before him and their struggle to go around the Cape of Good Hope. Their fear was greater because they thought the world was flat. They didn't want their ship to sail over the edge of the earth."[75] This is not the first account to mistakenly alter the circumstances of Slocum's involvement with antipodean Flat Earthers. In a 1953 biography based on the recollections of his daughter, Jessie Slocum Joyce, the encounter is framed as

> Josh was introduced to three scholars who were working on a scientific paper for the Boer government. "We are preparing a pamphlet to prove that the world is flat," they said. "We would like your help."
> "But how can I help," Josh asked, "when I have been sailing *around* the world?"[76]

The idea that, to the Boers, (pro) Flat Earth writing was scientific is a reflection of the same science versus religion issues that tinge the Flat Error.

Perhaps the most telling departure of Slocum's actual account of the Flat Earthers is provided via unsourced quotes in South African novelist Stuart Cloete (1897–1976)'s biography. In Cloete's depiction of the meeting between Slocum and Kruger, the latter combined his Flat Earth and Israelist views explicitly, as "the President told him with conviction that 'the earth, with certain minor exceptions, such as high mountains, was flat and had always been so.' He appeared to consider that the Transvaal was the chosen land of the globe and the eyeball of the world.

[73] Rowbotham, *Zetetic Astronomy*, 223–28.

[74] Kruger, *Memoirs*; Garwood, *Flat Earth*.

[75] Myra A. Lopes, *Captain Joshua Slocum: A Centennial Tribute* (New Bedford, MA: RPI Graphics, 1994), 117.

[76] Day, *Joshua Slocum, Sailor*, 229.

Slocum says, rightly or wrongly, the President gave him 'the impression that Pretoria was practically the centre of the earth.'"[77] The meaning of the British condescension toward Kruger is clear: the Boers were backwards relics of the Middle Ages, and not only could not stand against modern British military industry, but desperately needed the progress that British administration would bring. Indeed, one account of Kruger, written in 1984 by British Fortean John Michell (1933–2009), goes so far as to state that

> There are very few instances where a person with decided views on whether the earth is flat or spherical has ever been converted to the opposite belief. A notable case was President Paul Kruger of the Transvaal Republic. In 1900, while on his way into exile on board a Dutch man-of-war, he was invited by the captain to step onto the bridge and watch the process of navigation. As a member of the fundamentalist Dutch Reformed Church, Kruger believed literally in the pillars and corners of the flat earth. He learnt from the captain that the ship was navigated on the basis of the earth being spherical, inspected the sextant and other instruments, and then went below for his Bible and threw it into the sea. If the earth was really a globe, he said, then the Book was untrue and of no further interest to him.[78]

Michell's account is not completely accurate, as he seems to conflate Kruger's 1900 exile with his 1877 voyage to London to negotiate the independence of the Transvaal. During that voyage, Kruger heard his fellow deputation members, the Dutch-born E. J. P. Jorissen (1829–1912) and Willem Eduard Bok (1846–1904), discuss the changing stars overhead as the ship voyaged into the northern hemisphere. Kruger's response was to reject the evidence overhead and declare that *if* it was true, he might as well throw his Bible overboard.[79] Along with Slocum's book, Jorissen's memoir of the event, published in 1897 (albeit in Dutch), seem to be the basis for most modern awareness of Kruger's Flat Earth

[77] Cloete, *Against These Three*, 252–53.

[78] Michell, *Eccentric Lives*, 32.

[79] Meintjes, *President Paul Kruger*, 87.

views, through their citation in a 1974 English-language biography of Kruger.[80]

By repositioning the incident as taking place during the voyage to exile, and with Kruger actually throwing the Bible overboard when confronted with the truth of modern naval navigation, Michell perfectly embodies the endpoint of the British propaganda wave that began with Slocum's account, and which echoes Irving's *Knickerbocker* satire of the New York Dutch. Indeed, when recounting the event in a biography of his father, Victor Slocum (1872–1949) states that Joshua had first heard about Kruger's Flat Earth views on landing in Natal, but before the arrival of the three advocates had simply dismissed it as British propaganda against the Boers.[81] In his thesis on Slocum's voyage, the historian Donald A. Laskey compares Slocum with Joseph Conrad (1857–1924) and Mark Twain (1835–1910), two other Victorian sailors-turned-authors who wrote about Africa. In contrast with the former two, Slocum was not concerned with imperialism and the oppression it brings. While Laskey finds Slocum's view of the Boers as respectful but dismissive, regardless of his intent in recounting the Kruger story, the ultimate outcome "clearly has great propaganda value for humiliating one of the great Boer leaders," and as such Slocum's account was reprinted in newspapers across the British Empire during the Second Boer War.[82] Slocum's depiction of Kruger, and whether it might be libelous, resulted in a minor wave of public debate when Slocum's book was published.[83] Notably, the more combative details of Slocum's encounter with Kruger's Flat Earther friends were removed from initial printings of the book version of *Sailing Alone Around the World*.[84]

In Michell's reframing of Kruger's views, the defeat of the Boers is paired with the defeat of their Flat Earth beliefs; the tool for sustaining Britain's empire, the modern naval navigation, is the tool which convinces Kruger to reject his medieval religious views. This also is a clear reflection

[80] Meintjes, *President Paul Kruger*, 272n101; E. J. P. Jorissen, *Transvaalsche Herinneringen 1876–1896* (Pretoria: J. H. de Busy, 1897), 17.

[81] Victor Slocum, *Capt. Joshua Slocum: The Life and Voyages of America's Best Known Sailor* (New York: Sheridan House, 1950), 355–56.

[82] Donald A. Laskey, "A Passage of Time: Joshua Slocum's Circumnavigation and the World Behind It" (MA diss., Central Michigan University, 2015), 15, 34–35.

[83] Grayson, *A Man for All Oceans*, 275.

[84] Teller, *Search*, 142.

of the degeneracy fear mentioned above. Like the Portuguese settlers in their prazos, the Dutch had been in southern Africa for centuries longer than the British. When given the chance to become incorporated into the British Empire, the Dutch had instead chosen to flee into the interior, to maintain their pastoralist lifestyle among the various African kingdoms that the British had not yet subjected. Only after Boer sovereignty was extinguished by the British in 1902 could they become fully seen as fellow whites united against the African majority. Of course, it should be noted that Kruger's argument against Slocum's global navigation claim reflects the same argument made by the Anglo-American Flat Earther William Carpenter against the 1872–1876 HMS *Challenger* expedition, as mentioned in Chapter 2.[85] In light of the association of Carpenter's adopted hometown of Baltimore with Flat Earth beliefs, the city was a center of pro-Boer sentiment in the United States; the local *Baltimore American* newspaper printed several pro-Boer accounts during the war, and the Maryland state legislature debated a pro-Boer resolution in 1900 (proposed by the delegate from Charles County, with a heavily Black population).[86]

However, in light of this, it is worth noting that the Boer Doppers were not the only Flat Earthers in southern Africa—and this did not include African natives. As mentioned throughout earlier chapters, opinions on the shape of the Earth naturally varied among the many diverse cultures across the African continent. For example, before European arrival in Central Africa, the Banyarwanda people of the region believed that the entirety of the world could not be seen, because the world is flat and its distant borders consist of fences like those surrounding kraals. Those fences support a large rock, which is the sky overhead. Above that rock is another world much like our own, just as below our world is another, but unlike in Christianity, there are no moral characteristics assigned to either. Some people and animals came from the world above, and at least one man went and returned from the world below. The Moon and the stars are types of glow-worms which remain in the sky. The sun daily goes from east to west, at which point it is captured by a powerful man who kills it,

[85] Carpenter, *One Hundred Proofs*, 20.

[86] New-York Historical Society Patricia D. Klingenstein Library: Pamph DT1896.C65 1899 Oversize Non-circulating, "Collection of clippings of newspaper articles pertaining to the South African War, 1899–1902."

cuts it into pieces, and throws it back east for the process to begin again.[87] However, for the Lobedu, a Bantu people who inhabited what would become the Transvaal region before the arrival of the Boers—although their royal house traced their origins from the collapsing Monomatapa state centered around Great Zimbabwe—there was no great myth about the creation of the world, its shape, or even conceptions of sky gods. Indeed, curiosity about cosmology and creation was seen as impractical.[88] Instead, the fellow Flat Earthers in southern Africa were British—and not only British, but distant members of the imperial metropole's Universal Zetetic Society.

In the 1880s, prior to his involvement in the UZS, the Black Jamaican minister Emanuel Vaun Mulgrave worked as a missionary in the Transvaal, as described in the last chapter.[89] The 27 February 1900 issue of the *Mid Sussex Times* reported Mulgrave speaking at a public meeting in the Corn Exchange at Hurstpierpoint on "'Life in the Transvaal', 'an account of the South African peoples, their past state, and future prospects', and the lecturer's experiences of Boer character 'gained both by travel and residence in that country to which the eyes of all civilised people are turned at the present."[90] This indicates that Mulgrave had at least some interaction with Boers during his time in southern Africa. One wonders whether theological debates with them started Mulgrave down the path to Zeteticism; if so, this would be perhaps the only evidence of Dopper theology converting Anglophone people to Flat Earth belief.

As also mentioned in the previous chapter, through the figure of John Edward Quinlan and his Society for the Protection of Dark Races, the Zetetics in London were also interested in promoting the rights of native Africans in southern Africa. However, Quinlan's society was established in 1906, after the end of Boer independence and Mulgrave's return from missionary work. In the previous decade, prior to both the Second Boer

[87] J. J. Maquet, "The Kingdom of Ruanda," in Daryll Forde, ed., *African Worlds: Studies in the Cosmological Ideas and Social Values of African Peoples* (Oxford: James Currey Publishers, 1999 [1954]), 166.

[88] J. D. Krige and E. J. Krige, "The Lovedu of the Transvaal," in Daryll Forde, ed., *African Worlds: Studies in the Cosmological Ideas and Social Values of African Peoples* (Oxford: James Currey Publishers, 1999 [1954]), 55, 59–66.

[89] David Killingray, "Black Baptists and Pan-Africanism in Britain, 1890–1913," *Immigrants & Minorities*, Vol. 35, No. 1–2 (Sep., 2020), 115.

[90] Killingray, "Black Baptists," 128n60.

War and the joining of the Society by Mulgrave and Quinlan, there is evidence that at least some Zetetics were less racially progressive. From the earliest period of Dutch settlement of the Cape, the term "Hottentot" was used to describe the indigenous non-Bantu Khoisan people, with the term quickly entering English. By the eighteenth century, Hottentot had become a general term for Africans, with derogatory connotations related to lack of civilization, or even lack of humanity. The term is perhaps most famously associated with Sarah Baartman (1789–1815), a Khoi woman exhibited as a circus freak show under the name "Hottentot Venus."[91]

With the term "Hottentot" firmly cemented in English as a racist term depicting Africans as culturally and biologically primitive, it is therefore notable that a frequent commenter to Zetetic publications in London in the 1890s went by the pseudonym "A. Hottentot." A. Hottentot's identity is unknown, and Robert Schadewald has speculated there may have been a Cape connection, which is plausible but not necessary for one to be familiar with the term.[92] If anything, the connection seems to be the idea that evidence for the Flat Earth was so widespread and simple that even "a Hottentot" could grasp it—as opposed to the idea that Africans were so uneducated that they would gravitate toward a Flat Earth view, as later American advocates would claim. Among the available remaining archival materials, A. Hottentot's first appearance was in the July 1893 issue of the UZS publication, *The Earth not a Globe Review*, in a letter to the editor defending Rowbotham, which Albert Smith proclaimed as being too long. Of note, Hottentot states that the 20 May issue of the journal *Science Shiftings* proclaimed that "the Nile has a fall of only six inches in 1,000 miles!," a sign of the Flat Earther interest in the long African river mentioned in Chapter 1.[93] In the March 1894 issue, Hottentot wrote a letter to the editor complaining about another critique on Rowbotham

[91] Britt Rusert, *Fugitive Science: Empiricism and Freedom in Early African American Culture* (New York: New York University Press, 2017), 220.

[92] Schadewald, *The Plane Truth*, 76.

[93] A. Hottentot, "'Parallax' Triumphant," *The Earth not a Globe Review*, No. 3 (Jul., 1893), 19.

by a reader identified as "Enquirer."[94] Hottentot can be found in several other letters to the editor throughout 1894.[95]

Although A. Hottentot may or may not have been a resident of the Cape, the Cape played a role in metropolitan Zetetic speculation. The aforementioned undated *Chart & Compass* pamphlet argued that

> If their [spherical] system was as correct as it professes to be, ought not a fully trained and perfectly educated officer be able to take a ship, say, to Malta and back, on second or third voyage? Instead of which, when his theoretical education is over, he has to put his theories in his sea chest and learn from actual observation, for many long years, how to take a vessel, say, to the Cape of Good Hope, without finding himself, some foggy morning, on the coast of Mexico?[96]

The January 1894 issue of *The Earth not a Globe Review* contained a letter to the editor from an "H. C. Bowker, MA" informing the paper that

> Dear Sir,—I saw the other evening a gentleman from New Zealand, who declared they had a considerable amount of twilight there, and further that the outward voyage was made round the Cape of Good Hope to the east, and the homeward voyage east to Cape Horn, and then to the west coast of Africa before turning north, to get the benefit of easterly currents, and thus the outward and homeward voyages circumnavigated the "globe" in S. latitudes in a time which would be impossible if the degrees of longitude were so much larger as they should be were the earth a plane. Have you any sellable information on this point, or can you refer me to any books that give it. Birley's [Rowbotham's] book (Parallax) says p. 96, last line, the direct distance Valencia to Cape Town is miles—this must be a mistake. The S. African Steamers [*sic*] from Southampton cannot do it under 15 days. I am sure that one of the most pressing matters that requires settlement in this controversy in the measurement of a degree of longitude at two different latitudes south of the equator: or some reliable

[94] A. Hottentot, "Audi Alteram Partem," *The Earth not a Globe Review*, No. 6 (Mar., 1894), 135.

[95] Albert Smith, "Notes," *The Earth not a Globe Review*, No. 5 (Jan., 1894), 111; John Bradley and A. Hottentot, "Letters to the Editor," *The Earth not a Globe Review*, Vol. IV (Jul./Sep., 1894), 78.

[96] Zetetic Society, *Chart & Compass*, 3.

> information from a sea-captain who has sailed a degree there, and can give some reasonably good estimate of of [*sic*] the distance.[97]

In the same issue, Smith republished an article from "Henry G. Schlichter, D.Sc. in the *Geographical Journal*, Vol. 2, November 1893" discussing the difficulty of establishing longitudinal measurements in the African interior, with the end result being "interior of Africa is to-day almost as much a *terra incognita* as it was a hundred years ago. The reason of this complete failure is evidently twofold—firstly that the reliable methods are too complicated or difficult for the majority of travellers, and secondly that several of the instruments as well as of the methods employed are not accurate enough."[98] Africa itself, in other words, was a locale where the seeming impossibility of conducting modern, Western science reflected the deeper truth of Zetetic astronomy. In 1906, amidst his activist work for Black laborers in the region, Quinlan wrote how

> A sea-going ship could sail round the navigable world on portions of a uniform course, say like the Equator, in the following manner. It could start from the east coast of Africa at the Equator, and keep along the easterly course till it reached the many islands of the East Indies. It could sail round each of the islands till it got back to the Equator on their east side; it could then proceed on its uniform course across the Pacific till the west coast of South America was reached. Then by sailing southerly and around Cape Horn, and northerly till it reached the Equator near the mouth of the Amazon, it could pursue its uniform course across the Atlantic Ocean and the Gulf of Guinea (where it would cross the meridian of Greenwich), to the west coast of Africa, and by again sailing southerly

[97] H. C. Bowker, "Letter to the Editor," *The Earth not a Globe Review*, No. 5 (Jan., 1894), 112.

[98] Albert Smith, "Geographical Longitudes," *The Earth not a Globe Review*, No. 5 (Jan., 1894), 106. For further information on the difficulties of British efforts to chart the African interior in the nineteenth century, see Dane Kennedy, *The Last Blank Spaces: Exploring Africa and Australia* (Cambridge: Harvard University Press, 2013) and Robbie McLaughlan, *Re-imagining the 'Dark Continent' in fin de siècle Literature* (Edinburgh: Edinburgh University Press, 2012), 99–122. For more general difficulties on early European Southern Hemisphere exploration and charting, see Kenneth Morgan, *Navigating by the Southern Cross: A History of the European Discovery and Exploration of Australia* (London: Bloomsbury Academic, 2021), 1–20. For development of the grid system of cartography and its application to land maps, see Mick Ashworth, *Why North is Up: Map Conventions and Where They Came From* (Oxford: Bodleian Library Publishing, 2019), 29–34.

> and around the Cape of Good Hope and then northerly it would get back to the point from which it started; but the ship would have been sailing all the time along a flat ocean, and never along a convex or globular one, and its captain would have been consulting a flat chart, and never a globe to sail it along the desired course.[99]

Proper colonial navigation therefore, far from being contingent on proper Zetetic astronomical techniques, also served to validate the truth of Zetetic astronomy. This might be seen as explaining the reception of the Boers to Slocum's claims about sailing around the Earth—save for the fact that, along with the three Boer friends of Kruger, Slocum had another Flat Earther encounter while docked in Durban.

Outside of the Cape and Transvaal, Natal had smaller, but still notable, connections to the Zetetic movement. Joseph Steer Christopher (1805–1894) was an English merchant who resided in Natal for several years on either side of 1850; in that year, he published in London the book *Natal, Cape of Good Hope*, which included several maps of the colony made by himself. In the 1880s, Christopher seems to have become a Zetetic convert, drawing a map of the Flat Earth which became widely reproduced on either side of the Atlantic.[100] As mentioned in the last chapter, the 1906 Bambatha Rebellion in the colony may have inspired Quinlan's formation of the National Society for the Protection of the Dark Races. But in between Christopher and Quinlan, the colony was home to an actual British member of the UZS, and one with a far different take on Slocum than his Dopper co-planists: a man named Thomas Winship.

Not much is known about Winship, save that he was an English-speaking settler in southern Africa on either side of the fin de siècle, and a UZS member. Among existing correspondence, Winship first appears in the September 1894 *The Earth not a Globe Review*, where Lady Blount responds to him, "Thos. Winship.— Thanks for your letter and cuttings from *Cape Argus*. Hope you have received letter and

[99] Quinlan, *The Earth a Plane*, 4–5.

[100] Schadewald, *The Plane Truth*, 87; Michael Buehler, "Extraordinarily rare 1892 flat Earth map by Alexander Gleason," *Boston Rare Maps*, n.d., accessed August 4, 2022, https://bostonraremaps.com/inventory/alexander-gleason-flat-earth-1892/; Joseph Steer Christopher, *Natal, Cape of Good Hope* (London: Effingham Wilson, 1850).

pamphlets, forwarded."[101] The *Cape Argus* is an English-language newspaper established in Cape Town in 1857 and still existent as of 2025. This presumably means that Winship was resident in Cape Town as of 1894. While it is unknown what contents of the *Cape Argus* Winship forwarded to his comrades in London, it is notable that the newspaper long had an editorial stance of criticizing the Cape Colony government's racist political and social policies.

Winship next appears in the January 1895 issue of *The Earth not a Globe Review*, where he wrote about attending a lecture titled "Glimpses of the Solar System" at the Cape Observatory. The lecturer, who stated that he had been employed at the Royal Observatory in Greenwich when younger, showed a supposed picture of a lunar volcano that was actually Mount Vesuvius, and made a disparaging comment about Flat Earthers. When Winship wrote a letter to try and debate him, he received only a curt reply.[102] Here is a startling predecessor to the Apollo Moon landing hoaxes that emerged seventy years later, and will be discussed in the next chapter. The well-meaning astronomer, intending to bring the solar system to life for his audience, used a terrestrial mountain as a stand-in for an astronomical object; this deception was then used by a Flat Earther to cast doubt on the entirety of the astronomer's comments, the field of astronomy overall, and the notion of trust in scientific expertise—a lesson to modern science communicators.

Winship continued his letter by writing how

> Being an amateur navigator I am much interested in the truth of the earth's planarity and would much like to get a chart on the natural principle if there are any to be had. I am at present living right east of Table Mountain [a flat mountain then at the south of Cape Town], and therefore cannot determine where the Southern Cross sets; but should I go to live in the city I shall have great pleasure in observing it and reporting to you.[103]

[101] Elizabeth Blount, "Answers to Correspondents," *The Earth not a Globe Review*, No. 1, New Series (Sep., 1894), 23.

[102] Thomas Winship, "The Contents of our Letter Box," *The Earth not a Globe Review*, No. 2, New Series (Jan., 1895), 44.

[103] Winship, "The Contents of our Letter Box," 44–45.

Living directly east of Table Mountain would place Winship in close proximity to the Royal Observatory. But more to the point, it establishes that Winship's belief in the Flat Earth is connected to his interest in sea navigation, and the need to create a usable map which syncretized both hobbies. Further, between both his interest in creating a Flat Earth sea chart and helping to establish where the Southern Cross sets for his northern hemisphere correspondents, he applies the same willingness to use experiments to determine the truth—or rather, validate his own beliefs—that were evident in the Flat Earther movement since the Bedford Level Experiment, representing their ARIG status.

Winship's next appearance is again in the form of a response by Lad Blount in the July 1895 issue, where she writes, "T. WINSHIP.—Thanks for *The Cape Magazine*. We hope you will continue to write to it on 'Natural Cosmogony.' We wish you God speed in the propogation [*sic*] of truth. We shall be glad to see Mr. Dunn's reply, if he attempts to answer such practical facts as you have brought before the readers of so racy a magazine."[104] *The Cape Magazine* appears to be the *Cape Monthly Magazine*, established in the Cape Colony in 1851 by Roderick Noble (1829–1875), who in 1859 would go on to become a professor of physical science at the South African College, predecessor to the University of Cape Town. Noble would also for a time edit the *Cape Argus*, and the *Cape Monthly* shared its generally critical stance on racial policies, a consistency that is indicative, if circumstantial, of Winship's racial views. Although Noble had died by 1895, that Winship had chosen a magazine founded by a physical scientist to attempt to proselytize his Zetetic worldview also indicates a combativeness on behalf of Winship, typical of Flat Earthers then and now. The Dunn that Winship had been responding to was perhaps Edward John Dunn (1844–1937), the English-born geologist who from 1871 to 1873 had been the government geologist of the Cape Colony. During that time he published several articles in the *Cape Monthly Magazine*. Although living in Australia as of 1895, the Dunn article could have been a reprint or a submission from abroad, especially as he continued to write on southern Africa long after leaving service in the Cape.

As mentioned above, since the start of the Zetetic movement in the mid nineteenth century, followers had taken on pseudonyms, following

[104] Elizabeth Blount, "Answers to Correspondents," *The Earth not a Globe Review*, No. 4, New Series (Jul., 1895), 96.

the example of Samuel "Parallax" Rowbotham, with Lady Blount taking "Zeteo" and Albert Smith taking "Zetetes." The October 1895 issue of *The Earth not a Globe Review* contained a letter from someone identified only as "Yachtsman"—but who was likely Winship, in his first attempt at committing deeper to the Zetetic Society's culture, much like a modern Flat Earther slowly getting drawn into the internet culture of a Facebook group. The letter, titled "Navigation Proves the World a Plane!," is worth exploring large segments of:

> When at school, I was taught that the most conclusive proof that the Earth is a Globe, was found in the fact that the upper masts and sails of ships at sea are seen first, and as the vessels approach the observer, the lower masts and sails become visible, and finally the hull.... I am in possession of one of the most powerful glasses [binoculars] ever invented, or made by man. But to go into details. The iron barque, "La Querida," of Liverpool, left Capetown [*sic*] for Australia some time ago. I watched her from an elevation of 200 feet. She was in ballast, and thus presented a hull high out of water. As she sailed away, the entire vessel, masts, sails and hull *gradually* became less, until the hull was scarcely visible. I applied the "Emperor" binocular glass to my eyes, and could see the hull as plain as the sails and masts. The ship went on until I could no longer see the hull, and only the masts and sails indistinctly. Again I applied the glass, and again saw the hull, masts and sails *all* very small, *but all distinctly visible*, although about 10 miles distant. Hence, I came to the conclusion that either my binocular glass has the power of piercing a segment of water miles in thickness, or that the earth is a plane and therefore not a globe.... On a fine afternoon I took up my usual position (200 feet above sea level) to watch the schooner "Lilia," of Capetown, sail away about due North for Saldanha Bay [northwest of Cape Town in Cape Colony]. As she left the land behind she appeared to climb the water in front of her until she came up to the line of vision of the horizon and my eye, and then *gradually disappeared* without making the least attempt to get *below* the horizon. When I could no longer see her hull, which was painted black, I applied my glass to my eyes, and the whole vessel was restored to sight. This continued until she appeared as a black spot on the water and the glass failed to distinguish between hull and sails. At vanishing point, she was still in the same line of sight that she appeared to ascend to when first leaving the harbour. At my usual position again one very fine and clear morning, I saw a black spot far away on the horizon. As it drew nearer I could indistinctly see a funnel and two masts, but nothing whatever of the ship's body. Now, thought I, here's a fine chance to test the "Emperor" and the

> scientists [*sic*] globe teaching at one and the same time, and suiting the action to the thought, I applied the glass, when lo! and behold! What did I see? I saw a funnel, two masts, and the body of the steamer, and also the white wave being dashed from her bow by her speed through the water!!! Now if any Observatory man can be found to attempt to prove the Earth a Globe, I am ready to meet him.[105]

Yachtsman's name being a reference to yachting, the location in Cape Town, and the use of amateur navigational tools (the Emperor binoculars) align with this being Winship. And when the time came "to test the 'Emperor' and the scientists['] globe teaching at one and the same time," Winship's support of his own equipment and experience over scientific authority is in line with someone who had been disillusioned by the earlier lecture at the Royal Observatory, as well as the pattern of Zetetic science established in the Bedford Level Experiment. But also of note is Yachtsman's use of the classic proof of the globe Earth, of ships' masts vanishing after and appearing before the hull of the ship on the horizon. A settler on a colonial port, watching ships go to and from the metropole and the next settler colony, and using a binocular whose brand was named Emperor, no less—the traffic needed to sustain a settler colony and connect it to other outposts of the Empire was connected to the observational proofs of the supposed Flat Earth.

Assuming Yachtsman was indeed Winship, the letter indicates several things. First, it is likely that Yachtsman's "Emperor" was an Emperor field glass made by the firm of Thomas Armstrong & Brother of Manchester, England. A 1902 advertisement proclaimed their Emperor to be "a very fine Binocular for Land or Marine use, from £2 15s to £3 15s [roughly £200–300 in 2025], in best leather sling case."[106] If Yachtsman was Winship, it indicates that he was from Manchester, the city at the heart of the Industrial Revolution in Britain and therefore the spiritual antithesis to the Boer philosophical and economic mindset. More directly, the letter marked a transition for the author. It was Winship's first time using a pseudonym, and last time writing from Cape Town. After a few years

[105] Yachtsman, "Navigation Proves the World a Plane!," *The Earth not a Globe Review*, No. 5, New Series (Oct., 1895), 115–16.

[106] "Advertisements," *The Journal of the Manchester Geographical Society*, Vol. 18, Nos. 10–12 (Oct.-Dec., 1902), 246. With thanks to Horace A. Smith for suggestions on identification.

of absence from the remaining Zetetic record, Winship would return in 1899, with a new name, new location, and now author of no mere letter, but a book of his own: the amply named *Zetetic Cosmogony; or Conclusive Evidence That the World Is Not a Rotating Revolving Globe but a Stationary Plane Circle*. At 192 pages, this was an expanded version of an "an unpretentious pamphlet of 48 pages" he had written earlier; the new version was published in his new abode of Durban in November 1899, just after the start of the Second Boer War.[107] Further, Winship published *Zetetic Cosmogony* under both his name and his (new) pseudonym, "Rectangle." Winship would use the Rectangle byline going forward in his Zetetic career.

Winship's book contains several Zetetic standards. In line with the phantom balloon sightings of the contemporary war, he quotes James Glaisher, French astronomer and Spiritualist Camille Flammarion (1842–1925), and other aeronauts he had met who say that the horizon looks level even from the heights of balloon ascent.[108] Because east–west-oriented rivers like the Congo and Amazon (the former in an extant colony, the latter in a settler colonial state) flow in directions that would be "uphill" according to the rotation of the Earth at various times, a spherical Earth is impossible. Surprisingly, he does not cite the Zetetic favorite of the Nile flowing "up."[109] Of course, as a Flat Earther Winship rejects both Darwin's theory of evolution and the idea of deep time, in favor of Young-Earth Creationism.[110] More intriguing, however, was Winship's evidence for a young Earth, which came from two seemingly opposed sources.[111] The first was an 1895 debate between Irish physicist Lord Kelvin (1824–1907) and Irish engineer John Perry (1850–1920), centered around the fact that physicists could not account for the timescale of the Earth determined by geologists prior to the discovery

[107] Thomas Winship, *Zetetic Cosmogony; or Conclusive Evidence that the World is not a Rotating Revolving Globe but a Stationary Plane Circle* (Durban: T. L. Cullingworth, 1899), i–ii.

[108] Winship, *Zetetic Cosmogony*, 7–8.

[109] Winship, *Zetetic Cosmogony*, 110.

[110] Winship, *Zetetic Cosmogony*, 4–6, 154–55, 165–66.

[111] Winship, *Zetetic Cosmogony*, 162–65.

of radioactivity (the next year), providing an explanation for the energy source for both the Earth's internal heat and the Sun's luminescence.[112]

The second piece of evidence was the use of deep time—time scales of hundreds of thousands of years—in the cosmology of the Hindu and Buddhist religions, something which ironically had been used by both British Christians and Indian Muslims to build support for the time scales necessary for Darwinian evolution.[113] As mentioned in Chapter 1, English philosophers had been using Hindu astronomy to support the spherical shape of the Earth in England since at least the early thirteenth century. To Winship, as Hinduism and Buddhism were obviously incorrect due to their opposition to the true faith of Christianity, the large time scales of those religions were obviously also incompatible with the mere few thousand years of his YEC cosmology. Natal, and Durban specifically, was the center of the Indian population in southern Africa, ever since the first Indian—mainly Hindu—laborers were brought to the region by the British in 1836 to work the sugarcane industry, after the law abolishing slavery in the Cape Colony went into effect in 1834. The first Hindu temple in southern Africa was built in Durban in 1869, and the city remains one of the highest concentrations of both Indians and Hindus outside of India.[114]

[112] Philip C. England, Peter Moinar, and Frank M. Richter, "Kelvin, Perry and the Age of the Earth," *American Scientist*, Vol. 95, No. 4 (Jul., 2007), 342–49; Gillian Beer, "'The Death of the Sun': Victorian Solar Physics and Solar Myth," in *The Sun is God: Painting, Literature and Mythology in the Nineteenth Century*, ed. J. D. Bullen (Oxford: Clarendon Press, 1989), 174–76.

[113] Thomas Simpson, "Historicizing Humans in Colonial India," in Efram Sera-Shriar, ed., *Historicizing Humans: Deep Time, Evolution, and Race in Nineteenth-Century British Sciences* (Pittsburgh: University of Pittsburgh Press, 2018), 113–37; Sarah Ahmed Qidwai, "Sir Syed (1817–1898) and Science: Popularization in Nineteenth Century India" (PhD diss., University of Toronto, 2021), 60–61; Sarah A. Qidwai, "Reexamining Complexity: Sayyid Ahmad Khan's Interpretation of "Science" in Islam," in Bernard Lightman, ed., *Rethinking History, Science, and Religion: An Exploration of Conflict and the Complexity Principle* (Pittsburgh: University of Pittsburgh Press, 2019), 58–59; C. Mackenzie Brown, *Hindu Perspectives on Evolution: Darwin, Dharma, and Design* (Abingdon: Routledge, 2012), 7.

[114] P. Pratap Kumar, "Hinduism in South Africa," in *The Wiley-Blackwell Companion to African Religions*, ed. Elias Kifon Bongmba (Hoboken, NJ: Wiley-Blackwell, 2012), 389–90; Thomas R. Metcalf, *Imperial Connections: India in the Indian Ocean Arena, 1860–1920* (Berkeley, CA: University of California Press, 2007), 136–64; Paul Younger, *New Homelands: Hindu Communities in Mauritius, Guyana, Trinidad, South Africa, Fiji, and East Africa* (Oxford: Oxford University Press, 2010), 126–60; Jon Soske, *Internal*

Winship's dismissal of Hinduism and Buddhism as false religions may seem like a typical colonialist viewpoint, which it may be. However, given his residence in Durban and specific reference to Hinduism and Buddhism may indicate that Winship had actually gone out of his way to learn at least a little about the non-white population of his new city, even if he disagreed with them. Such interracial and interreligious friendships were not impossible in colonial Natal, but as often reinforced the colonial hierarchies of power as challenged them.[115]

However, from a Zetetic standpoint, the most interesting part of Winship's book was the paragraph where he notes that

> In December, 1897, I met Captain Slocum on board the "Spray". *This navigator told me that he had sailed his little craft 33,000 miles [53,000 kilometers] by plane sailing*. Rather a LONG voyage, it must be admitted. A PLANE or LEVEL SURFACE for 33,000 miles, and yet the world a globe? To the pre-historic "man of science" at the North Pole, and the Darwinian Ape at the South Pole of the astronomers [*sic*] imaginary globe, with such a delusion.[116]

This is a notable passage for several reasons. First, Winship takes the exact opposite view of Kruger: Slocum's account *confirms*, rather than *opposes*, the notion of a Flat Earth, as Slocum never had to adjust the *Spray* to remain on the surface of the globe (demonstrating the typical lack of understanding of the nature of gravity among Zetetics then and now). Given that Slocum's visit and Winship's writing happened in the immediate prelude to the Boer War, it makes sense that the English Winship would have a more positive view of the Nova Scotian Slocum than the Boer Kruger did. Perhaps this positive reception is why Slocum did not mention Winship in his own book; although there is also the potential

Frontiers: African Nationalism and the Indian Diaspora in Twentieth-Century South Africa (Athens, OH: Ohio University Press, 2017), 145–48; Goolam Vahed, "Gokhale, Polak, and the end of Indian indenture in South Africa, 1860–1911," in *South Asian Migrations in Global History: Labor, Law, and Wayward Lives*, ed. Neilesh Bose (London: Bloomsbury Academic, 2022), 37–62.

[115] T. J. Tallie, "The Impossible Handshake: The Fault Lines of Friendship in Colonial Natal, 1850–1910," in Shannon Walsh and Jon Soske, eds., *Ties that Bind: Race and the Politics of Friendship in South Africa* (Johannesburg: Wits University Press, 2016), 100–24.

[116] Winship, *Zetetic Cosmogony*, 89.

that Winship never revealed his views to Slocum, or that a common (and Anglophone) settler having Flat Earth views was less interesting to Slocum than those views being held by the State President of the South African Republic.

But of more interest is Winship's contrast of "the pre-historic "man of science" at the North Pole, and the Darwinian Ape at the South Pole." The association of scientists with the northern hemisphere, and "Darwinian Apes" with the southern hemisphere, maps onto contemporary views of the Global North as developed, and Global South as undeveloped. It also reflects the notion that non-whites, particularly Africans, were less developed evolutionarily than Europeans, a byproduct of the scientific racism which emerged out of Darwinian thought (albeit without Darwin's input or approval).[117] For example, Alfred Russell Wallace, the enemy of the Zetetics from the Bedford Level Experiment, promoted polygenism, the idea that the various human races evolved separately from different simian ancestors.[118] In the US South, while supporters of anti-Black racism latched onto Darwinism to shore up their argument, there was no firm pro- or anti-Darwinian line among secular or religious Black thinkers from 1859 into the first decades of the twentieth century.[119] In an ironic twist, during Reconstruction, anti-Black politicians like Missouri's Liberal Republican senator Carl Schurz embraced Darwinian evolution to justify the defeat of an 1871 proposal to annex

[117] Guy Burrows, *The Land of the Pygmies* (London: C. Arthur Pearson, 1898), 182; Chris Ballard, "Strange Alliance: Pygmies in the Colonial Imaginary," *World Archaeology*, Vol. 38, No. 1, Race, Racism and Archaeology (Mar., 2006), 131–51.

[118] John R. Durant, "Scientific Naturalism and Social Reform in the Thought of Alfred Russel Wallace," *The British Journal for the History of Science*, Vol. 12, No. 1 (Mar., 1979), 39–41; Eric D. Anderson, "Black responses to Darwinism, 1859–1915," in Ronald L. Numbers and John Stenhouse, eds., *Disseminating Darwinism: The Role of Place, Race, Religion, and Gender* (Cambridge: Cambridge University Press, 1999), 253–59; Rusert, *Fugitive Science*, 70–73.

[119] Ronald L. Numbers and Lester D. Stephens, "Darwinism in the American South," in Ronald L. Numbers and John Stenhouse, eds., *Disseminating Darwinism: The Role of Place, Race, Religion, and Gender* (Cambridge: Cambridge University Press, 1999), 123–44; Eric D. Anderson, "Black responses to Darwinism, 1859–1915," in Ronald L. Numbers and John Stenhouse, eds., *Disseminating Darwinism: The Role of Place, Race, Religion, and Gender* (Cambridge: Cambridge University Press, 1999), 247–66; Jeffrey P. Moran, *The Scopes Trial: A Brief History with Documents* (Boston: Bedford/St. Martin's, 2002), 66–69, 180–88; Jeffrey P. Moran, "The Scopes Trial and Southern Fundamentalism in Black and White: Race, Region, and Religion," *The Journal of Southern History*, Vol. 70, No. 1 (Feb., 2004), 111–15.

the sole independent Black republic of Haiti.[120] By the second decade of the twenty-first century, this Black agnosticism toward evolution had changed, with the US Black support for teaching Creationism in schools much higher than white support, at 58% to 44%, and with similarly higher rates of both church attendance and belief in Biblical literalism.[121]

One of the major exemplars of the debate between the teaching of Creationism and evolution is the so-called Scopes Monkey Trial of 10–21 July 1925, in which advocates for teaching Darwinian evolution were found to be in violation of a Tennessee state law endorsing the Book of Genesis. In the aftermath of the trial, the Society for Science & the Public claimed that religious-backed anti-evolution laws being debated by the legislatures of such southern US states as North Carolina, South Carolina, Missouri, Tennessee, and Alabama would also require the teaching of Flat Earth thought in addition to Creationism in state schools. An example of this can be found in the Rollin Kirby (1875–1952) cartoon from the 18 July 1925 issue of the *New York World* newspaper, printed during the middle of the trial.[122] The most prominent Flat Earther in the United States at the time, the Illinois cult leader Wilbur Glenn Voliva (1870–1942), went to Tennessee in an attempt to get called as a witness for the prosecution at the Scopes Trial, though he was unsuccessful in that endeavor.[123]

Just as general knowledge of the 1692–1693 Salem Witch Trials tends to be mediated through Arthur Miller's (1915–2005) 1953 play *The Crucible*, widespread knowledge of the Scopes Monkey Trial tends to be mediated through Jerome Lawrence (1915–2004) and Robert E. Lee's (1918–1994) 1955 play *Inherit the Wind*, particularly its 1960 film adaptation. Like *The Crucible*, *Inherit the Wind* highly fictionalizes its source

[120] Brenda Wineapple, *Ecstatic Nation: Confidence, Crisis, and Compromise, 1848–1877* (New York: HarperCollins, 2013), 488.

[121] Esmeralda Sánchez Salazar, Brandon Vaidyanathan, Elaine Howard Ecklund, and Adriana Garcia, "Challenging Evolution in Public Schools: Race, Religion, and Attitudes toward Teaching Creationism," *Socius: Sociological Research for a Dynamic World*, Vol. 5 (Sep., 2019), 1–13.

[122] "Flat Earth and Anti-Evolution," *The Science News-Letter*, Vol. 11, No. 305 (Feb. 12, 1927), 95–96; Ronald L. Numbers, *The Creationists: The Evolution of Scientific Creationism* (New York: Alfred A. Knopf, 1992), 98–101.

[123] Garwood, *Flat Earth*, 202.

events and serves as a critique of contemporary McCarthyism, but in addition it has helped shape the perception and teaching of both law and science in the United States.[124] The Book of Joshua plays a major role in the climactic scene of the trial. Defense attorney Henry Drummond (stand-in for the real Clarence Darrow, 1857–1938) pushes prosecuting attorney Matthew Harrison Brady (stand-in for the real William Jennings Bryan, 1860–1925) on whether he thinks the stillness of the Sun in Joshua actually happened, and if it did, if Copernicus should be banned along with Darwin.[125] Although expanded from the actual words spoken, the back and forth on the historicity of Joshua is based on transcribed dialogue between Darrow and Bryan that took place on the trial's seventh day, with the Copernicus element taken from statements by defense attorney Arthur Garfield Hays (1881–1954) on the second day.[126]

The role of Joshua in both the actual Scopes Trial, and its additional emphasis in *Inherit the Wind*, perhaps reflects the lingering influence of John Jasper. Several individuals in the defense team of the Scopes Trial—Darrow, attorney Dudley Field Malone (1882–1950), and witness Winterton C. Curtis (1875–1965), zoology professor at the University of Missouri—equated the rejection of evolution with the endorsement of Flat Earth (though perhaps tellingly, none of the prosecution did).[127] *Baltimore Sun* journalist H. L. Mencken (1880–1956, depicted as E. K. Hornbeck in *Inherit the Wind*), who covered the trial very sympathetically to the defense, was an admirer—albeit not uncritical—of esotericist Charles Fort, described in Chapter 5.[128] Indeed, in 1997, historian of science Edward J. Larson (b. 1953) published a book on the Scopes Trial, *Summer for the Gods*, which won him the Pulitzer Prize for History

[124] Randy Moore, "Creationism in the United States: VII. The Lingering Impact of *Inherit the Wind*," *The American Biology Teacher*, Vol. 61, No. 4 (Apr., 1999), 246–50; Gad Guterman, "Field Tripping: The Power of *Inherit the Wind*," *Theatre Journal*, Vol. 60, No. 4 (Dec., 2008), 563–83; Edward J. Larson, *Summer for the Gods: The Scopes Trial and America's Continuing Debate Over Science and Religion* (New York: Basic Books, 1997), 225–66.

[125] Jerome Lawrence and Robert E. Lee, ed. Alan Woods, *The Selected Plays of Jerome Lawrence and Robert E. Lee* (Columbus, OH: Ohio State University Press, 1995), 51–52.

[126] *The World's Most Famous Court Trial: Tennessee Evolution Case* (Cincinnati: National Book Company, 1925), 56–57, 285–86.

[127] *The World's Most Famous Court Trial*, 76–77, 116, 183, 256, 265.

[128] Joshua Blu Buhs, *Think to New Worlds: The Cultural History of Charles Fort and His Followers* (Chicago: University of Chicago Press, 2024), 28–30.

the next year. In it, Larson notes that "Relatively little comment about the trial survives from African Americans. A few black evangelists, such as Virginia's John Jasper, endorsed Bryan's position, while the [National Association for the Advancement of Colored People], which worked regularly with the ACLU [American Civil Liberties Union], participated in some of the ACLU's New York meetings on the trial."[129] It would be an impressive feat for Jasper to endorse the attack on evolution twenty-four years after his death—although as mentioned earlier, modern parishioners of the Sixth Mount Zion Church believe his spirit to haunt it, so Jasper could perhaps provide feedback in that manner.

That this error could appear in a Pulitzer Prize-winning work is perhaps somewhat less shocking given the aforementioned nadir of Jasper's popularity among scholars in the 1990s. It also could have come from a conflation of Jasper with a figure at the time associating himself with Jasper. In March 1925, several months before the Scopes Trial, James S. Hatcher, Reverend of the Third Street Bethel African Methodist Episcopal Church in Richmond, preached a sermon titled "The Sun Do Move and the Earth Has Corners," a conscious attempt to echo Jasper. Hatcher was Black, but his congregation of nearly a thousand was mixed race, as was the audience when he repeated his sermon by request the next month. Hatcher's sermon emphasized the Joshua story, but also incorporated Jasper himself into it, elevating his predecessor to the level of Moses and saying that ignoring the flatness of the Earth would be to turn away from both of those great prophets.

Unlike Jasper with his lack of formal education, Hatcher was a graduate of Alabama State Normal School (now Alabama State University) and had served as secretary to Booker T. Washington. Such a figure preaching Flat Earth in the 1920s was accorded a less positive reception than Jasper preaching it in forty years earlier. The *Baltimore Afro-American* newspaper claimed it could only find one Black person to agree with Hatcher, and the Black journalist Ernest Rice McKinney (1886–1984) suggested that to give such a sermon, Hatcher must have fallen "under the spell of Macbeth's witches, the head hunters of Borneo or some long distance African medicine man"—or possibly that "friend Hatcher may have been talking with William Jennings Bryan," linking Flat Earth not only to the beliefs of indigenous peoples and magic from around the world, but

[129] Larson, *Summer for the Gods*, 122.

with anti-evolution sentiment. Indeed, in the aftermath of the Scopes Trial, Hatcher was called upon to give several anti-evolution sermons in Richmond.[130] While the fear that the Scopes Trial's endorsement of Creationism might lead to Flat Earth spreading was perhaps overstated, it does seem that at least in one case, the trial's outcome led one Flat Earther to Creationism.

This connection between Flat Earth and Creationist thought will be explored more in Chapters 6 and 7. However, in the South African context of Thomas Winship's usage of the hemispheric contrast of men of science and Darwinian Apes, it is important to remember that Winship rejected both the idea of hemispheres and of evolution, and certainly would have been inclined to reject any scientific idea promoted by Wallace. By labeling the "men of science" like Perry as "pre-historic," and associating the southern hemisphere "apes" with the modern term of "Darwinian," Winship deliberately blended the categories that would have been used by his mainstream opponents, and the use of "Darwinian Ape" is meant to reflect, rather than endorse, his opponents' views. After all, while rejecting the beliefs of the Hindus and Buddhists, Winship still placed them on equal footing with European scientists.

It may be illuminative to briefly discuss John Colenso (1814–1883), who from 1853 until his death served as the first Anglican Bishop of Natal. Colenso combined advocacy of African causes with critical views on Biblical statements including not only polygenism and the literalness of the Noachic flood, but also the accuracy of the Book of Joshua narrative of the Sun standing still.[131] Colenso's endorsement of Biblical errancy was partially inspired by critiques from his Zulu assistants and parishioners, and his belief in the accuracy of Egyptian records predating the YEC timeframe.[132] Colenso's views of Christianity were at odds with the Victorian-era Anglican mainstream just as Winship's were, and both

[130] Moran, "The Scopes Trial and Southern Fundamentalism," 104–11. There was seemingly no connection between James Hatcher and the John Jasper biographer and preacher William Hatcher.

[131] John William Colenso, *The Pentateuch and Book of Joshua Critically Examined, Volume 1* (London: Longman, Green, Longman, Roberts, & Green, 1862), x–xi.

[132] Jeff Guy, "The Religious Thinking of J. W. Colenso: The Theology of a Heretic," *Religion in Southern Africa*, Vol. 4, No. 2 (Jul., 1983), 4, 7; Jeff Guy, "Class, Imperialism and Literary Criticism: William Ngidi, John Colenso and Matthew Arnold," *Journal of Southern African Studies*, Vol. 23, No. 2 (Jun., 1997), 220–21, 230; Tallie, "Impossible Handshake," 100–01.

drew (in differing ways) on both ancient and contemporary non-Christian views to support their claims, but it was Winship's Flat Earth views which stressed the unity of all human races.

While it may be a stretch to call Winship enlightened, even by the standards of fin de siècle settlers in southern Africa, his writings reflect how his Flat Earth beliefs played into his holding more nuanced takes on non-European peoples and their ideas, certainly when compared to those of Kruger and the Boer Flat Earthers. Conversely, by all accounts, Winship appears to have existed comfortably within the British Empire, and used Flat Earth beliefs in a defense of the navigational systems used to maintain it, while the Boers' Flat Earth beliefs played into their anti-imperialism (albeit not any kind of critical view of settler colonialism). It also adds to the question of how much the Boer War and its aftermath radicalized the Zetetic Society to support pro-African movements. While the joining of Quinlan and the 1902 Mount Pelée eruption undoubtedly played a part, if Winship was broadly representative, a section of the membership might already have been so inclined. It is perhaps instructive to compare the (known) membership of the UZS with the membership patterns of the Aborigines' Protection Society (which Quinlan also was a member of) and other British abolitionist groups earlier in the nineteenth century, which tended to be dominated by industrialists, Nonconformists, and Liberal Party members.[133]

Zetetic Cosmogony was not the last appearance by Winship in the record. He returns, again using the Rectangle name, in the June 1902 issue of Lady Blount's new UZS publication, *The Earth*, where he authored a multi-part treatise, "A Defence of Philosophic Doubt." The article states that it is a continuation from the previous volume of *The Earth*, an issue which is not extant, but this installment of the article contains his thoughts on the Noachic flood, Young-Earth Creationism, and fossils as remnants of the Deluge.[134] Among Winship's thoughts are the injunction that

[133] Richard Huzzey, "Minding Civilisation and Humanity in 1867: A Case Study in British Imperial Culture and Victorian Anti-Slavery," *The Journal of Imperial and Commonwealth History*, Vol. 40, No. 5 (Nov., 2012), 807–25; Darren Reid, "British Humanitarianism, Indigenous Rights, and Imperial Crises: Assessing the Membership Base of the Aborigines' Protection Society, 1840–73," *Journal of British Studies*, Vol. 63, No. 3 (Jul., 2024), 625–42.

[134] Thomas Winship, "A Defence of Philosophic Doubt," *The Earth*, Vol. III, Nos. 25/26 (Jun., 1902), 30.

> Be it remembered that mankind in the old world dwelt in Asiatic Turkey, speaking the same language, and it was not till after Noah's death that the dispersion from Babel, over the earth, took place. Asiatic Turkey contains at present fifteen million human beings, and there only could fossilized man be found. To what extent, if at all, has that country been geologically examined.[135]

Here is made clear both his rejection of scientific polygenism in favor of the unity of all human races from the offspring of Noah, but also an implicit call to begin the process of "geological" expeditions to the heart of the Ottoman Empire to determine the truth of the Noachic flood in the fossil record. As mentioned above, this reflects the synergy between fin de siècle scientific expeditions and imperial power, but also the unity of using science to prove religious outcomes, a key aspect of Flat Earth scientific views.

The next four installments of "A Defence of Philosophic Doubt," presumably parts three through six, lack materials relevant to this work.[136] However, the final entry, from the October/November 1903 issue, contains much to unpack.[137] It begins with the exclamation that

> But what about the Turanian languages? Do they show any such resemblances to the Semitic? Do the Negro and the Hottentot and the savage of Terra [*sic*] del Fuego make good their claim to be reckoned both men and brethren? Here the answer is a still more STARTLING AFFIRMATIVE.... There is a very close connection between the various African languages. Dr. Latham says: "That the uniformity of languages throughout Africa is greater than it is either in Asia or in Europe, is a statement to which I have not the least hesitation in committing myself." Among these languages we have to number the chief of them—the ancient Egyptian. Now this we are able to inspect in its most ancient forms. The inscriptions enable us to trace it for more than 2,000 years before the Christian era. The result is that the

[135] Winship, "Defence," Nos. 25/26, 32.

[136] Thomas Winship, "A Defence of Philosophic Doubt," *The Earth*, Vol. III, Nos. 27/28 (Jul., 1902), 49–52; Thomas Winship, "A Defence of Philosophic Doubt," *The Earth*, Vol. III, Nos. 31/32 (Feb./Mar., 1903), 140–43; Thomas Winship, "A Defence of Philosophic Doubt," *The Earth*, Vol. III, Nos. 33/34 (Apr./May, 1903), 173–77; Thomas Winship, "A Defence of Philosophic Doubt," *The Earth*, Vol. III, Nos. 35/36 (Jun./Jul., 1903), 214–16.

[137] Thomas Winship, "A Defence of Philosophic Doubt," *The Earth*, Vol. III, Nos. 39/40 (Oct./Nov, 1903), 280–83.

> Scripture account is most marvellously vindicated in more than one particular. The Egyptian presents us with a language intermediate between the three families of human speech. Latham and Bunsen agree in regarding it as intermediary and have called it "sub-Semitic." This means that the further we go back the more do the languages of the three families approach each other, and the more do they show the traces of their common origin.[138]

Of note is the use of the term "Hottentot" near the beginning of the quote, but as stated above, it was widely enough in use to not be evidence that Winship was A. Hottentot. Further, while a derogatory term, as with his earlier use of "Darwinian Ape," there is a clear rejection of the attempts to determine Africans to be anything other than "both men and brethren." Additionally, Winship's connection of ancient Egyptian civilization to that of Black Africa, at a time when the Victorian imagination generally viewed ancient Egypt as a white civilization, makes him more akin to later twentieth-century Afrocentrists like Ivan Van Sertima.[139]

Also worth identifying in Winship's quote are "Dr. Latham" and "Bunsen." "Dr. Latham" was Robert Gordon Latham (1812–1888), a member of the Ethnological Society of London and critic of polygenism; the quote Winship uses is taken from his 1851 book *Man and His Migrations*.[140] "Bunsen" was the Prussian diplomat and amateur scholar Christian Charles Josias von Bunsen (1791–1860), whose major work, *Egypt's Place in Universal History*, was published in five volumes from 1845 to 1857, and attempted to harmonize ancient Egyptian dating with Biblical chronology.

[138] Winship, "Defence," Nos. 39/40, 280.

[139] Ironically, as Afrocentric theories tend to focus extensively on ancient Egypt, or to ahistorically depict ancient Egyptians as modern sub-Saharan Africans, they unintentionally sideline actual sub-Saharan Africans and their achievements in a similar way to the earlier Eurocentric narratives. Mia Bay, "The Historical Origins of Afrocentrism," *Amerikastudien / American Studies*, Vol. 45, No. 4, Time and the African-American Experience (2000), 501–12; François-Xavier Fauvelle, trans. Troy Tice, *The Golden Rhinoceros: Histories of the African Middle Ages* (Princeton: Princeton University Press, 2018), 160–68; David S. Anderson, "Black Olmecs and White Egyptians: A Parable for Professional Archaeological Responses to Pseudoarchaeology," in Jeb J. Card and David S. Anderson, eds., *Lost City, Found Pyramid: Understanding Alternative Archaeologies and Pseudoscientific Practices* (Tuscaloosa: University of Alabama Press, 2016), 68–80; Wyatt MacGaffey, "Who Owns Ancient Egypt?," *The Journal of African History*, Vol. 32, No. 3 (1991), 515–19.

[140] Robert Gordon Latham, *Man and His Migrations* (London: John Van Voorst, 1851), 152.

However, while Egyptian civilization was similar to other African civilizations in Winship's eyes, it was not native to the continent itself. Winship continues to state that

> If we regard physical characteristics, manners and customs, and mythologies, as well as mere languages, it is much easier thus to link together nearly all the populations of the globe.... That is to say, that the Bible has clearly stated from the beginning what Philology is now dimly feeling to be the truth! Will any "advanced" friend tell us how this has come to pass if Genesis is not fully inspired? *How has it been filled with the light of eternal truth, and not with the darkness of the age of Moses or of Ezra?* One more result of investigation will complete this testimony. Professor Hommel, of Munich, has lately announced that there can no longer be any doubt that Egyptian civilization PROCEEDED FROM BABYLON. This is another tribute paid by science to the truth of Genesis, which has made men heirs for thirty-three centuries of the knowledge that the human race dwelt together in its early days in the valley of the Euphrates.[141]

"Professor Hommel" was Fritz Hommel (1854–1936), a historian and linguist of the ancient Near East at the University of Munich. The work of Hommel and other social scientists, much like the physics of those like Lord Kelvin, served as legitimate science to Winship, in that they confirmed the Biblical truths: in this case, that the origins of humanity were consisted geographically and chronologically with the Book of Genesis. As such, by arguing that African civilizations came from Egypt, and Egyptian civilization came from the heartland of Genesis, it is not a denunciation or rejection of African civilization, but evidence that the continent's peoples share a direct lineage from those of the Old Testament.

Winship ends the final entry in his series with a summary of his opinions:

> Science boldly avers that: — 1. The earth is a globe, the product of evolution. 2. Man is the highest species of evolution, from lower forms through incalculable periods of time. 3. The sun and planets are superior to this earth, which is as a mere speck compared to the myriad (possibly inhabited) worlds around us. 4. The moon is a reflector, and not a giver of light. 5. The stars are millions of times larger than the earth. 6. All life

[141] Winship, "Defence," Nos. 39/40, 281.

> having started by spontaneous generation, there is no room for a God in Nature (see *Modern Review*, October, 1880,[142]) either to create or direct its forces. On the contrary, the Bible consistently states from the first page to the last, that; 1. The world is a stretched-out structure built on foundations, and the result of Special Divine Creation. 2. Man is the result of Special Creation and was made perfect at one time. 3. The sun and stars are inferior to this (the only) world, which they were created to serve. 4. The moon is self-luminous—a giver of light. 5. The stars are small objects. Some of them will at a future time fall upon this earth, and it is manifestly impossible for bodies millions of times larger than the earth to fall upon it. 6. All life was created by Divine Power and made perfect at the start, and that God creates and controls all the "forces of nature," which without His sustaining hand would cease to be.[143]

It would be hard to come up with a more concise precis of fin de siècle Zetetic views than that—and as seen, in Winship's case, they are directly connected to his experiences as a settler in southern Africa.

It is fitting for a Zetetic in southern Africa that, while *Zetetic Cosmogony* was published within weeks of the start of the Second Boer War, Winship's publication of "A Defence of Philosophic Doubt" spanned the ending of the conflict on 31 May 1902. The South African Republic and Orange Free State ceased to exist, respectively becoming the Transvaal Colony and Orange River Colony under British control. In 1910, the two former Boer republics were united with the Cape and Natal colonies to create the self-governing (under a white minority) Union of South Africa, which became the modern Republic of South Africa in 1961. Partly as a result of the Boer defeat and annexation, after 1902 there was a major push to make the more common Afrikaans—the Boer offshoot of the more formal Dutch—an officially recognized language, which happened in 1925.[144] After its recognition, and especially after the 1948 implementation of apartheid and resulting increasing international isolation of South Africa, the South African government invested enormous resources into making

[142] This appears to be a reference to William B. Carpenter, "Nature and Law," *The Modern Review*, Vol. 1, No. 4 (Oct., 1880), 748–70. It is important to note that this William Carpenter is William Benjamin Carpenter (1813–85), a Unitarian zoologist and critic of Spiritualism associated with the *Challenger* expeditions, and not the Flat Earther William Carpenter featured in the previous chapter.

[143] Winship, "Defence," Nos. 39/40, 282–83.

[144] Duvenage, "Afrikaner intellectual history," 104–09.

Afrikaans a language of science to the same degree as English, translating works into the language as well as developing new terms within it.[145]

The transformation of Afrikaans into a language of science by a government still dominated by Boers belonging to the Dutch Reformed Church likely had Kruger spinning in his grave (and his body had been repatriated to Pretoria after his 1904 death in exile). But even by the official recognition of Afrikaans in 1925, Flat Earth support in South Africa had faded amongst both English and Afrikaans speakers. In 1961, facing rising nationalist sentiment in its African colonies, Britain committed itself to the "Wind of Change" that was decolonization and majority African rule. However, the white settlers of South Africa and Rhodesia decided this was a step too far and attempted to go their own ways. The Rhodesians, not as numerous as their comrades to the south and facing a more militant resistance from their African majority, folded first, with their settler colony becoming Zimbabwe in 1980. In South Africa, majority rule came with Nelson Mandela (1918–2013)'s election as president in 1994.

To the few Flat Earthers remaining in the UK and US by the 1990s, this was not a surprise. If the white settlers of southern Africa had abandoned the Biblical truths of Zetetic cosmogony, it only made sense that their punishment would be the loss of their temporal authority. But the start of the twenty-first century would see the rise of a more dominant Flat Earther religious movement in a former British African colony—one which quite explicitly based its Flat Earth beliefs on opposition to the postcolonial legacy Britain had left behind.

[145] Sibusiso Biyela, "Decolonizing Science Writing in South Africa," *The Open Notebook*, February 12, 2019, accessed August 18, 2022, https://www.theopennotebook.com/2019/02/12/decolonizing-science-writing-in-south-africa/.

CHAPTER 6

"Western Education is Impermissible"

On 15 November 2014, during his closing speech at the first summit of Latin American Muslim leaders in Istanbul, Recep Tayyip Erdoğan (b. 1954), newly inaugurated President of the Republic of Turkey, made a strange assertion: "Muslim sailors had arrived in the shores of America in 1178. In his diaries, Christopher Columbus referred to the presence of a mosque on top of a mountain in Cuba."[1] Erdoğan appears to have been citing a 1996 claim by the Lebanese-American scholar Youssef Mroueh, who drew from both the Afrocentric hyperdiffusionist work of Leo Wiener and Mansa Musa's claims of the Malian Transatlantic expedition, described in chapter one.[2] In reality, this appears to be a (perhaps deliberately) distorted takeaway from the description given by Columbus in his 29 October 1492 journal entry, where (as paraphrased by Bartolomé de las Casas), "Remarking on the position of the river and

[1] Mustafa Küçük, "Erdoğan says America was discovered by Muslims, not Columbus," *Hürriyet Daily News*, November 17, 2014, accessed December 10, 2023, https://www.hurriyetdailynews.com/erdogan-says-america-was-discovered-by-muslims-not-columbus-74371.

[2] Youssef Mroueh, "Precolumbian Muslims in the Americas," *Beautiful Islam*, 1996, accessed December 10, 2023, http://www.beautifulislam.net/history/muslims_americas.htm.

E. Guimont, *The Power of the Flat Earth Idea*,
Palgrave Studies in the History of Science and Technology,
https://doi.org/10.1007/978-3-032-23981-5_6

port, to which he gave the name of San Salvador, he describes its mountains as lofty and beautiful, like the Peña de los Enamorados [of Spain], and one of them has another little hill on its summit, like a graceful mosque."[3] This metaphorical description of a hill as resembling a mosque was taken literally by Mroueh, Erdoğan, and other Muslim critics of the West from the late twentieth century on—and in Erdoğan's case, was part of his drive to erode the post-Ottoman secularization of Turkish society as part of his personal consolidation of power.[4]

Claims of pre-Columbian travel to the Americas did circulate in the Ottoman Empire in the sixteenth century; one prominent claim was that Alexander the Great, folklorically associated with the Qur'anic figure Dhu al-Qarnayn, had voyaged to what was termed the "New Country" and carved two stone idols, one hundred cubits high, as a warning to sail no further.[5] A mountaintop mosque seems like a reasonable extrapolation from these legendary towering idols. Of course, there really was a very early West African Muslim presence in the Americas, but it did not come from medieval Malian or Arab expeditions. Prior to his voyage across the Atlantic, Columbus had been involved in the West African slave trade, visiting Portuguese slave ports on the Guinea coast. But while instituting the practice of European enslavement of the Indigenous Americans, he did not initiate the Transatlantic slave trade. That dubious dishonor falls to a Juan de Córdoba of Seville, who in 1502 was the first to transport several of his African slaves from Spain to Hispaniola. The first direct transport of slaves from West Africa to the Caribbean was done by the Spanish in 1517. It is beyond doubt that at least some, and perhaps most, of those slaves would have been Muslim. It can therefore be said with certainty that North and West African Muslims were present in the

[3] Bartolomé de las Casas, "Journal of the First Voyage of Columbus," Early Americas Digital Archive, 2003, accessed December 10, 2023, http://eada.lib.umd.edu/text-entries/journal/.

[4] Gideon Rachman, *The Age of the Strongman: How the Cult of the Leader Threatens Democracy Around the World* (New York: Other Press, 2022), 41–53; Pankaj Mishra, *Age of Anger: A History of the Present* (New York: Farrar, Straus and Giroux, 2017), 268; Jason Colavito, "Did Columbus Find an Ancient Mosque in Cuba?," JasonColavito.com, June 22, 2013, accessed December 10, 2023, https://www.jasoncolavito.com/blog/did-columbus-find-an-ancient-mosque-in-cuba.

[5] Giancarlo Casale, "Did Alexander the Great Discover America? Debating Space and Time in Renaissance Istanbul," *Renaissance Quarterly*, Vol. 72, No. 3 (Fall, 2019), 895–901.

Americas before the religion of Protestant Christianity even existed. It should also be noted that Columbus' third voyage of 1498–1500 overlapped with the first European circumnavigation of Africa, the 1497–1499 voyage of Portuguese sailor Vasco da Gama (c. 1460–1524). For Europeans, sub-Saharan Africa was just as much a New World as the Americas that were being explored at the exact same time.[6]

Although Muslim empires did not take part in the conquest of the Americas from the late fifteenth century, they were well aware of the European expansion, including the military and academic officials of the Republic of Turkey's predecessor state, the Ottoman Empire. This is most evident from the career of Ahmed Muhiddin Piri (1465–1553), better known as Piri Reis (Reis was an Ottoman title equivalent to admiral, which he was granted in 1547). In 1513, Piri compiled a map of the known world at the time for Sultan Selim I (1470–1520; r. 1512–1520); the map vanished from history until approximately a third of it was discovered in Istanbul's Topkapı Palace in 1929. Now known simply as the Piri Reis Map, it was synthesized from several sources, including Greek, Arabic and Indian, but also four Portuguese maps from their recent expeditions to Asia—and most curiously, a map of the Caribbean which Piri stated had been drawn and given to him by Columbus himself. Unfortunately, no more information on this Columbus-Piri connection is known, but it does demonstrate that even without being involved in the contemporary European exploration of the Americas (or sub-Saharan Africa, or East Asia), the power of the Ottoman imperial state—or at least certain ranking actors wielding its state power—could be used to develop cutting-edge cartography.[7] So detailed was the Piri Reis Map that since

[6] P. E. H. Hair, "Columbus from Guinea to America," *History in Africa*, Vol. 17 (1990), 121–23; Ruth Pike, "Sevillian Society in the Sixteenth Century: Slaves and Freedmen," *The Hispanic American Historical Review*, Vol. 47, No. 3 (Aug., 1967), 346–47, 354; Albert J. Raboteau, *Slave Religion: The "Invisible Institution" in the Antebellum South* (New York: Oxford University Press, 1978), 5–7, 46–47.

[7] John Freely, *Light from the East: How the Science of Medieval Islam Helped to Shape the Western World* (London: I. B. Tauris & Co., 2011), 120; Karen C. Pinto, *Medieval Islamic Maps: An Exploration* (Chicago: University of Chicago Press, 2016), 25–26; Abbas Hamdani, "Ottoman Response to the Discovery of America and the New Route to India," *Journal of the American Oriental Society*, Vol. 101, No. 3 (Jul.-Sep., 1981), 323–30; Kemal Özdemir, trans. Priscilla Mary Işın, *Ottoman Nautical Charts and the Atlas of Ali Macar Reis* (Istanbul: Creative Yayıncılık ve Tanitim, 1992); Casale, "Did Alexander the Great Discover America?," 871–77.

the 1950s—coincidentally anticipating the new development of Earth-orbiting satellites—conspiracy theorists have claimed that Piri had access to the space maps of ancient aliens, or of some pre-Ice Age ATAC such as the one associated with Atlantis.[8]

But Piri and his works (including books and maps other than the world map) were rare exceptions to the general state of Ottoman cartography being far less sophisticated and comprehensive than the European field.[9] Indeed, the recognition of the deficiencies of Ottoman cartography goes back centuries, and was the source for a Turkish claim about the discovery of the Americas very different, but centuries earlier, than that of Mroueh and Erdoğan. In his *Seyāḥatnāme* (*Book of Travels*), Ottoman traveler Evliya Çelebi (1611–1682) recounts his version of the discovery of the Americas. According to Evliya, the New World was discovered by two priests, the Spaniard Padre and Portuguese Kolon. In 1484, they appeared before the Sultan Bayezid II (1447–1512; r. 1481–1512) during his siege of Bilhorod-Dnistrovskyi, and offered him the new continent they had discovered for him to rule. Bayezid turned them down, saying that Mecca and Medina were enough for a sultan to conquer. In response to his rejection, Padre and Kolon went to the Pope and the Spanish crown, leading to Europeans settling the Americas. A partial source for Evliya's strange history seems to be several Native Americans he met and spoke to via German interpreters in the Atlantic coastal city of "Lonçar" (possibly Rotterdam) following the Ottoman capture of Nové Zámky in October 1663. Evliya seems to have intended this account of Bayezid turning down the conquest of the Americas to be a satire of the shortsightedness of Ottoman elites. In line with this, the first Ottoman book about the Americas was written not in 1484 or even 1492, but around 1580, in the form of a cosmography which limited its geographical section to the new continent.[10]

[8] Hamdani, "Ottoman Response," 327-28n24a; Kenneth L. Feder, *Encyclopedia of Dubious Archaeology: From Atlantis to the Walam Olum* (Santa Barbara, CA: ABC-CLIO, 2010), 215–16.

[9] S. Frederick Starr, *Lost Enlightenment: Central Asia's Golden Age from the Arab Conquest to Tamerlane* (Princeton: Princeton University Press, 2013), 509.

[10] Robert Dankoff, *An Ottoman Mentality: The World of Evliya Çelebi* (Leiden: Brill, 2006), 62–63, 221. Kolon is a transliteration of Colón (Columbus), and Padre is the Spanish word for "priest."

Cartography was not the only science which lacked respect in the Ottoman Empire; astronomy was similarly undervalued. The Ottomans captured Constantinople and made it their capitol in 1453. The first Ottoman observatory there was not built until 1576 under the astronomer Taqi ad-Din (1526–1685). However, after only four years, the Sultan Murad III (1546–1595; r. 1574–1595) had the observatory destroyed under the influence of the Grand Vizier Koca Sinan Pasha (c. 1506–1596). After ad-Din's observatory was destroyed, there was a continuing lack of interest in astronomy in the Ottoman Empire. The first translation of a European work into Turkish discussing Copernicanism was not published until 1660, over two centuries after the publication of *On the Revolutions of the Heavenly Spheres*.[11]

As such, it is perhaps not surprising that at least some Flat Earth sentiment developed in the Ottoman Empire by the early twentieth century, on the eve of the First World War that would lead to its collapse. One of these adherents was John George Abizaid. An Arab Christian, Abizaid was born c. 1868 in the city of Douma in what is now Lebanon but at the time was part of the Ottoman province of Syria. In the late nineteenth century, Abizaid emigrated to Boston in the United States, where he married and had eleven children. Lacking a formal education, Abizaid attended the Quincy Evening School on Tyler Street, near his residence, for three terms. Its principal, Bertha Hazard, praised him as "one of my most valued pupils" in a letter. In 1906, Abizaid wrote his own Arabic Flat Earth booklet, *The Enlightenment of the World*, publishing a revised English version in 1910. Charles W. Morse of the Boston suburb of Brookline—author of several Flat Earth books and official Representative Member and Lecturer of Lady Blount's Universal Zetetic Society—read the first English version and, while endorsing the ideas, believed it needed editing and revising, which Abizaid did before publishing a revised second edition in 1912. A dramatically expanded third English edition would follow in 1935. By then Abizaid continued to give interviews on the topic in local newspapers and would take his children to preach Flat Earth truth to passers-by on Boston Common on Sundays. Abizaid would eventually return to Douma without his family, dying there in his late 70s.[12]

[11] Starr, *Lost Enlightenment*, 509–10.

[12] John G. Abizaid, *The Enlightenment of the World, Second Edition* (Boston: John G. Abizaid, 1912), 5, 32; Robert J. Schadewald, *The Plane Truth: A History of the Flat Earth Movement* (Wendy S. Schadewald, 2015), 126–27; Charlie Abizaid, "Re: The Enlightment

In Boston, Abizaid lived in South Cove, a neighborhood of Chinatown now called Bay Village and which also had the nicknames "Little Syria" and "Syriatown," due to its large immigrant population from that region.[13] South Cove was also the location of Denison House, a woman-run facility aimed at helping provide various social services to the neighborhood's immigrants. For the editing of his second edition, Abizaid turned to help from Denison House, and the facility assigned one of its personnel, Eleanor S. Upton, to handle the task.

Upton was likely chosen by the Denison staff because of her father. Winslow Upton (1853–1914) was an astronomer and the first director of Brown University's Ladd Observatory, holding that position from its opening in 1891 until his death.[14] In the short period between the publication of Abizaid's second edition and Winslow's death, he acquired a copy of Abizaid's work, likely through his daughter. From a 1952 letter Eleanor left on Upton's copy in his personal papers at Brown, some more background details on Abizaid emerge. Eleanor recalls that Abizaid had developed "quite a reputation" among the Arab immigrants of South Cove for his Flat Earth advocacy, and that the Denison House staff were split on "whether the old man should be encouraged in his erroneous beliefs, or have his heart broken by being given an unanswerable argument against them." This demonstrates not only a patronizing tone perhaps not surprising to find in middle-class liberal reformers discussing immigrants, but also a complete unfamiliarity with the resilience of the conspiracist mindset when faced with facts of any sort. In any case, Eleanor states that she did neither, and stuck to the editorially sound approach of merely correcting Abizaid's English. However, both Eleanor and Winslow seemed to have had a fondness for Abizaid, appreciating

of the World," The Flat Earth Society, January 3, 2009, accessed February 8, 2024, https://www.theflatearthsociety.org/forum/index.php?topic=25630.0; Charles W. Morse, "Letters," *The Earth*, Vol. V, Nos. 52 & 53 (Nov./Dec., 1904), 55–56; Elizabeth Blount and Albert Smith, *Flat or Spherical?* (Kingston Hill: Lady Blount, 1905), i. For Abizaid's model of the Earth, see John G. Abizaid, "New correct map of the flat surface, stationary earth," Library of Congress, c. 1920, accessed January 6, 2024, https://www.loc.gov/item/2013585077/.

[13] Kathering Giraldo, "Boston's Syriatown: Preserving Cultural Heritage Through Religious Centers," *Medium*, June 5, 2019, accessed January 7, 2024, https://medium.com/@kgiraldo/bostons-syriatown-94cca93626cc.

[14] Edward Guimont and Horace A. Smith, *When the Stars Are Right*: H. P. Lovecraft and Astronomy (New York: Hippocampus Press, 2023), 57–60.

his responses to patronizing comments. Eleanor noted that as late as 1940, she had seen Abizaid continue to be active in promoting Flat Earth among the South Cove Arabs.[15] Notably, this was the second time the Uptons had crossed paths with fringe science. By this point, Winslow had already served as a scientific mentor to the future author H. P. Lovecraft, and the astronomical knowledge that Upton imparted on Lovecraft would shape the latter's pioneering fantasy and science fiction, which in turn would influence modern ancient alien conspiracies (Fig. 6.1).[16]

Although Abizaid stated in his introduction that his belief in Flat Earth stemmed from his interpretation of the Bible, he knew there were many for whom Scripture did not hold that central role. Therefore, Abizaid made his publication an entirely secular work relying on commonsense arguments on the Earth's shape.[17] Many of these are familiar from some of the figures discussed earlier in this book. In line with Ebenezer Breach's "Level Seas" argument from the last chapter, Abizaid noted that steamships, always sailing straight and not having to adjust to a curve, proved the Earth was flat.[18] As with the Zetetics and their interest in the evidence provided by James Glaisher and other aeronauts, Abizaid argues that the ability of balloons and airships to fly proves that gravity does not exist.[19] Abizaid even endorses the Flat Error, stating that "500 years ago" people believed the Earth was flat, but modestly claims they stopped believing because they lacked the proofs of his book.[20]

These are all fairly by-the-book Flat Earth claims, though it is worth noting that they were commonplace by then as well. Robert Schadewald noted that while Abizaid became quite familiar with Flat Earth history and literature by the 1930s, it is not clear what inspired his interest in the movement in the early 1900s.[21] Although Abizaid does not mention John Jasper (or any other Flat Earther), writing his first edition less than a

[15] Brown University, John Hay Library, MS-12AS-1, Box 72: 30,897 Upton—notes on eclipses of 1878, 87, 83 Correspondence 1886–1913, Notebooks, offprints, graphs—unalphabetized Ladd Observatory, folder 7:16, 15 July 1952 letter.

[16] Guimont and Smith, *When the Stars Are Right*, 60–66, 170.

[17] Abizaid, *The Enlightenment of the World*, 5.

[18] Abizaid, *The Enlightenment of the World*, 11–12, 21–22.

[19] Abizaid, *The Enlightenment of the World*, 14.

[20] Abizaid, *The Enlightenment of the World*, 21.

[21] Schadewald, *The Plane Truth*, 125.

DESCRIPTIVE KEY.

JOHN G. ABIZAID, 3 Malden St., Boston, Mass. U. S. A.

Fig. 6.1 Pamphlet advertising Abizaid and his booklet in both English and Arabic, with a photograph of him and a diagram of his world model, c. 1920[22]

decade after Jasper's death, he could not have failed to have been aware of his Flat Earther predecessor. Several historians have noted that the treatment of free Blacks by white authorities in the post-Reconstruction South was not too different from the contemporary treatment of Christians by

[22] John G. Abizaid, "New correct map of the flat surface, stationary earth," Library of Congress, c. 1920, accessed January 6, 2024, https://www.loc.gov/resource/g9930.ct003852v/?r=-0.832,-0.034,2.664,1.241,0.

Muslim authorities in Ottoman Lebanon.[23] It is not inconceivable that even with only a general knowledge of Jasper, Abizaid might have seen a kinship in him even beyond religious and Flat Earth ties.

More concrete connections come from the last few pages of the book, where Abizaid lists correspondence from individuals he sent his argument to. Unsurprisingly, a number of those individuals were also Arabs. Daniel Joseph Faour and Brothers, major Syrian merchants and representatives of the Bank of Syria in New York City, were sent a copy to which they diplomatically replied, "from our point we found it built on an excellent foundation."[24] Philip K. Naoufal, Secretary of Boston's Hudson Club, found it "indeed very interesting."[25] However, not all of Boston's Arab population were merely polite; another Lebanese immigrant to the city, Segean Joseph Geageah, wrote to strongly endorse Abizaid's arguments.[26] Nor also were his correspondents limited to immigrants; a Habeeb Cury Thrubuy of Tannourine, back in Ottoman Lebanon, wrote to state "That the earth is flat and stationary was believed for centuries and has not as yet been absolutely proved to be otherwise. I do not believe the earth is round and the fallacious theory should be exploited."[27]

But the most interesting of his Arab correspondents was Naoum Anthony Mokarzel (1864–1932) of New York, a fellow Lebanese Christian immigrant and owner and publisher of *Al-Hoda* (*The Guidance*), the largest Arabic newspaper in the United States at the time. *Al-Hoda*, as with similar newspapers at the time, attempted to depict the community it was aimed at as Syrians, not Arabs, Turks, or Ottomans, as a project of nation-building for independence for the region from Ottoman

[23] Ussama Makdisi, *Age of Coexistence: The Ecumenical Frame and the Making of the Modern Arab World* (Oakland, CA: University of California Press, 2019), 4–6, 55; Bruce Masters, *Christians and Jews in the Ottoman Arab World: The Roots of Sectarianism* (Cambridge: Cambridge University Press, 2001), 132.

[24] Abizaid, *The Enlightenment of the World*, 27; "D. J. Faour & Bros. Importers Exporters & Commission Merchants," James Cummins, Bookseller, n. d., accessed January 7, 2024, https://www.jamescumminsbookseller.com/pages/books/352074/arab-american/d-j-faour-bros-importers-exporters-commission-merchants-in-notions-dry-goods-cutlery-jewelry-c.

[25] Abizaid, *The Enlightenment of the World*, 28.

[26] Abizaid, *The Enlightenment of the World*, 29.

[27] Abizaid, *The Enlightenment of the World*, 28. "Exploited" is perhaps an error in translation.

rule. This anti-imperial nationalist pretension also had a domestic counterpart: by creating a new category of "Syrian," the community could be interpreted as white (and the newspaper aimed to inculcate middle-class white sensibilities into its immigrant readership), hoping to circumvent legal and social racial categories.[28] In the second of two letters, Mokarzel stated his interest in Abizaid's arguments and that "You deserve to have more success for your useful book." Nor was this idle flattery; after this letter, Abizaid included an aside stating that Mokarzel published a long and supportive article on Abizaid's book in the 25 June 1910 issue of *Al-Hoda*.[29] The largest Arabic newspaper in the United States printing Flat Earth material in 1910 brings to mind that at that same time, the only Black-owned newspaper in the United Kingdom was owned and published by a Flat Earther, as mentioned in chapter four.

However, also of interest is that Abizaid did not limit his outreach to Arabs only. Among those he sent a copy to was Cardinal John Murphy Farley (1842–1918), Archbishop of New York from 1902 until his death; Farley's secretary sent a politely worded congratulatory response on the work.[30] Helena S. Dudley (1858–1932), director of Denison House as well as a dedicated activist for women's rights, trade unionism, and the League of Nations, congratulated Abizaid on the work he put into the pamphlet and the reception it received.[31] In turn, full-fledged support was given to Abizaid's theories by a "Professor H. B. Newton" of Liverpool, England; William J. Hutchinson of the New York Academy of Sciences; Lewis Pennini, a steamship agent, justice of the peace, and notary public in Boston; and Lewis Hahn, an evangelical Christian in Salem, Oregon, who promised to send a copy of *The Enlightenment of the World* on to Oswald West (1873–1960), Governor of Oregon (1911–1915).[32]

Also voicing his full support was Aurin F. Hill (1853-c. 1930), General Secretary and Treasurer of the radical-leaning American Press-Writers' Association. A Boston gadfly, Hill was a working-class carpenter; trade

[28] Ussama Makdisi, *Faith Misplaced: The Broken Promise of U.S.-Arab Relations: 1820–2001* (New York: PublicAffairs, 2010), 95–96.

[29] Abizaid, *The Enlightenment of the World*, 26.

[30] Abizaid, *The Enlightenment of the World*, 25.

[31] Abizaid, *The Enlightenment of the World*, 27, 29.

[32] Abizaid, *The Enlightenment of the World*, 25–28, 32.

union activist; promoter of civil and women's rights; and opponent of lynching—along with being a Spiritualist, phrenologist, and a Flat Earth theorist of his own right.[33] Hill's apartment in Boston was 13 Isabella Street, half a mile from Abizaid's residence. In 1905, Hill had written a pamphlet titled *The Earth is Flat*, whose primary inspirations were a reference to Copernicus' heliocentrism as being only a theory in a book on religious freedom, and a literalist reading of ancient Egyptian mythology works at the Boston Public Library, particularly the 1904 text *The Gods of the Egyptians* by British Egyptologist E. A. Wallis Budge (1857–1934). Hill incorporated this cosmology into a wider claim that the Heaven of the ancient Egyptians was the same as the Garden of Eden; that the physical location of Heaven-Eden was in the delta of the Amazon River in South America; and that Egyptian, Greek, Hebrew, and Native American cultures preserved elements from that original cultural homeland.[34] However, as will be described later, Hill would later claim to have been interested in Flat Earth as early as 1892.

On the other hand, one group of people who did not send Abizaid even polite endorsements was Harvard University; Abizaid claimed that a Professor Davis of Harvard had tried to silence him by telling him it would be too expensive to print the manuscript, while Jay B. Woodworth (1865–1925), a Harvard professor of geology and pioneering seismologist, rejected Abizaid's arguments in a back and forth series of letters.[35]

When the United States entered World War I, *Al-Hoda* promoted the war effort among the Lebanese-American community as a means of liberating Lebanon from the Ottomans (although the US never declared war on the Ottoman Empire), and Mokarzel would serve as a Lebanese delegate to the Paris Peace Conference of 1919–1920, once more connecting,

[33] Abizaid, *The Enlightenment of the World*, 25–26; Schadewald, *The Plane Truth*, 127.

[34] Aurin F. Hill, *Elysian Fields of the Egyptians—The earth is flat—Heaven and Hell, 1905–1907*, 5–7, Aurin F. Hill Papers (MS 579), Box 3, Special Collections and University Archives, University of Massachusetts Amherst Libraries; *Points of History: The Inquisition; The Albigenses and the Waldenses; The Massacre of St Bartholomew; The Fifth of November or the Gunpowder Plot; Galileo and the Inquisition; Religious Toleration a Question of First Principles* (Boston: Marlier & Co., 1901), 191–92; E. A. Wallis Budge, *The Gods of the Egyptians, or, Studies in Egyptian Mythology, Vol. I* (London: Methuen & Co., 1904), 216.

[35] Abizaid, *The Enlightenment of the World*, 30–31.

however loosely, Flat Earth advocacy with anti-colonialist politics.[36] But despite opportunistic promises made by the Allies during the war, the collapse of the Ottoman Empire—like that of the German colonial empire—would not lead to various nationalist groups establishing their own states. Instead, Ottoman and German territories would be carved up by the French and British, further expanding their empires' holdings in Asia and Africa—a process that was not new, but merely the latest stage of colonial expansion whose acceleration had been unleashed by the Berlin Conference thirty years earlier. But of course, European colonialism did not begin with the General Act of Berlin in 1885, and as seen, Britain had played a leading role. The development of longitude needed to navigate and map a spherical Earth stemmed from the eighteenth-century Board of Longitude and the need of the British state and merchants to provide better navigation to the warships and merchants of empire.[37]

Even the Universal Zetetic Society, with its criticism of Britain's empire described in previous chapters, did not always have a positive involvement in southern Africa. In 1908, an Englishwoman named Mrs. Horne was put on trial at the King's Bench over an alleged scam by which she had raised £30,000 (over £3 million in 2025) to invest in a land concession in Madagascar she claimed to have inherited containing large sums of gold, diamonds, and timber. Madagascar had been a French protectorate since 1882, and a formal colony since 1897 (achieving independence in 1960). According to a report in *The Times*, Mrs. Horne was a strictly religious woman who was a close acquaintance of both Lady Blount and Edward Blyden.[38]

Interestingly, this opens the possibility that the Liberian national Blyden, known for his educational activism and insistence that Africans

[36] Makdisi, *Faith Misplaced*, 147–52; "Al-Hoda Newspapers, الهدى," Khayrallah Center for Lebanese Diaspora Studies, n. d., accessed January 7, 2024, https://lebanesestudies.omeka.chass.ncsu.edu/collections/show/58.

[37] Dava Sobel, *Longitude: The True Story of a Lone Genius Who Solved the Greatest Scientific Problem of His Time* (New York: MJF Books, 1995), 51–60; Joshua Nall, "Seafaring, enslavement, and the longitude problem," in Victoria Avery and Jake Subryan Richards, eds., *Black Atlantic: Power, People, Resistance* (London: Philip Wilson Publishers, 2023), 91–93.

[38] As covered in several 1911 news publications, including *The Times* (2 Mar., 3; 3 Mar., 3; 4 Mar., 3; 8 Mar., 3; 13 Apr., 3); *Lloyd's Weekly News* (5 Mar,. 6); and the *Appeal* (23 Nov., 3). With thanks to David Killingray for the details on the case and its newspaper coverage.

look to African instead of European culture, may have been involved, however peripherally, with some of the British Flat Earthers.[39] Just as the Dutch astronomer Christiaan Huygens (1629–1695) helped in the development of longitude by building clocks carried on ships to the Cape Verde islands off the coast of West Africa, it was Blyden's Liberia that served as a place for observing the solar eclipse of 29 May 1919—despite British objections to the use of what was effectively their protectorate—which helped confirmed Einstein's general theory of relativity, a revolution as radical as Copernicanism had been, and just as opposed by the Flat Earthers of the twenty-first century.[40]

By 1919, the British presence in West Africa was already centuries old. Since the nineteenth century, the British expansion into Muslim-majority regions of North and West Africa had contributed to an Islamic reform movement in the area, in which a fundamentalist interpretation of Sunni Islam was seen as an ideology helping motivate resistance to both Western imperialism and the Westernizing Muslim governments of the Ottoman Empire and Egypt. An example of this was the Shilluk Kingdom in what is now the south of Sudan, on the border of the Islamic world. Given this frontier status, Shilluk culture was a syncretic mix of not only Muslim but Monophysite Christian, Egyptian, and Funj notions—reflected in their cosmology, which is something akin to a combination of a Flat and Hollow Earth, with our world on the interior bottom half of a sphere, and the sky as the dome overhead.[41] In 1861, the Shilluk were conquered by the Ottomans, their territory incorporated into the government of Egypt, which had become increasingly autonomous from Istanbul's rule. From

[39] Edward Wilmot Blyden, *Christianity, Islam and the Negro Race* (Baltimore, MD: Black Classic Press, 1994 [1888]), 82–109; William J. Simmons, *Men of Mark: Eminent, Progressive and Rising* (Cleveland: George M. Rewell & Co., 1887), 916–21; Allen Dwight Callahan, *The Talking Book: African Americans and the Bible* (New Haven: Yale University Press, 2006), 152–54.

[40] Sobel, *Longitude*, 34–40; Daniel Kennefick, *No Shadow of a Doubt: The 1919 Eclipse That Confirmed Einstein's Theory of Relativity* (Princeton, NJ: Princeton University Press, 2019), 239.

[41] Godfrey Lien-Hardt, "The Shilluk of the Upper Nile," in Daryll Forde, ed., *African Worlds: Studies in the Cosmological Ideas and Social Values of African Peoples* (Oxford: James Currey Publishers, 1999 [1954]), 138, 144–45. In contrast to the rejection of evolution by modern Flat Earthers, Shilluk cosmology also recognizes that humans and animals were alike in the distant past, but differentiated over time.

1876, Egypt came under the domination of British and French, and ultimately just the British, further blurring the distinction between province, kingdom, and protectorate. By the time the British formally established a protectorate over the technically-Ottoman province of Egypt in 1882, a joint Anglo-Egyptian military campaign was already being undertaken in the former Shilluk territory of the Sudan. This was the Mahdist War, initiated the year before by Muhammad Ahmad (1843–1885), a Sudanese Sunni Sufi sheikh who claimed descent from the Prophet Muhammad and in 1881 proclaimed himself the Mahdi, a messianic figure in Islam.

The Mahdi's forces succeeded in establishing an independent Islamist state after the defeat of the Anglo-Egyptian forces at Khartoum in 1885. Although the Mahdi died from typhus shortly thereafter, Sudan continued to be ruled independently by his successor, Abdallahi ibn Muhammad (c. 1846–1899), who proclaimed himself the new Caliph and was only defeated when the British invaded again and recaptured Sudan in 1899, in a war whose final two months overlapped with the start of the Second Boer War against Paul Kruger's forces in the southern end of the continent. Nor would the Mahdist War would be the last time that the British fought Africans who used messianic Islam as an organizing force of resistance.

In 1903, the British conquered the Sokoto Caliphate, an Islamic state in West Africa that was formed through jihad against the existing Hausa kingdoms in the region from 1804–1808. Sokoto would be the largest West African Muslim entity annexed into the British Empire. In contrast to the duration of the Mahdist War, Sokoto was conquered after a relatively brief military campaign which saw the final sultan, Muhammadu Attahiru I, killed in battle and his government disbanded. The ease of Sokoto's conquest was seen by the British as confirming both the decadence of Islam and the backwardness of West Africans, but this interpretation unsurprisingly ignored the deeper complexities and tangled political nature of the caliphate. In particular, the Sokoto elite were willing to give up political independence in order to stave off a more intense British attack which might destroy the foundations of Islamic society in their territory. The irony of the conquest is that while Attahiru's government had tried to stem the increasing support of Mahdist millenarianism amongst the caliphate's subjects, the British killing of him and dismantling of his government only increased support to the Mahdist movement, leading to the 1906 Satiru uprising by Mahdists outside the city of Sokoto. The caliphate itself would be integrated into Britain's new

Northern Nigeria Protectorate, which would be united with its southern counterpart to form the colony of Nigeria in 1914.[42]

The Mahdist and Sokoto movements, combining Sunni fundamentalist reform with resistance to Western colonialism, were trends of wider developments in Sunni Islam, called Salafism, which continued to spread into the twentieth century. Among those who embraced this broad trend were the members of the House of Saud, who established Saudi Arabia after World War I through the rump Arab territories leftover from the British and French seizure of lands from the collapsing Ottoman Empire, abrogating the promise of Arab liberation supported by such venues as *Al-Hoda*. In Saudi Arabia, Salafism merged with Wahhabism, a related Sunni movement with eighteenth-century origins, such that the terms are generally (though technically inaccurately) used interchangeably. Salafism/Wahhabism gave the new Saudi royal family a ground of public support, especially as they became the new guardians of the holy cities of Mecca and Medina. The wealth afforded to the Saudi state through the production of oil, combined with the later Cold War alliance with the United States as a fellow anti-communist power, afforded the Saudi state a unique, if somewhat delicate, position in the wider Muslim world in terms of funding sympathetic Islamic movements and clerics, particularly through the large number of foreign students educated at the Salafist Islamic University of Medina (IUM) established by the Saudi government in 1961. But this uneasy balance extended also to clerics within the state, whose interpretations of Islam could be at odds with the kingdom's strategic foreign relations.[43]

In 1966, Abd al-Aziz ibn Baz (1912–1999), a Salafist scholar who was then Vice President of IUM, argued that the Qur'an stated that the Earth is flat and the sun orbits the Earth. Three years later, he used

[42] T. C. McCaskie, "Cultural Encounters: Britain and Africa in the Nineteenth Century," in Philip D. Morgan and Sean Hawkins, eds., *Black Experience and the Empire* (Oxford: Oxford University Press, 2004), 183–84; Paul E. Lovejoy, *Jihād in West Africa during the Age of Revolutions* (Athens, OH: Ohio University Press, 2016), 36–67.

[43] Robert Lacey, *Inside the Kingdom: Kings, Clerics, Modernists, Terrorists, and the Struggle for Saudi Arabia* (London: Penguin Books, 2009), 9–11, 131; Natana J. DeLong-Bas, *Wahhabi Islam: From Revival and Reform to Global Jihad* (Oxford: Oxford University Press, 2016), 7–40, 227–79; Shiraz Maher, *Salafi-Jihadism: The History of an Idea* (Oxford: Oxford University Press, 2016), 3–27, 114–34, 169–85; Pankaj Mishra, *From the Ruins of Empire: The Intellectuals Who Remade Asia* (New York: Farrar, Straus and Giroux, 2012), 120–22.

these Qur'anic interpretations to state that the Apollo 11 Moon landing was a hoax.[44] Ibn Baez was not the only Muslim scholar to develop interpretations around spaceflight in general or the Apollo missions specifically. As with the African American folklore around flight, the Islamic world, particularly the Ottomans, had a lineage of claimed flights, including Abbas ibn Firnas (c. 809–887)'s attempts to fly in al-Andalus; Hezârfen Ahmed Çelebi (1609–1640)'s alleged flight in a glider across the Bosphorus; and his brother Lagâri Hasan Çelebi's even more alleged construction of a gunpowder-powered rocket craft. As with Erdoğan's attempts to reinterpret the early Muslim connection with the Americas, twentieth-century Turkish authors attempted to rehabilitate some of these efforts to justify not just a Turkish claim to early aviation, but a scientific basis for the Turkish nation as an alternative to Islamism. For example, the Byzantine chronicler Niketas Choniates (c. 1155–1217) wrote of an unidentified Muslim who tried to fly across the Constantinople Hippodrome in 1162, but instead plummeted to his death. In the twentieth century, Turkish authors gave him the name Siraj al-Din, identified him as a Turkish associate of Seljuk Sultan Kilij Arslan II, and claimed he had developed a sound method of flight that had almost worked.[45]

At the same time ibn Baez rejected the Moon landing, Turkish press reported a claim that Neil Armstrong (1930–2012) had heard a radio broadcast of the Islamic call to prayer and converted to Islam on the Moon's surface, an urban legend which has remained circulating in the Muslim world since.[46] So widespread has this claim become that Turkish Islamic creationist Adnan Oktar (b. 1956) argued that Qur'an 54:1 actually predicts the Apollo 11 landing, based on him claiming the abjad values of the words in the first verse indicate the year 1969.[47] However, others in the Muslim world used the popularity of the Apollo program to promote the ways in which the Qur'an was compatible with modern

[44] Lacey, *Inside the Kingdom*, 89–90; Maher, *Salafi-Jihadism*, 127–28.

[45] Gary Leiser, "Birdmen of the Middle East: Early Attempts at Human Flight," *Aerospace Historian*, Vol. 35, No. 3 (Fall, 1988), 173–78. For a broader exploration of discussions of flying machines in premodern Asian, Hopi, and Oceania myth, see Benjamin B. Olshin, *Lost Knowledge: The Concept of Vanished Technologies and Other Human Histories* (Leiden: Brill, 2019), 40–110.

[46] Ian Richard Netton, *Islam, Christianity and the Realms of the Miraculous: A Comparative Exploration* (Edinburgh: Edinburgh University Press, 2020), 22–23.

[47] Netton, *Islam, Christianity and the Realms of the Miraculous*, 178.

science. One such group was Egypt's Supreme Council for Islamic Affairs, which in 1967 was inspired by the Space Race to publish a book exploring how the Qur'an's passages on the natural world, particularly celestial matters, were compatible with modern science. This included sections on the Sun and Moon, and explicit defenses of the Earth being a sphere.[48]

Just as ibn Baez was not the only Islamic scholar who had an interpretation of the Apollo program, he also was far from the only Flat Earther to challenge the landing. Apollo Moon landing hoaxers are often jokingly associated with Flat Earthers by debunkers.[49] However, in reality, they are closer than might be assumed. As mentioned in chapter two, the leading Flat Earther of the mid-twentieth century was the Briton Samuel Shenton. In 1966, far ahead of Apollo 11, Shenton arguably pioneered Apollo skepticism by claiming that any future NASA Moon landing would be a hoax.[50] Shenton's successor as the head of the International Flat Earth Research Society, the American Charles K. Johnson, would find a new, unlikely way to integrate Apollo denialism and Flat Earth: the 1994 murder trial of Black athlete-turned-actor O. J. Simpson (1947–2024). In 1977, Simpson had starred in a film called *Capricorn One*, which features a NASA plot to stage a fake space mission, and then kill the astronauts involved in the hoax to cover it up after the uncrewed dummy capsule is destroyed on re-entry. Although the destination in the film was Mars, *Capricorn One* both built off of and reinforced the Moon landing hoaxes, as had earlier films *2001: A Space Odyssey* (1968) and the James Bond entry *Diamonds Are Forever* (1971). Of note also is that *Capricorn One* was written and directed by Peter Hyams (b. 1943), who in 1984 would also write and direct *2010: The Year We Make Contact*, the sequel to *2001*.

When Simpson was charged with murder, Johnson was convinced it was because *Capricorn One* had been Simpson's attempt to reveal the truth of the Apollo hoax to the public, and NASA had framed him for

[48] Muhammed Jamaludddin El-Fandy, trans. Ismail Cashmiry, revised by Shawki Sukkary, *On Cosmic Verses in the Quran* (Cairo: Supreme Council for Islamic Affairs, 1967), 5, 47–49, 69–75.

[49] Christopher D. Bader, Joseph O. Baker, L. Edward Day, and Ann Gordon, *Fear Itself: The Causes and Consequences of Fear in America* (New York: NYU Press, 2020), 31–32.

[50] Christine Garwood, *Flat Earth: The History of an Infamous Idea* (New York: Thomas Dunne Books, 2007), 219–79; Kelly Weill, *Off the Edge: Flat Earthers, Conspiracy Culture, and Why People Will Believe Anything* (Chapel Hill: Algonquin Books, 2022), 64–67.

murder in an attempt to cover up Simpson's noble efforts at disclosure. Johnson heavily promoted the innocence of Simpson; as polling at the time indicated that only 21% of white Americans believed in Simpson's innocence (in contrast to 66% of Blacks), this made Johnson one of the few whites to support Simpson, all due to his Flat Earth dedication. Likewise, in 1994, while only 9% of Americans in general doubted the Apollo missions, the highest demographic of doubters were Blacks, meaning Apollo skepticism may have been a route to evangelize Flat Earth among Black people, as with endorsing Simpson's innocence.[51] Well into the modern period of Flat Earth revivalism, Simpson's innocence due to the *Capricorn One* connection continues to be a tenet of faith amongst American Flat Earthers; Hyam himself also believed Simpson's innocence originally before becoming convinced of his guilt.[52] It should also be noted that Simpson was not the only controversial Black actor to be tied to Flat Earth; in 1971, the leaders of the Flat Earth Society of Canada debated trying to recruit Bill Cosby (b. 1937) (Fig. 6.2).[53]

Even outside of the belief that Apollo was a hoax, the Moon plays a major role amongst modern Flat Earthers, who believe that it is self-luminous and that moonlight lowers temperature, just as sunlight raises temperature.[54] This is widespread enough that even traditional Young-Earth Creationists seek to disprove it, afraid that such a ridiculous notion

[51] "Races disagree on impact of Simpson trial," CNN, October 6, 1995, accessed January 24, 2024, http://www.cnn.com/US/OJ/daily/9510/10-06/poll_race/oj_poll_txt.html; "Footballer arrested to hide faked moonwalk," *The Independent*, July 21, 1994, accessed February 18, 2024, https://www.independent.co.uk/news/world/footballer-arrested-to-hide-faked-moonwalk-1415442.html.

[52] Garwood, *Flat Earth*, 347–48; Weill, *Off the Edge*, 73–74; Donald R. Prothero, *Weird Earth: Debunking Strange Ideas about Our Planet* (Bloomington: Red Lightning Books, 2020), 27–28; Robert J. Schadewald, "The Flat-out Truth: Earth Orbits? Moon Landings? A Fraud! Says This Prophet," *Science Digest*, Vol. 88, No. 1 (Jul., 1980), 62–63; Edward Hendrie, *The Greatest Lie on Earth: Proof That Our World Is Not a Moving Globe* (Great Mountain Publishing, 2016), 156–265; Zen Garcia, *The Firmament: Vaulted Dome of the Earth* (Augusta, GA: Sacred Word Publishing, 2016), 13–16; It should be noted that one aspect of Simpson's life hasn't been covered by conspiracy theorists: that he was born 9 July 1947, the day after the US Army announced it had captured a flying saucer at Roswell. Coincidence?

[53] Garwood, *Flat Earth*, 290–91.

[54] Thomas Winship, "A Defence of Philosophic Doubt," *The Earth*, Vol. III, Nos. 39/40 (Oct./Nov, 1903), 282–83; Hendrie, *The Greatest Lie on Earth*, 166–71.

Fig. 6.2 Samuel Shenton's 1972 illustration of how the flight path of the Apollo missions would have to be shaped over the surface of the Flat Earth[55]

will challenge their own heterodox Christian interpretations.[56] Beyond the Apollo program specifically, as mentioned in chapter two, the balloon aeronauts of the nineteenth century held the same cultural space as pioneers of the high frontier that astronauts would hold in the twentieth century, and their observations of the Earth just as scrutinized by Flat Earthers as that of the subsequent astronauts. As such, it is perhaps appropriate that something of a precursor to the *Capricorn One* theory emerged from the nineteenth-century aeronauts. On 4 July 1892, as part of Independence Day celebrations, aeronaut George Augustus Rogers ascended in his balloon from Boston Common with his assistant Thomas Fenton and reporter Delos Goldsmith. When the wind began to shift and blow his balloon into the Atlantic, Rogers decided to ditch in Boston Harbor, ironically close to modern Logan Airport. Rogers drowned; Fenton and Goldsmith were rescued, but the former soon died from

[55] Samuel Shenton, "Mr Shenton's Map with Apollo Flight Path," Flat Earth Society, 1972, accessed February 13, 2024, https://www.theflatearthsociety.org/home/index.php/featured/maps.

[56] Danny Faulkner, "Testing a Flat-Earth Prediction: Is the Moon's Light Cooling?," *Answers Research Journal*, Vol. 12 (2019), 1–11. See also Robert J. Schadewald, "Scientific Creationism, Geocentricity and the Flat Earth," *The Skeptical Inquirer*, Vol. 6, No. 2 (Winter, 1981–82), 41–48 and Terry Mortenson, "The Firmament: What Did God Create on Day 2?," *Answers Research Journal*, Vol. 13 (2020), 126–30.

having inhaled poisonous gas from the deflating balloon.[57] Writing in an autobiography from sometime in the years 1911–1913, Aurin Hill claimed that he had shared his ideas on the shape of the Earth with Rogers prior to his death, and that his crash was extremely suspicious given Rogers' decades of experience.[58] This is perhaps the first evidence of the Flat Earther trope of claiming a conspiracy to silence explorers of the upper air who might dare to disclose the Earth's shape.

In the modern era, it should be noted that Saudi and Turkish clerics were not the only nationalists who held skeptical views of the Apollo program. Apollo skepticism is also prominent among members of the avowedly secular, social democratic Scottish National Party seeking Scottish independence from the United Kingdom; even in the United States, a 2022 study indicated that Apollo conspiracism was more prevalent among liberals than conservatives.[59] It should also be noted that Apollo astronauts have, on occasion, invited paranormal speculation onto themselves. In 1973, Armstrong himself was tricked by Erich von Däniken into joining an expedition he had organized to the Cueva de los Tayos in Ecuador to search for a vast underground library of golden books left by ancient aliens; Armstrong reportedly was not pleased when he discovered the goals of the expedition.[60] Armstrong's crewmate Buzz Aldrin (b. 1930) would later coauthor an ancient alien novel (and also infamously punched a leading Apollo denier, Bart Sibrel, in 2002).[61] Apollo 14 astronaut Edgar Mitchell (1930–2016) believed he had psychic powers and conducted private tests of his alleged abilities while alone in lunar orbit; he also later engaged in psychic remote healing and strongly endorsed

[57] "George Augustus Rogers Balloonist," New England Aviation History, April 3, 2015, accessed February 10, 2024, https://newenglandaviationhistory.com/tag/george-augustus-rogers-balloonist/.

[58] Aurin F. Hill, Ledger A, 71, Aurin F. Hill Papers (MS 579), Box 3, Special Collections and University Archives, University of Massachusetts Amherst Libraries.

[59] Gabriela Manley, "A Scottish Kind of Conspiracy," *Anthropology News*, July 12, 2019, accessed January 24, 2024, https://www.anthropology-news.org/articles/a-scottish-kind-of-conspiracy/; Dan Ariely, *Misbelief: What Makes Rational People Believe Irrational Things* (New York: HarperCollins, 2023), 29.

[60] Erich von Däniken, *The Gold of the Gods* (London: Souvenir Press, 1973), 1–45; Erich von Däniken, trans. Michael Heron, *According to the Evidence: My Proof of Man's Extraterrestrial Origins* (London: Souvenir Press, 1977), 291–96.

[61] Buzz Aldrin and John Barnes, *Encounter with Tiber* (London: Hodder & Stoughton, 1996).

his belief that the US government was covering up recovered UFOs and alien bodies.[62] Closer to the beliefs of Flat Earthers, Apollo 15 astronaut James Irwin (1930–1991) became a YEC advocate after returning from his Moon landing, and led several expeditions to Mount Ararat in Turkey to attempt to locate the remains of Noah's Ark.[63] Apollo 16 pilot Charles Duke (b. 1935) likewise converted to evangelical Christianity and embraced YEC after walking on the Moon.[64]

Due to the technological achievements needed for it and its accelerated timescale, it is not a surprise that the Apollo program has fueled a number of conspiratorial beliefs. Its successor, the space shuttle program, seemingly led to fewer such beliefs, perhaps due to its reduced ambition—but in one case, it led to a prominent Flat Earther abandoning his views. In 1985, thanks to the strategic importance of Saudi Arabia to US foreign policy, Saudi prince Sultan bin Salman Al Saud (b. 1956) flew on the STS-51-G mission of the space shuttle *Discovery*, becoming the first Muslim, first Saudi, and first Arab in space. As a result of this mission, ibn Baz publicly denounced his belief in the Flat Earth and accepted that the world was a sphere around which a member of the ruling family—his patrons—had circled. However, while his public belief in a Flat Earth changed, ibn Baz apparently remained unconvinced of heliocentrism. This did not prevent ibn Baz from being appointed Grand Mufti of Saudi Arabia—the leading Islamic figure in the kingdom, re-established after being dissolved in 1969—by King Fahd in 1993, a position he held until his 1999 death.[65] Even after his death, ibn Baz remained influential. In 2015, a YouTube video of Sheikh Bandar al-Khaibari—a cleric of the Saudi Ministry of Islamic Affairs—lecturing university students in the UAE went viral on Arabic social media because of his endorsement of geocentrism and rejection of the Apollo landings. Al-Khaibari began

[62] Alex Pasternack, "The Moon-Walking, Alien-Hunting, Psychic Astronaut Who Got Sued By NASA," *Vice*, May 14, 2016, accessed January 24, 2024, https://www.vice.com/en/article/aek7ez/astronaut-edgar-mitchell-outer-space-inner-space-and-aliens.

[63] Jerry Bergman, "Colonel James Irwin: Creationist Astronaut," Institute for Creation Research, October 31, 2013, accessed January 24, 2024, https://www.icr.org/article/7724/.

[64] Ronald L. Numbers, *The Creationists: The Evolution of Scientific Creationism* (New York: Alfred A. Knopf, 1992), 209.

[65] Jörg Matthias Determann, *Space Science and the Arab World: Astronauts, Observatories and Nationalism in the Middle East* (London: I. B. Tauris, 2018), 116–17.

his talk citing both ibn Baz and Saleh Al-Fawzan (b. 1933), an Islamic scholar who became Grand Mufti of Saudi Arabia in 2025, who has also interpreted the Qu'ran as endorsing slavery and opposing all you can eat buffets. Al-Khaibari's university talk also included a brief denunciation of Mohammad al-Massari, a Saudi physicist and political dissident who left the country in 1993 and has critiqued ibn Baz's scientific claims as part of his opposition to the Saudi government.[66]

While Saudi Arabia as a state emerged independently, though relatedly, to the European colonial project, many other Islamic states in Asia and Africa were the byproducts of European colonization in the nineteenth century and the decolonization struggles of the twentieth century. This includes Nigeria, the former British colony on the West African coast which achieved independence in 1960, and severed final ties with the British monarchy three years later. Nigeria is the country with the largest Black population in the world, along with the largest city in Africa, the former capitol of Lagos, which itself is home to Aliko Dangote (b. 1957), both the richest Black person and richest African.[67] In 1998, the Nigerian government established the National Space Research and Development Agency, which led to Nigeria becoming a leader in not only astronomy on the African continent, but also space exploration, with the construction of Africa's first communications satellite in 2007, a point of major national pride.[68]

However, for all of Nigeria's economic, scientific, and political achievements, its post-independence history has had its share of conflict.

[66] "Saudi Preacher Bandar Al-Khaybari: The Earth Does not Revolve around Itself - Extended Version," Middle East Media Research Institute, January 15, 2015, accessed January 25, 2024, https://www.memri.org/tv/saudi-preacher-bandar-al-khaybari-earth-does-not-revolve-around-itself-extended-version; Sarah A. Qidwai, "Reexamining Complexity: Sayyid Ahmad Khan's Interpretation of "Science" in Islam," in Bernard Lightman, ed., *Rethinking History, Science, and Religion: An Exploration of Conflict and the Complexity Principle* (Pittsburgh: University of Pittsburgh Press, 2019), 52–53; Nidhal Guessoum, "Lessons from the 'Earth does not rotate' debate," *Gulf News Opinion*, February 23, 2015, accessed February 5, 2024, https://gulfnews.com/opinion/op-eds/lessons-from-the-earth-does-not-rotate-debate-1.1461490.

[67] Kevin Ochieng Okoth, *Red Africa: Reclaiming Revolutionary Black Politics* (London: Verso, 2023), 59, 153n1.

[68] A. A. Ubachukwu and P. N. Okeke, "Astronomy in Nigeria: Recent Developments," *African Skies*, No. 6 (Dec., 2001), 17–18; Peter Martinez, "Nigcomsat-1 Launch Approaches," *African Skies*, No. 10 (Oct., 2006), 3. See also Oloyede S. Oyelekan and Julius B. Omiwale, "Trends in Nigeria's Science Education within the Global Context," in

Northern Nigeria is heavily Muslim, due to its history dating to the Sokoto Caliphate, while the southern part of the country is largely Christian. This regional religious division is further overlaid on multiple local ethnic, linguistic, and political divisions. A civil war erupted from 1967–1970, and several military and quasi-military coups have occurred since independence, although by the 2010s a solidification of multiparty democracy and peaceful transition of power has occurred. On top of these issues is that modern Nigeria has, in the words of one scholar, "long been an archetypal neo-colony, with an economy characterised by extractive capitalism and the massive outflow of surplus value in the form of corporate profits abroad" which has resulted in 40% of the population living in abject poverty, in contrast to the small number of urban elites. This has resulted in occasional large-scale protests against both the Nigerian state and foreign extractive corporations, which have been put down with brutality even by democratic civilian governments.[69]

Peruvian sociologist Aníbal Quijano (1930–2018) created the term coloniality to refer to "the structures of power, control, and hegemony that have emerged during the modernist era, the era of colonialism, which stretches from the conquest of the Americas to the present."[70] The lingering British architecture in the educational, economic, and social structures of the politically independent Nigeria reflects coloniality. Its education system was originally implemented by the British upon the federation of the Nigerian colonies in 1914. Although the new Nigerian state took over the British education system in 1960, it failed to adequately resolve the imperfections it inherited, including a division whereby schools were more prominent and Western-oriented in the south, and sparser and Islamic-based in the north. The slow pace of

Femi S. Otulaja and Meshach B. Ogunniyi, eds., *The World of Science Education: Handbook of Research in Science Education in Sub-Saharan Africa* (Rotterdam: Sense Publishers, 2017), 43–64.

69 Okoth, *Red Africa*, 123–25, 153n1.

70 Walter D. Mignolo, "DELINKING: The rhetoric of modernity, the logic of coloniality and the grammar of decoloniality," *Cultural Studies*, Vol. 21, Nos. 2–3 (2007), 449–514.

educational advancement in both regions has been termed a policy of "de-emancipation" which limits the ability of Nigerians to fully end coloniality and exercise true sovereignty.[71]

The history of Sunni Islam as a motivating force for northern Nigerians to both resist the federal government and the Western-derived neocolonial education system would find a potent outlet from what was at first a modest start: an Islamic school in the northeastern city of Maiduguri, founded in 2002 by Mohammed Yusuf (1970–2009), a Salafist preacher of the Hausa ethnic group—composing 70% of the population of northern Nigeria—who had received his education in an Islamic university. Yusuf's followers originally were referred to as the Taliban, a word meaning "Students" in the Pashto language of Afghanistan. Yusuf's Taliban appeared to have derived some inspiration from the Afghan organization of the same name, but had no major connection, outside of both groups receiving support from attendees of IUM. However, Yusuf's group would soon become known by a more permanent name: Boko Haram.[72]

Major scholarship into the origins of Boko Haram was not carried out until a decade into the group's existence, making some early details murky—not least of which is the meaning of the name, which is contested amongst both locals in Maiduguri and scholars. One of the most common claims for the name is that it comes from the Hausa word *boko*, meaning "Western education" (from the English "book") and the Arabic word *haram*, meaning "forbidden"—in other words, Boko Haram means "Western education is forbidden." An alternative origin was proposed in 2013, that *boko* was an indigenous pre-colonial Hausa word meaning "deceitful knowledge." Other Hausa speakers in northern

[71] Grace Oluremi Akanbi and Alice Arinlade Jekayinfa, "Education and emancipation, educational policies and «de-emancipation»: A history of the Nigerian education system from 1914 to 2014," *Tiempo y Educación*, Vol. 6, No. 2 (Jul.–Dec., 2019), 178–83; Alexander Thurston, *Boko Haram: The History of an African Jihadist Movement* (Princeton: Princeton University Press, 2018), 36–37, 72–81; A. Babs Fafunwa, *History of Education in Nigeria* (London: George Allen and Unwin, 1974), 92–181.

[72] Joe Boyle, "Nigeria's 'Taliban' enigma," BBC News, July 31, 2009, accessed August 18, 2020, http://news.bbc.co.uk/2/hi/8172270.stm; Ousmane Oumar Kane, *Beyond Timbuktu: An Intellectual History of Muslim West Africa* (Cambridge, MA: Harvard University Press, 2016), 183.

Nigeria believe that *boko* has no specific meaning in the way used by the group.[73]

Until its terrorist campaign escalated in 2009, Boko Haram, while promoting Sunni extremism, was mainly a regionalist group, supporting the autonomy of the Hausa of northern Nigeria against the expansion of federal authority. This included the continuation of efforts dating from the 1970s, and accelerating since the resumption of democracy in 1999, to expand the secular, British-modeled federal education system to the north, which would challenge the more independent, Islamic schooling system present there since the pre-colonial period. To their view, Western-style education was not merely *un*-Islamic, it was *anti*-Islamic, as it was the source of what Yusuf viewed as such fundamentally anti-Islamic ideas as secularism, democratic pluralism, feminism, and scientific notions he viewed as opposing the Qu'ran's teachings.[74] One of the main inspirations Yusuf took for this view was the book *Global, Foreign and Colonial Schools: Their History and Danger* by Bakr bin Abdullah bin Abi Zayd (c. 1944–2008). A Saudi Salafist scholar and former librarian at IUM, Zayd argued that Western secular education was a plot to destroy Islam.[75]

From 2004–2008, Yusuf held a series of debates with other Nigerian Salafist scholars. The oldest surviving recording of this series is a 26 June 2006 debate with Isa Ali Pantami (b. 1971). In 2014, Pantami would join IUM's faculty, and would later serve as Nigeria's Minister of Communications and Digital Economy from 2019–2023. The Western-educated Pantami was widely seen as being more prepared and skilled in the debate than Yusuf. In it, Yusuf illuminates several core tenets of his—and by extension, Boko Haram's—views on Western education and its coloniality, asking

[73] Thurston, *Boko Haram*, 1–2; Kane, *Beyond Timbuktu*, 182–83; Fred Femi Akinfala, Gabriel Aunde Akinbode, and Ibi-Agiobu Kemmer, "Boko Haram and Terrorism in Northern Nigeria: A Psychological Analysis," *British Journal of Arts and Social Sciences*, Vol. 17, No. 1 (2014), 116.

[74] Akinfala, Akinbode, and Kem, "Boko Haram and Terrorism," 116–18; Muhammad Mann Shaaba, "Absolutism in Religion and Global Peace: Boko-Haram Factor in Nigeria and Its Educational Imperatives," *World Journal of Education*, Vol. 7, No. 2 (Apr., 2017), 11–18.

[75] Kane, *Beyond Timbuktu*, 185.

> what is Western education? Western education is contemporary knowledge that came to us through European colonialism and it includes medicine, engineering, agriculture, geography, physics, biology, chemistry, and so on. All these disciplines together, and even the English language itself, once it does not contradict the Sunna of the Prophet, in its capacity as knowledge, there is no blame [in learning it]. If it contradicts the Sunna of the Prophet or the Qur'ān, it would be rejected. If it does not contradict or conflict with the Qur'ān then it is acceptable. If we have our schools, we can adopt those disciplines and teach them in our schools. The reason they cannot go to those schools, as I said, is not because of the disciplines, but because of the system.[76]

In between rejections of both evolution and the Big Bang theory in favor of Islamic creationism, Yusuf also provides his endorsement of geocentrism:

> Secondly, there is a topic in *Geography*, the theory of revolution. In their perspective, the world moves in a way they understand, not in the way that Allah created the world, with the sun. They have what they call the "ten planets," just as Ibn Taymiyya narrated, or nine planets—with the addition of the sun they become ten. These ideas, after they invented them, in their own research, do not conform to the sayings of the Qur'ān and *ḥadīth*. Furthermore, they gave them the names of the Greek idols such as Mercury, Venus, Earth, Mars, Jupiter, Saturn, and Uranus and so on. This way of thinking contradicts the Qur'ān and Sunna. The study of these ideas cannot be made permissible based on this contradiction.[77]

Yusuf's appeal to Ibn Taymiyya (1263–1328) is also revealing. A Sunni theologian of mixed Arabic and Kurdish ancestry in what is now southern Turkey, Ibn Taymiyya is considered by many Salafists to be a predecessor to their theology. He campaigned against various philosophical fields including astrology but was not widely respected in his time.[78]

[76] Muhammed Yusuf and Isa Pantami, "Debate on the Status of Western Education and Working for the Nigerian Government Between Mallam Isa Ali Ibrahim Pantami and Mallam Muhammad Yusuf Maiduguri," in Abdulbasit Kassim and Michael Nwankpa, eds., *The Boko Haram Reader: From Nigerian Preachers to the Islamic State* (Oxford: Oxford University Press, 2018), 12.

[77] Yusuf and Pantami, "Debate on the Status of Western Education," 14.

[78] Yusuf and Pantami, "Debate on the Status of Western Education," 26n4; Ikram Hawramani, "The Quran and the Shape of the Earth: Is It Round or Flat?," Ikram

Yusuf and Pantami then debate whether or not the Qur'an accepts nine planets (including the then-classification of Pluto) and whether those are the same as the seven heavens of the Qur'an; and if the Qur'an is silent on an issue, whether scientific observations can be adapted into those silences. Yusuf ultimately concedes, "If it does not contradict it, yes."[79] This concession to the more educated, Westernized Pantami seems to symbolize Yusuf's standing in the overall debate. In response to the perception of him losing to Pantami, sometime in late 2006 or early 2007, Yusuf attempted to save face by writing a 168-page, Arabic manifesto titled *This is our Creed and Method of Proclamation*, which elaborated on some of the positions he took with Pantami. In a list of criticisms of Western-style education, Yusuf first makes an explicit Flat Earth endorsement, in between rejections of Marxism, Freud, evolution, deep time, and—perhaps most incoherently and bizarrely—seasonal weather patterns:

> These foreign secular colonialist schools contain elements that are opposed to sharīʻa, which forbids practicing, announcing, studying and teaching in them, as in what follows.... Fourth, the belief that the sun is stationary, not moving, and does not orbit, but that it is the earth which orbits around the sun. This is what they call the "globality of the world," because the world in their belief is a globe. Glory be to You, this is a monstrous falsehood, [as] the Almighty states: "And the sun runs to its fixed destination. That is the decree of the All-mighty, the Allknowing." (Q36:38).[80]

The rejection of the Western style of education endorsed by the Nigerian state was central to Yusuf and Boko Haram's endorsement of both the Salafist interpretation of Islam and Hausa regionalist politics. Although not the sole aspect of the rejection of education, Flat Earth—based on strict interpretations of passages of the Qur'an—were key to

Hawramani's Website, November 30, 2018, accessed August 14, 2020, https://hawramani.com/the-quran-and-the-shape-of-the-earth-is-it-round-or-flat/; DeLong-Bas, *Wahhabi Islam*, 247–56.

[79] Yusuf and Pantami, "Debate on the Status of Western Education," 15–16.

[80] Muhammed Yusuf, "Translation of Selections from Muhammed Yusuf: Hādhihi ʻAqīdatunā WaMinhāj Da'watinā (This is our Creed and Method of Proclamation)," in Abdulbasit Kassim and Michael Nwankpa, eds., *The Boko Haram Reader: From Nigerian Preachers to the Islamic State* (Oxford: Oxford University Press, 2018), 31.

Yusuf's views. However, Yusuf would not have the chance to implement most of these ideas.

Until 2009, Boko Haram followers engaged in the occasional murder of moderate Muslim scholars who supported Western-style education. But that year, a dramatic escalation would happen. Tensions between Boko Haram and police had escalated, initially because of the police attempting to enforce a law requiring motorcyclists to wear helmets leading to the arrest of several Boko Haram couriers. This led to a cycle of clashes between Boko Haram and the police, culminating in shootings and arrests. On 26 July, Boko Haram members stormed the police station where their members were being held, resulting in a widespread uprising by Boko Haram members drawing in federal military forces. On 30 July, after a two-day siege of his house, Yusuf was taken into custody by the Nigerian army. The army handed Yusuf over to local police, who subsequently shot him dead in the street. However, the police murder of Yusuf did not stop Boko Haram; leadership fell to his lieutenant, Abubakar Shekau (c. 1975–2021), and the uprising only escalated in both scale and violence.[81] As of this writing, Boko Haram continues to exist, though its peak was the period of 2012–2017. From 2015–2016, Boko Haram was further allied with the Islamic State of Iraq and Syria (ISIS) group of the Middle East. The groups would eventually split, and Shekau would eventually kill himself during a battle between the two organizations.

Boko Haram, despite its origins as group promoting Hausa regionalism, has always had a measure of internationalism to it. Yusuf had been inspired by Bakr Zayd but Abdullah el-Faisal (b. 1963; né Trevor William Forrest), a Jamaican convert to Islam then living in the United Kingdom, and had connections to Hausa and Islamic institutions of learning in neighboring countries as well.[82] Under Shekau's leadership, Boko Haram increased these regional ties, including the use of social media to promote itself among Western Muslims, but its efforts paled before the ISIS efforts to propagandize and proselytize on social media. YouTube, Twitter, and Facebook were extremely effective for ISIS to spread its message to Western audiences. These were the same venues that were being used to not only spread Flat Earth and other pseudoscience beliefs (and in

[81] Akinfala, Akinbode, and Kem, "Boko Haram and Terrorism," 117–18, 122–23; Kane, *Beyond Timbuktu*, 186–88.

[82] Kane, *Beyond Timbuktu*, 184–85.

Arabic language channels and groups, alongside ISIS and Boko Haram recruitment propaganda) but also far-right racist and terrorist recruitment messages in the English language.[83] Efforts to moderate social media to deal with the spread of Islamic ideology in the early 2010s could have been turned to Western promoters of right-wing terrorism and pseudo-science who were growing their influence at the same time, partly in reaction to the fear generated by ISIS and Boko Haram. That those social media strategies were not applied equally to the Western figures would come to have a major impact in politics and health policy by the end of the decade.[84]

These are not issues exclusive to Boko Haram, ISIS, or Islamic terrorism. An example from the United States is Kurt Cofano (b. 1987), a Flat Earther from Pittsburgh who in May 2020 threatened to kill protesters in the Black Lives Matter movement and two months later was arrested driving to Washington, D.C. with over thirty homemade bombs in his trunk that he planned to use on government buildings. Sentenced to five years in prison in 2021, he was let out less than three years later.[85] In 2023, Neo-Nazi Flat Earther Hardy Lloyd (b. 1978) was sentenced to six years in prison for threatening jurors in the trial of the man who killed eleven people in the 2018 mass shooting of the Tree of Life synagogue

[83] H. Colleen Sinclair, Andrew R. Burns, and Brett S. Burton, "Taking Conspiracies to Extremes: How Extremist Groups Use Conspiracy Narratives to Stoke Violence," in Luke Ritter, ed., *American Conspiracism: An Interdisciplinary Exploration* (London: Routledge, 2024), 137–158.

[84] Bader, Baker, Day, and Gordon, *Fear Itself*, 52–54, 76–78; Max Fisher, *The Chaos Machine: The Inside Story of How Social Media Rewired Our Minds and Our World* (New York: Little, Brown, and Company, 2022), 229; P. W. Singer and Emerson T. Brooking, *LikeWar: The Weaponization of Social Media* (Boston: Houghton Mifflin Harcourt, 2018), 65–66, 159–60, 212–23; Asma Shakir Khawaja and Asma Hussain Khan, "Media Strategy of ISIS," *Strategic Studies*, Vol. 36, No. 2 (Summer, 2016), 104–21; Anne Speckhard, Ardian Shajkovci, Claire Wooster and Neima Izadi, "Mounting a Facebook Brand Awareness and Safety Ad Campaign to Break the ISIS Brand in Iraq," *Perspectives on Terrorism*, Vol. 12, No. 3 (Jun., 2018), 50–66; Rita Katz, *Saints and Soldiers: Inside Internet-Age Terrorism, From Syria to the Capitol Siege* (New York: Columbia University Press, 2022), 148–67, 217–30; Sharon A. Hill, *Scientifical Americans: The Culture of Amateur Paranormal Researchers* (Jefferson, NC: McFarland & Company, 2017), 42–44; Mishra, *Age of Anger*, 343–45.

[85] Weill, *Off the Edge*, 162.

in Pittsburgh.[86] Social scientist Becca Lewis has termed this connection between conspiracy theory and extremism the "Alternative Influence Network" (AIN) while archaeologist Jeb Card (b. 1974) developed the related term of "Paranormal Unified Field Theory" (PUFT) to categorize the overlap of not only multiple paranormal and conspiracy views, but of paranormalism and radical right politics.[87] There is a long history of the blending of this "supernatural imaginary" with far-right politics; this was especially common in European movements after World War I.[88] In 2022, a poll by Fairleigh Dickinson University indicated that believers of 2020 election conspiracies in the United States were also more likely to believe in anti-vaccination conspiracies, Bigfoot, and Flat Earth. Poll director Dan Cassino explained it as, "Even the wildest theories about the 2020 election don't include Bigfoot. But people are really trying to be consistent, and so endorsing one set of false beliefs makes them more likely to endorse another, even if it's pretty far out there."[89]

A more specific contemporary US example of this can be found in the case of Mount Shasta in California. Since the early twentieth century, the mountain has become central to a number of distinct legends, including that it was an outpost of the lost continent of Lemuria, it is home to Bigfoots, it contains an entrance to the Hollow Earth, it covers a UFO landing base, and it is a nexus of crystal energy. As such, the area around

[86] Stephanie Halmhofer, "Manufacturing History: Atlantis, Aryans, and the Use of Pseudoarchaeology by the Far-Right," in Christopher T. Conner; Matthew N. Hannah and Nicholas J. MacMurray, eds., *Conspiracy Theories and Extremism in New Times* (Lanham, MD: Rowman & Littlefield, 2024), 66, 69.

[87] Jeb J. Card, "*America Before* as a Paranormal Charter," *The SAA Archaeological Record*, Vol. 19, No. 5 (Nov., 2019), 29; Rebecca Lewis, "Alternative Influence: Broadcasting the Reactionary Right on YouTube," Data & Society, September 18, 2018, accessed February 19, 2021, https://datasociety.net/wp-content/uploads/2018/09/DS_Alternative_Influence.pdf, 33–35; Zachary J. Goldberg and Sean Richey, "Anti-Vaccination Beliefs and Unrelated Conspiracy Theories," *World Affairs*, Vol. 183, No. 2 (Summer, 2020), 105–24; Talia Lavin, *Culture Warlords: My Journey Into the Dark Web of White Supremacy* (New York: Hachette Books, 2020), 164–66; Sarah Krasnostein, *The Believer: Encounters with the Beginning, the End, and our Place in the Middle* (Portland, OR: Tin House, 2022).

[88] Eric Kurlander, *Hitler's Monsters: A Supernatural History of the Third Reich* (New Haven: Yale University Press, 2017), 3–61.

[89] Dan Cassino, "2020 Election Conspiracy Believers More Likely to Embrace Bigfoot, Flat Earth," FDU Poll, May 5, 2022 (accessed May 19, 2025), https://portal.fdu.edu/fdupoll-archive/220505/.

Mount Shasta has attracted New Age and paranormal proponents of every shade as residents.[90] Despite this, the elected offices of the surrounding Shasta County have long been dominated by conservative Republicans. However, the 2022 midterm elections saw even the traditional Republicans evicted in favor of members of a local militia group who see them as too liberal. The militia members were able to use covid conspiracies, and the backing of a local evangelical church opposed to almost all of the New Age Shasta-resident beliefs, to come to office on the promise of not only ending covid restrictions, but also gun control and local education, and to push for political secession. Shasta found itself to be simultaneously home to arguably the highest concentration of paranormal believers in the country, and the most radical right-wing political establishment.[91]

The AIN/PUFT issues demonstrated by Mount Shasta will be returned to in chapter seven. But a decade earlier in Nigeria, Boko Haram continued to promote an interpretation of Islam which included support for a Flat Earth. In a 16 February 2009 speech titled "Western Civilization is Atheism and Anti-Islam," from several months before he took over Boko Haram, Shekau stated that

> Yes, Western education is impermissible, but these electrical appliances are good. This light is good because it brightens this place. These speakers are good because you can all hear my voice. These appliances are good. However, Western education is impermissible because they have said the sun is stationary. Is it impermissible in your own view? Is it permissible to say that the sun is stationary? The sun is not the speakers. If the same view was said about the speakers, I can also leave them. This is not Western education.[92]

[90] Colin Dickey, *The Unidentified: Mythical Monsters, Alien Encounters, and Our Obsession with the Unexplained* (New York: Viking, 2020), 13–17; Donald R. Prothero, *Weird Earth: Debunking Strange Ideas about Our Planet* (Bloomington: Red Lightning Books, 2020), 167–77; Justin McHenry, *Lemuria: A True Story of a Fake Place* (Port Townsend, WA: Feral House, 2024), 105–28; John O'Connor, *The Secret History of Bigfoot: Field Notes on a North American Monster* (Naperville, IL: Sourcebooks, 2024), 212.

[91] Samuel McIlhagga, "The California militia ready for civil war," *UnHerd*, June 27, 2023, accessed February 18, 2024, https://unherd.com/2023/06/the-california-militia-ready-for-civil-war/.

[92] Abubakar Shekau, "Western Civilization is Atheism and Anti-Islam," in Abdulbasit Kassim and Michael Nwankpa, eds., *The Boko Haram Reader: From Nigerian Preachers to the Islamic State* (Oxford: Oxford University Press, 2018), 127.

Several months later, Shekau would claim more generally that

> even those who have obtained Western education and have gone deep into its principles still express doubts about the morality of Western education. Truthfully, they know that the teachings of Western education go against the *tawḥīd*, the oneness of Allah. This includes not only subjects as sociology, but all other course of human sciences. That is why you will hear them use concepts as nature in place of Allah.... That is why today you find the promotion of absolute freedom and liberty has become commonplace. Illicit sexual relations are permitted, gay marriage (homosexuality) is allowed and lesbianism is promoted. You will hear unbelievers tell you, geography is this, and geography is that. Such sayings are among the many hundreds of vices these unbelievers and promoters of faithlessness rail against Allah.... This is our point of dispute that we said they do not believe in Allah. In this education, they make reference to mathematics, geography, physics, biology, geology and many other sciences. From the beginning to the end of their program, there is no reference to Allah. Allah is excluded from the litany of science. No reference is made to the Prophet and his tradition.[93]

While not as explicit as Yusuf was, it is clear that Shekau still harbored doubts on established notions of geography and astronomy, all due to his belief that Western education and science were both colonialist and anti-Islamic in nature.

While Shekau might not have been as explicit about his Flat Earth beliefs as Yusuf, they still informed Boko Haram policies. While from 2003–2009, the victims of the group tended to be moderate Muslim scholars, from 2009 its victims tended to be schoolteachers and students. The group's targeting of female students for murder and kidnapping due to their interpretation of an Islamic injunction against female education has been well publicized in Western media. Less publicized is the fact that Boko Haram specifically targeted geography teachers, specifically because they instructed that the world was spherical. According to a 2016 report by Human Rights Watch, over the previous seven years, Boko Haram had directly killed 600 school staff, and led to the resignation out of fear of 19,000 more. Eyewitnesses reported Boko Haram gunmen entering

[93] Abubakar Shekau, "Hādhihi 'Aqīdatunā (This is Our Creed)," in Abdulbasit Kassim and Michael Nwankpa, eds., *The Boko Haram Reader: From Nigerian Preachers to the Islamic State* (Oxford: Oxford University Press, 2018), 125.

schools to specifically kill the geography teachers, while not harming anyone else. Also targeted were science teachers, for their views on both astronomy and evolution.[94]

Although Yusuf and Shekau drew their Flat Earth belief from their particularly Salafist strict interpretation of the Qur'an, and likely ultimately from the interpretation of ibn Baz and others at IUM, these views are not universal even amongst Salafists—take for example Yusuf's debate opponent Pantami, who likewise was a Salafist who would join the IUM faculty. Even other notable Salafist groups who set up state organizations, such as the Taliban and the onetime Boko Haram ally ISIS, did not incorporate those views into their education system. In Afghanistan, aside from the destruction and collapse of infrastructure caused by decades of total war, there is a wider Taliban neglect of "worldly" higher education due to the belief that it lacks value. But while some of the Taliban's school districts ban teaching of such subjects as English, culture, and history, and the use of books containing the images of humans or animals, there does not seem to be any official revision to geographical or astronomical concepts.[95]

In the areas of Iraq and Syria that fell under ISIS control from 2014–2019, textbooks were distributed in the form of PDFs over social media channels, as a means of providing new doctrines quickly and avoiding the war's destruction of physical textbook, schools, and libraries—including by ISIS itself—as in Afghanistan. These ISIS textbooks incorporated the pursuit of jihad into everyday lessons, such as posing a math lesson around how many unbelievers a bomb attack could kill. In this, the ISIS textbooks followed the format of Afghanistan's Taliban textbooks, initially provided by the CIA as part of the US support for the mujahedeen during the 1980s war against the Soviet Union.[96]

[94] Colin Freeman, "Geography teachers targeted by Boko Haram," *The Daily Telegraph* (12 April, 2016), 12; Mausi Segun, *"They Set the Classrooms on Fire": Attacks on Education in Northeast Nigeria* (New York City: Human Rights Watch, 2016), 47–49.

[95] Misbah Abdulbaqi, "Higher Education in Afghanistan," *Policy Perspectives*, Vol. 6, No. 2 (Jul.-Dec., 2009), 102–03; Maija Liuhto, "The Schools of the Taliban," *The Diplomat*, November 3, 2016, accessed February 10, 2021, https://thediplomat.com/2016/11/the-schools-of-the-taliban/.

[96] Kyle Almond, "How ISIS changed Iraqi schools," CNN, July, 2017, accessed February 19, 2021, https://www.cnn.com/interactive/2017/07/world/iraq-schools-cnn photos/; Kinana Qaddour, "Inside ISIS' Dysfunctional Schools: The Failure of the Group's Educational System," *Foreign Affairs*, October 13, 2017, accessed February

On 2 August 2014, the newly appointed ISIS Minister of Teaching and Education Reda Seyam (1960–2017, nom de guerre Dhu al-Qarnayn) issued a set of twelve guidelines for education, which included the removal of the subjects of "Art and music, civics, art and sculpting, social studies and history, sports, philosophy, sociology, psychology, Islam, and Christianity" along with "Anything that has to do with Darwin's theory, creation by nature or creation from nothingness must be removed from science classes; all creation must be attributed to God Almighty." While physics and chemistry were allowed subjects, "it must be constantly repeated that all laws of physics and chemistry are laws of divine origin inspired by God in his creation." Science (along with separate, more advanced classes on physics and chemistry), mathematics, computers, and geography remained in the allowed curriculum.[97] However, even these nonreligious subjects were required to have religious elements in them, including the science and geography courses, the latter of which included the use of the Qur'an to identify planets and stars. This was meant to counter the "theories put forth by atheists such as communists" that developments in the natural world happened without the divine intervention of God.[98]

Despite all of these efforts to introduce a Salafist interpretation of Sunni Islam into even secular topics, and to remove as many Western elements from the pre-existing secular education of both Iraq and Syria, the ISIS schools did not prohibit astronomy or geography, or introduce Flat Earth, or even geocentrist, notions. Indeed, the fourth-grade ISIS science textbook, introduced in 2015, explicitly describes how to prove that the Earth is a sphere, by pointing out that "We can notice this when we look at ships. The first thing we see of the ship is its sail. If the Earth

21, 2021, https://www.foreignaffairs.com/articles/syria/2017-10-13/inside-isis-dysfunctional-schools; Craig Davis, ""A" Is for Allah, "J" Is for Jihad," *World Policy Journal*, Vol. 19, No. 1 (Spring, 2002), 90–94; DeLong-Bas, *Wahhabi Islam*, 266.

[97] Olivier Arvisais and Mathieu Guidère, "Education in conflict: how Islamic State established its curriculum," *Journal of Curriculum Studies*, Vol. 52, No. 4 (2020), 499, 507; Olivier Arvisais, Marie-Hélène Bruyère, Chirine Chamsine, and Mohamed Amine Mahhou, "The educational intentions of the Islamic State through its textbooks," *International Journal of Educational Development*, Vol. 87 (Nov., 2021), 2.

[98] Arvisais and Guidère, "Education in conflict," 511; Arvisais et al., "The educational intentions of the Islamic State," 6; Olivier Arvisais and Mathieu Guidère, "The Integration of Religious Elements into ISIS Textbooks," *Religion & Education*, Vol. 47, No. 2 (2020), 194, 198.

were flat, we would have seen the whole ship."[99] The same book cites the Qur'an (Surah Ya-Sine, verse 40) to provide justification for a heliocentric model of the solar system.[100] Clearly, Boko Haram's insistence on Flat Earth as a motivating factor for their own movement owed more to their leaders' personal interpretations of Muslim theology and its political application to the Nigerian context than to wider Salafist acceptance (Fig. 6.3).

While Boko Haram may have been unique among Salafist movements of the 2010s in using the Flat Earth as a metaphor for revolutionary resistance to coloniality, they were not alone in doing so in movements—or at least members of those movements—in Muslim societies of northern Africa. What is notable is that the context of those examples is almost completely different from the motivations and methods of Boko Haram.

On 17 December 2010 in Tunisia—the modern region of the ancient Roman province of Numidia from whence came the original Flat Earther Lactantius, as described in chapter one—a street vendor named Mohamed Bouazizi (1884–2011) lit himself on fire to object to the illegal confiscation of his wares by local police. The immediate protests which formed in support of Bouazizi and against the authoritarian regime of Tunisian president Zine El Abidine Ben Ali (1936–2019), who had been in power since 1984, formed the basis for not only what became known as the Tunisian Revolution, but the wider Arab Spring movement across North Africa and the Middle East. The Arab Spring would result in a number of revolutions and counter-revolutions in the region, including the civil war in Syria which would allow for the consolidation of power by ISIS in that country. On 14 January 2011, ten days after Bouazizi died from his wounds, Ben Ali stepped down from office. On 23 October of that year, a Constituent Assembly was elected to draft a new constitution, which went into effect in January 2014. The new constitution guaranteed freedom of speech, a commitment to elections and gender equality in the

[99] Patrice Potvin, Marianne Bissonnette, Chirine Chamsine, Marie-Hélène Bruyère, Mohamed Amine Mahhou, Olivier Arvisais, Patrick Charland, Stéphane Cyr, and Vivek Venkatesh, "Science Education under a Totalitarian Theocracy: Analyzing the ISIS Primary Curriculum," *Journal of Research in Science, Mathematics and Technology Education*, Vol. 2, No. 3 (Sep., 2019), 190.

[100] Potvin et al., "Science Education," 191. See also the examples in El-Fandy, *On Cosmic Verses in the Quran*, 47–49.

Fig. 6.3 Two-dimensional image of the Earth as a globe, from the second-year ISIS science textbook PDF. Image courtesy of Olivier Arvisais and Patrice Potvin

legislature, and promoted both political neutrality of mosques, support for Islam while guaranteeing freedom of conscience, and stated support for Pan-Arab unification and the liberation of Palestine.[101] These, while

[101] "Tunisia's Constitution of 2014," Constitute Project, June 17, 2023, accessed February 4, 2024, https://www.constituteproject.org/constitution/Tunisia_2014.pdf.

giving weight to the role of Islam in Tunisia's governance, are all clearly more liberal than the ideologies of Boko Haram or Saudi Arabia.

Also present in the 2014 constitution was a requirement of education of all Tunisians until the age of 16, the right to free public education at all levels, and the guarantee of freedom from political interference in all centers of learning. This commitment to education was a reflection that a number of educators were elected as delegates to the Constituent Assembly. This included Jamel Touir, a professor of earth sciences at the University of Sfax in southern Tunisia. Touir was one of twenty delegates elected as part of the social democratic Democratic Forum for Labour and Liberties (abbreviated Ettakatol, after the Arabic word for 'political bloc'), a left-wing opposition party under Ben Ali's reign. After the 2011 elections, until the 2014 elections under the new constitution, Ettakatol was part of the government coalition as a junior partner to both a larger center-left secular party and a moderate Islamic party. Touir was chair of the Constitutional Bodies Committee, involved in such decisions as balancing the regulation of social media with freedom of speech.[102]

During his time as a social democratic member of Tunisia's revolutionary government, Touir maintained his academic work at Sfax, including serving as the advisor to a doctoral student, Amira Kharroubi. Starting in the 2011–2012 academic year, under the supervision of Touir, Kharroubi began work on her doctoral thesis: an argument that the Earth is flat. In August 2016, the two published a coauthored paper arguing for geocentrism in an open-access, ostensibly peer-reviewed "pay to publish" journal whose editor-in-chief specialized in homeopathy. At this time, their article only argued for geocentrism, and not the shape of the Earth, perhaps a result of them testing the waters.[103] Of curiosity is that their arguments for a geocentric Earth were entirely secular with not reliance on religion of any sort, something which made it almost unique in either the Muslim world or the West—where "pure" geocentrism tends to be

[102] Tofigh Maboudi, "Reconstituting Tunisia: Participation, Deliberation, and the Content of Constitution," *Political Research Quarterly*, Vol. 73, No. 4 (Dec., 2020), 783; Yaël Nazé, "A Doctoral Dissertation on a Geocentric Flat Earth: 'Zetetic' Astronomy at the University Level," *Skeptical Inquirer*, Vol. 42, No. 3 (May/Jun., 2018), 13.

[103] Amira Kharroubi and Jamel Touir, "The Geocentric Model of the Earth: Physics and Astronomy Arguments," *The International Journal of Science & Technoledge*, Vol. 4, No. 8 (Aug., 2016), 57–62.

associated with Traditionalist Catholicism, just as Flat Earth tends to be associated with evangelical Protestantism.[104]

Mention before has been made of Gerardus Bouw, a Traditionalist Catholic who received a PhD in astronomy from Case Western Reserve University and has written several works promoting geocentrism and attacking Flat Earth.[105] Another prominent case is Robert Sungenis (b. 1955), a Traditionalist Catholic Young-Earth Creationist and Holocaust denier who was converted to geocentrism by reading Bouw's work, and in 2014 produced *The Principle*, a pro-geocentric documentary notable for tricking many of its participants to appear under false pretenses.[106] In 2018, Sungenis both self-published his attack on Flat Earthers and appeared at the Second Flat Earth International Conference to debate prominent Flat Earther Rob Skiba (and where YEC astronomer Danny Faulkner was also present).[107] There is perhaps something to be said for the concept of the Flat Earth in the West as a bridge between Traditionalist Catholic-dominated neo-geocentrism, and evangelical Protestant-dominated Young-Earth Creationism. This mix may have been especially potent in Brazil, as discussed in the next chapter.

Despite the fact that Touir and Kharroubi's paper was open access, written in English, and with their names and academic credentials listed at the top, it drew little attention until the next year. On 1 April 2017, Hafedh Ateb, founder of the Astronomical Society of Tunisia, posted on his Facebook page critiquing Kharroubi's forthcoming Flat Earth thesis

[104] Nazé, "Doctoral Dissertation," 14n5; Paul Wallace, *Stars Beneath Us: Finding God in the Evolving Cosmos* (Minneapolis: Fortress Press, 2015), 125–40; Donald R. Prothero, *Weird Earth: Debunking Strange Ideas about Our Planet* (Bloomington: Red Lightning Books, 2020), 41–59.

[105] Gerardus D. Bouw, "Flat Earth Mythology and Fact," *The Biblical Astronomer*, Vol. 10, No. 94 (Fall, 2000), 23–30; Gerardus D. Bouw, "Panorama," *The Biblical Astronomer*, Vol. 19, No. 127 (Winter, 2009), 9; Gerardus D. Bouw, *Geocentricity: Christianity in the Woodshed* (Miamitown, OH: Daystar Publishing, 2013), 16–34.

[106] Lawrence M. Krauss, "I Have No Idea How I Ended Up in That Stupid Geocentrism Documentary," *Slate*, April 8, 2014, accessed November 23, 2020, https://slate.com/technology/2014/04/lawrence-krauss-on-ending-up-in-the-geocentrism-documentary-the-principle.html.

[107] Robert Sungenis, *Flat Earth Flat Wrong: An Historical, Biblical & Scientific Analysis* (State Line, PA: Catholic Apologetics International Publishing, 2018); Danny Faulkner, "Second Flat Earth International Conference," Answers in Genesis, December 14, 2018, accessed March 16, 2024, https://answersingenesis.org/blogs/danny-faulkner/2018/12/14/second-flat-earth-international-conference/.

(without identifying her), a copy of which he had been given. Ateb's post started an avalanche in the Tunisian science community, resulting in articles on regional French, Arabic, and English websites denouncing the development as a negative reflection of the state of both education and science in Tunisia and eventually leading to the identification of the university and individuals involved. Within a week, the Sfax administration announced it would both reject the thesis and launch an inquiry into the approval process. For his part, Touir defended the thesis on grounds of freedom of speech and religion, as while their 2016 article was purely secular, the dissertation—which as of 2025 has not made public—appears to have a conclusion rooted in a strict interpretation of the Qu'ran, according to Ateb's response. As has been pointed out, if Ateb had not initiated an international outcry with his post, it is likely that the University of Sfax would have approved the thesis with little fanfare.[108] However, as with Boko Haram among Salafist movements, it is worth emphasizing that the Sfax case does mean that Flat Earth belief is endemic among Western-style educational institutions in the Muslim world. In October 2017, in the aftermath of the Sfax issue coming to light, a study was conducted at a Muslim high school in Malang, Indonesia; of 92 students polled, all had heard of the Flat Earth theory, but eighty-eight rejected it outright and only four entertained the possibility.[109] It should also be noted that the vast majority of modern Muslims have no difficulty reconciling scientific cosmology with Islam, unlike evolution which is accepted by a minority of global Muslims.[110]

Although Kharroubi included an Islamic element within her dissertation, it is clearly of a different sort and a different context than the Salafist interpretations of ibn Baz, Yusuf, Shekau, and al-Khaibari. Kharroubi's Flat Earth dissertation was accepted at the start of the revolutionary era in Tunisia, and her supervisor was one of those leading the revolution from

[108] Nazé, "Doctoral Dissertation," 13–14; Nidhal Guessoum, "PhD thesis: The earth is flat," *Gulf News Opinion*, April 10, 2017, accessed March 12, 2019, https://gulfnews.com/opinion/op-eds/phd-thesis-the-earth-is-flat-1.2009202.

[109] Agus Widayoko, Praptika Septi Femilia, Albertus Djoko Lesmono, Heri Sudjatmi, Vicki Dian Prastiwi, and Ninik Munfarikha, "Description of Students' Scientific Literacy Competencies on the Scientific Issue of Flat Earth Theory," *Anatolian Journal of Education*, Vol. 4, No. 2 (Oct., 2019), 31–38.

[110] Nidhal Guessoum, *Islam's Quantum Question: Reconciling Muslim Tradition and Modern Science* (London: I. B. Tauris, 2011 [2009]), 179–218, 271–324.

above. Yusuf and Shekau saw the Flat Earth as a useful tool for organizing not just protest against, but proposing an organizational model different to, what they saw as the encroachment of the Nigerian federal government—with its educational coloniality—on the north. Likewise, John Jasper saw the Flat Earth as a similarly useful organizational tool against the hegemony of the Redemption-era Southern United States, and the Universal Zetetic Society and Boers saw Flat Earth as a way to critique the globe-spanning British Empire. As the decades-long authoritarian regime of Ben Ali disintegrated in a matter of weeks and was replaced by an alliance of Islamists and social democrats, Flat Earth—paradoxically framed as an Islamic interpretation guided by a social democratic politician—must have seemed a reflection of the forces that promised the development of the new, freer, more pluralistic Tunisia.

The education system was central to the strange case of the Tunisian Flat Earthers, just as the opposition to the education system was central to the Nigerian Flat Earthers. In both cases, Islamic interpretation played a role, just as it did in the absence of Flat Earth thought in the ISIS educational system. On these lines, it is worth noting one further element of the ISIS education program: the emphasis on rote memorization of facts over the development of advanced critical thinking, even in science curricula which ignore the scientific method in favor of memorizing details.[111] This is not too different in either methodology or approach to religion as the route taken by the educators of Timbuktu and other Malian cities of the golden age of Islamic West African universities, as described in chapter one. Malian opposition to the introduction of Western-style education in the late nineteenth century triggered similar reactions as Boko Haram's approach in the early twenty-first century.[112] This approach to education could help explain a curious aspect of Islamic terrorist movements: that the engineering profession is greatly overrepresented among their numbers. This is partly due to the training inherent in an engineer, which includes understanding equipment and how to repair it, but not

[111] Arvisais and Guidère, "Education in conflict," 512; Arvisais et al., "The educational intentions of the Islamic State," 6; Potvin et al., "Science Education," 182–83, 193–94.

[112] Elias N. Saad, *Social History of Timbuktu: The Role of Muslim Scholars and Notables 1400–1900* (Cambridge: Cambridge University Press, 1983), 58–93; Kane, *Beyond Timbuktu*, 124; Fafunwa, *History of Education in Nigeria*, 15–72.

the deeper scientific explanations for the process.[113] Skeptic Sharon A. Hill has called this the practice of being "scientifical"—trying to emulate science and the scientific method without knowing how to, through what she terms ARIGs—and that such scientifical notions pervade modern cryptozoology, ghost hunting, UFO watchers, and other paranormal groups, including Flat Earthers.[114]

This trend of scientifical belief and its related association of engineers with terrorism is also exhibited by how, as stated previously, mechanics tend to be overrepresented amongst leading Flat Earthers of the twentieth century, and others were interested in amateur airship design. They know how to fix and develop machines, but not the principles of flight, the misunderstanding of which drew Charles Johnson and others to suspect that standard notions of gravity and the Earth's rotation were false. A parallel can be found in US radical activist Lyndon LaRouche (1922–2019), whose movement attempted to integrate itself into both the pseudoscientific cold fusion research field and more mainstream (but still speculative) Strategic Defense Initiative design in the 1980s.[115] Another related trend is that while doctors tend to support vaccination, nurses tend to be disproportionately anti-vaccine.[116] Modern Flat Earthers tend to be anti-vaccination, a trend only accelerated by the covid pandemic, which resulted in a number of prominent Flat Earthers dying from covid, as well as at least one Flat Earther pharmacist destroying covid vaccine doses.[117]

[113] Diego Gambetta and Steffen Hertog, *Engineers of Jihad: The Curious Connection between Violent Extremism and Education* (Princeton: Princeton University Press, 2016), 11–16.

[114] Sharon A. Hill, *Scientifical Americans: The Culture of Amateur Paranormal Researchers* (Jefferson, NC: McFarland & Company, 2017), 13–26; Sharon A. Hill, "Flat-earthers as scientifical Americans: One message from 'Behind the Curve'," *I Doubt It*, March 21, 2019, accessed March 15, 2022, https://idoubtit.wordpress.com/2019/03/21/flat-earthers-as-scientifical-americans-one-message-from-behind-the-curve/.

[115] Dennis King, *Lyndon LaRouche and the New American Fascism* (New York: Doubleday, 1989), 61–81.

[116] Jagdish Khubchandani, Elizabeth Bustos, Sabrina Chowdhury, Nirbachita Biswas and Teresa Keller, "COVID-19 Vaccine Refusal among Nurses Worldwide: Review of Trends and Predictors," *Vaccines*, Vol. 10, No. 2 (Feb., 2022), 1–13.

[117] Meredith Mattlin, "Examining the Social Construction of Health, Illness, and Wellness in Anti-Science Communities," MA diss., Vanderbilt University, 2018, 20; Derek Beres, Matthew Remski, and Julian Walker, *Conspirituality: How New Age Conspiracy Theories Became a Health Threat* (New York: PublicAffairs, 2023), 87–89; Andrea

However, one group which did not reject covid vaccines was the Taliban, which organized quickly to distribute them. This effort contributed to building the legitimacy which let them take over the national government of Afghanistan in late 2021. The Taliban promotion of covid vaccination was a departure from an earlier anti-vaccination stance, partly due to the CIA's use of false vaccination campaigns as a pretext for espionage.[118] In this, unlike its science education, the Taliban did have some overlap with Boko Haram, as Shekau included invectives against the drug paracetamol, the main ingredient in Tylenol.[119]

Shekau likely drew not only from his Salafist interpretations of Islam, but a wider opposition to Western medicine in Africa. One key aspect that emerged in the former British colonies on the continent, especially Zimbabwe, Kenya, and South Africa, was the belief that AIDS was an American conspiracy for Black genocide. This belief was apparently first promoted in 1983 by the Soviet Union, but by the dawn of the new century, it was being promoted to Africans by white Americans in order to shore up their own anti-vaccination movement.[120] This movement was particularly embodied by Thabo Mbeki (b. 1942) during his 1999–2008 tenure as President of South Africa. Mbeki's health minister Manto Tshabalala-Msimang (1940–2009) drew influence for her AIDS denialism from the 1991 book *Behold a Pale Horse*, a cornerstone conspiracy text by American radio host William Cooper (1943–2001). Cooper's argument that AIDS was created by the CIA to kill Blacks—along with his criticism of the "War on Drugs," which disproportionately impacted Blacks, as an unconstitutional power grab—also won Cooper and his book widespread popularity among such Black rappers as Tupac Shakur (1971–1996) and

Salcedo, "Wisconsin pharmacist who destroyed more than 500 vaccine doses believes Earth is flat, FBI says," *The Washington Post*, February 1, 2021, accessed February 1, 2024, https://www.washingtonpost.com/nation/2021/02/01/wisconsin-pharmacist-vaccine-flatearth/; Weill, *Off the Edge*, 31–32.

[118] Roshni Kapur, "The Afghan Taliban and Covid-19: Leveraging the Crisis or a Change of Heart?," Middle East Institute, April 13, 2021, accessed January 31, 2024, https://www.mei.edu/publications/afghan-taliban-and-covid-19-leveraging-crisis-or-change-heart.

[119] Shekau, "Western Civilization is Atheism and Anti-Islam," 127–28.

[120] Anna Merlan, *Republic of Lies: American Conspiracy Theorists and Their Surprising Rise to Power* (New York: Metropolitan Books, 2019), 48–49.

Public Enemy.[121] On a slightly different tangent, President of the Gambia (1996–2017) Yahya Jammeh (b. 1965) accepted the existence of AIDS, but believed that it could be cured with natural herbs and bananas instead of Western medicine.[122]

The mix of AIDS denialism and suspicion of Western pharmaceutical companies has led various African government officials—specifically in South Africa, Cameroon, Kenya, Uganda, Tanzania, Sierra Leone, and Malawi—to embrace various homeopathic or alternative "medicines" promoted by Westerners. A key example is so-called "Miracle Mineral Solution" (MMS), a term given to chlorine dioxide (bleach) by Jim Humble—a US-based former Scientologist and ancient aliens proponent—in the mid-2000s and promoted by the cult he formed to avoid government food and drug regulations, the Genesis II Church of Health and Healing. Humble initially promoted MMS as an anti-malaria cure aimed at indigenous laborers while he was gold prospecting in Guyana, before he shifted to marketing it as an AIDS cure in Africa. In 2020, MMS further mutated when it was incorporated into anti-covid conspiracies and heavily promoted by QAnon followers in the United States and Australia.[123] So widely promoted was MMS by QAnon and its allies that at the start of the covid pandemic, US President Trump infamously

[121] William Cooper, *Behold a Pale Horse* (Flagstaff: Light Technology Publishing, 1991), 151–58, 379–81; Mark Jacobson, *Pale Horse Rider: William Cooper, the Rise of Conspiracy, and the Fall of Trust in America* (New York: Blue Rider Press, 2018), 175–203; Robert Evans and Jamie Loftus, "Part One: Bill Cooper: The Man Who Killed Truth," *Behind the Bastards*, July 14, 2020 (accessed July 20, 2020), https://podcasts.apple.com/us/podcast/part-one-bill-cooper-the-man-who-killed-truth/id1373812661?i=1000484862608.

[122] Daniel Jolley, Silvia Mari and Karen Douglas, "Consequences of conspiracy theories," in Michael Butter and Peter Knight, eds., *Routledge Handbook of Conspiracy Theories* (London: Routledge, 2020), 236–37; Theodore Powers, "Institutions, power and para-state alliances: a critical reassessment of HIV/AIDS politics in South Africa, 1999–2008," *The Journal of Modern African Studies*, Vol. 51, No. 4 (Dec., 2013), 605–26; Sarah L. Bosha, Michelle Adeniyi, Jenna Ivan, Roya Ghiaseddin, Fabakary Minteh, Lamin F. Barrow, and Rex Kuye, "The Impact of the Presidential Alternative Treatment Program on People Living with HIV and the Gambian HIV Response," *Health and Human Rights Journal*, Vol. 21, No. 1 (Jun., 2019), 239–52.

[123] Matthew Hongoltz-Hetling, *If It Sounds Like a Quack…: A Journey to the Fringes of American Medicine* (New York: PublicAffairs, 2023), 89–97, 156–61, 231–45; Mike Rothschild, *The Storm Is Upon Us: How QAnon Became a Movement, Cult, and Conspiracy Theory of Everything* (Brooklyn: Melville House Publishing, 2021), 127–28; Julian Feeld, Travis View, and Jake Rockatansky, "Episode 77: Drink Bleach To Own The Libs," *QAnon*

suggested on national TV that the virus could be cured by injecting bleach, a sign of MMS belief moving up the political latter.[124]

Outside of AIDS-related conspiracy theories, other African heads of state have been at the center of somewhat less serious claims. In 2018, President of Nigeria (2015–2023) Muhammadu Buhari (1942–2025) was victim of a conspiracy claiming that he had been replaced by a clone grown in Sudan.[125] Mbeki's predecessor as President of South Africa (1994–1999), Nelson Mandela (1918–2013), has an entire conspiracy theory named after him—the Mandela Effect—predicated on the false memory of many that he died in prison in the 1980s.[126] In reality, Mandela was let out of prison in 1990 and was elected South Africa's first Black president in 1994, ending the period of formal apartheid that had been in power since 1948, and the general system of white rule since the country was established in 1910.

The end of white rule in South Africa that Mandela's presidency inaugurated was the goal of John Edward Quinlan and the fear of Paul Kruger. By the time it happened, it had also been predicted by Charles Johnson, president of IFERS. As mentioned above, the same year as Mandela's election, Johnson promoted the innocence of O. J. Simpson, aligning him firmly with US Blacks and against majority-white opinion. However, his views on South Africa did not reflect that seeming racial progressiveness. In a March 1988 article titled "The fall of White South Africa" (reproduced verbatim), Johnson stated that

Anonymous, February 3, 2020, accessed February 4, 2020, https://soundcloud.com/qanonanonymous/episode-77-drink-bleach-to-own-the-libs.

[124] Tom Porter, "A fringe movement that claims bleach is a miracle cure is jubilant after Trump floated injecting disinfectant as a treatment for COVID-19," *Business Insider*, April 24, 2020, accessed February 23, 2024, https://www.businessinsider.com/mms-fans-celebrate-trump-claim-disinfectant-might-cure-coronavirus-2020-4.

[125] Peter Doerrie, "Insiders Insight: The serious side of the Buhari clone story," *African Arguments*, December 4, 2018, accessed February 7, 2024, https://africanarguments.org/2018/12/serious-side-buhari-clone-story/.

[126] William Shoki, "The 'Mandela Effect' describes the false memories many of us share. But why can't scientists explain Political Struggle Is the Answer—Not Conspiracy Theories," *Jacobin*, January 20, 2020, accessed February 3, 2020, https://jacobin.com/2020/01/conspiracy-theories-nelson-mandela-anc-south-africa; Jacopo Prisco, "The 'Mandela Effect' describes the false memories many of us share. But why can't scientists explain it?," CNN, September 18, 2023, accessed February 7, 2024, https://www.cnn.com/2023/09/18/world/mandela-effect-collective-false-memory-scn/index.html.

> Just a few years back we had lots of contact with South Africa, including a couple Universities, several members. In fact FLAT EARTH was strong there 90 and 100 years ago, faded out, till, might say... **the time of the end**. I remember in an issue of FLAT EARTH NEWS I said, "unless South Africa drops Greek ballism, the insane delusion of lunatics, the people from the jungles will have to come in and take it all over...as the so called Whites are unworthy to be leaders of ANYTHING, including the uncivilized black Africans." They didn't do it, neither the right nor the left...both reprobate. THE BIBLE says "don't just believe, but do listen to any who tells you something and/or warns you, and consider." Then, it says, **"IF THE THING COMES TO PASS... YOU WILL KNOW A PROPHET HAS BEEN AMONG YOU." What I said has come to pass**. South Africa is of course a white country, created by, run and lived in by white Africans. As the Bible and common sense says, "much was given them... and much expected" they failed DUE TO BELIEVING EARTH A PLANET. OH YEAH, THIS WAS THE DOWNFALL OF SF [*sic*], from that first step into INSANITY and it is INSANE for anyone, ESPECIALLY for a South African to believe earth a globe, for he CLAIMS HE IS UNDER THE WORLD, HANGING BY FEET HEAD DOWN! Is that NUTS or what? So God and common sense has driven the people in from the jungles, to come and TAKE OVER THE WHITES, COULD SAY WITH NETS, AS THEY THE BLACKS KNOW THE WHITES ARE CRAZY AND **MUST** be taken out of power!
>
> As you know by the newspapers, and I have before me a booklet by the South African Ambassador, Piet Koornhof, describing how SF is now given over to (the Communist-Socialist) Mandella [*sic*] and the millions of jungle people who have invaded SF, now a true democratic peoples [*sic*] republic, which translates, given over to the invaders from the jungles, so it's all over for the so called Whites. All their labor was in vain, creating the country, as they **rejected the FLAT EARTH FACTS and all that goes with that. Australia may be next...**will not, can not be allowed for a people to be such **MORONS** to claim to live "under the world" and still rule the NATIVES... the natives are superior to any Grease baller, reprobate white. So: my prophecy came to pass... in SA, what excuse do you have to disbelieve me?[127]

Johnson clearly states his view that South Africa was a country made by and for whites, at a time when many white South Africans believed in

[127] Charles K. Johnson, "The fall of White South Africa," *Flat Earth News*, No. 75 (Mar., 1988), 1.

Flat Earth—a memory, even if perhaps inaccurate in terms of the balance between Boer and Anglo settlers, of the time of Kruger. But the whites gradually abandoned their belief in the flatness of the Earth, in contrast to the "uncivilized black Africans" who understood the basic truth of the Earth's flatness.

In Johnson's view, the white acceptance of the Earth's sphericity was such an insane notion that it was this which galvanized Mandela's followers to "invade" from the "jungles" and take over South Africa, as the idea of being ruled over by white globularists was seemingly a step too far compared to the decades of apartheid. It should be noted that on writing this, Mandela's release from prison, let alone the end of apartheid and his election, were still years in the future. The fixation on anti-communism, and the left in general, indicates a coming rightward trend in future Flat Earthers, though it should be pointed out that Johnson's assertion that there were both left- and right-wing parties in apartheid South Africa is false. Under the apartheid era, South Africa was a one-party state under the right-wing National Party.

Johnson's reference to the occasional contact with South African Flat Earthers is borne out by the surviving correspondence of Samuel Shenton, Johnson's British predecessor as IFERS president. On 1 December 1968, Shenton wrote to a Mr. Hurley of Boksburg in the Transvaal province, responding to an inquiry by Hurley as to whether Shenton would ever lecture in South Africa. Shenton recommended the 1899 *Zetetic Cosmogony* pamphlet by Thomas Winship, as discussed in chapter five.[128] On 15 August 1970, Shenton wrote a reply to a 21 July letter from Graham H. Watt, Esq., of Cape Town, who had proposed an expedition to search for "water levels, strata changes, etc." in the Boland region of the Western Cape, apparently to find evidence of the Noachic flood. Shenton gave Watt his endorsement, membership in IFERS, and the title of "Honorary Patron."[129] Johnson's point of South Africans gradually abandoning Flat Earth is also supported by South African

[128] Samuel Shenton, "Outgoing Correspondence 5846," The Flat Earth Society Library, March 12, 2014, accessed November 30, 2019, https://www.theflatearthsociety.org/library/correspondence/Samuel%20Shenton%20-%20Outgoing%20Correspondence%205846.pdf.

[129] Samuel Shenton, "Outgoing Correspondence 4240," The Flat Earth Society Library, March 12, 2014, accessed November 30, 2019, https://www.theflatearthsociety.org/library/correspondence/Samuel%20Shenton%20-%20Outgoing%20Corresponcence%204240.pdf.

science educator Sibusiso Biyela, who noted that by 2020, what Flat Earth advocacy there was in South Africa came from American evangelical influence.[130]

In February 1990, Mandela and other political prisoners were freed, opposition parties were legalized, and several acts of apartheid legislation were repealed. In the September IFERS newsletter, Johnson would reprint what seemed to be a press release from the South African embassy (perhaps also from Ambassador Koornhof), followed by his own editorializing:

> Sure Great News! These native sons of AFRICA, the whites, have really been through a siege, demonstrated a wonderful tolerance and understanding, what wonderful "self control" these Native Africans (whites) have. We should always remember the Whites are NATIVE AFRICANS, just as anyone who is born there is! The long war conducted against the White people of South Africa, has been hard on all the Native Sons, both White and Black, and in between. Lets [*sic*] all WELCOME and BE FRIENDS with SOUTH AFRICA. Bible mentions Queen of Sheba came from South Africa, "the end of the earth." May God be with you.[131]

There is something of a whiplash reading the celebratory tone of the end of apartheid compared to Johnson's reporting on it from just two and a half years earlier. Nevertheless, there are some clear political constants here, not least of which is the assertion that the white South Africans are true "Native Africans," an inversion of Black Hebrew Israelites like Frank Cherry. Johnson's reference to the Biblical figure of the Queen of Sheba as being from South Africa, in contrast to the more typical placing of her in the Horn of Africa, reflects some of the nineteenth-century musings carried out by Kruger himself, in trying to connect Sheba and Solomon to the nearby ruins of Great Zimbabwe and the Lemba tribe.[132]

IFERS' membership shrunk dramatically in the 1990s, and Johnson's output nearly vanished after his wife's 1996 death. Much was additionally

[130] Zoom conversation with the author, 28 May, 2020.

[131] Charles K. Johnson, "Anti-Apartheid Act lifted in South Africa," *Flat Earth News*, No. 79 (Sep., 1990), 1.

[132] Edward Guimont, "From King Solomon to Ian Smith: Rhodesian Alternate Histories of Zimbabwe" (PhD diss., University of Connecticut, 2019), 19–20, 54–56. Ironically, the Boksburg address of the Mr. Hurley who corresponded with Shenton was 228 Solomon Road.

lost in a housefire prior to his own 2001 death. However, South Africa is mentioned briefly in one further article. Johnson lived in southern California near Edwards Air Force Base, approximately one hundred and sixty kilometers outside of Los Angeles. In spring 1992, Los Angeles was the center of large race riots springing from Black outrage over the police beating, and then acquittal for beating, a Black man. In the aftermath of the riots, Los Angeles Police Chief Daryl Gates (1926–2010) retired after fourteen years in office, as he was blamed for allowing a culture of racism to flourish in his department. The police brutality leading to the riots was widely credited with helping lead to Simpson's acquittal two years later.

In response to the riots, Johnson wrote an article tellingly titled "Police Chief Gates saved the United States," where he claimed that the riots were

> run and sponsored (same as South Africa) by African Methodist-Episcopal Church, located in downtown L.A. All day in this church, demigods, subversive anti-caucasian, anti-American, anti-decent preached and slobbered their black racist mad dog ravings.... This was the Nazi German Lenin, and Nazi Marx always planned the beginning of a revolution, from then on that's the cry MASSACRE (check S[outh] A[frica] right now).... **POLICE CHIEF GATES AND HIS THIN BLUE LINE OF KNIGHTS BEAT 'EM AT THEIR OWN GAME!**[133]

Gone was the tolerant view of South Africa's political evolution, returned were the anti-communist invectives. Not only was the end of apartheid the signal for a group of communist Blacks to massacre whites, but the anti-apartheid movement and the Black riots in Los Angeles were connected by the same nefarious forces. The 1992 riots were the sign that South Africans' liberation movement had spread to Blacks in the United States.

Not only does this show the significance of Flat Earth in convincing Johnson to support Simpson's innocence—given his attitude to the riots, it is highly unlikely Johnson would have done so otherwise—it also shows the development of a new international conspiratorial fear, where Black success in one part of the world was a sign of their eventual plans to oppress whites in the United States. The political turn to the conspiratorial right by Johnson in the 1990s was a harbinger of conspiracy culture

[133] Charles K. Johnson, "Police Chief Gates saved the United States," *Flat Earth News*, No. 83 (Sep., 1992), 2.

to come in the twenty-first century, including but not exclusively for Flat Earthers. John Jasper, the Universal Zetetic Society, John Abizaid and the related Flat Earthers of the late nineteenth and early twentieth centuries had what could be called progressive goals. Even Kruger and his Boers, while architects of white supremacy, had a political goal against the advance of the British Empire and its institution of industrial capitalism. The Tunisians might be seen as liberals of a sort as well. If their Salafist compatriot Flat Earthers were certainly not, then like Kruger, they did retain an element of anti-imperialism, and certainly anti-coloniality, in their Flat Earth worldview.

By the time of Johnson's death in 2001, the Flat Earth movement was seen as dead, only to be discussed as a term of mocking reference, or occasionally by academics. Though sparse, there was still occasional research done in this time. In 1997, Ruth Freitag (1924–2020), a librarian at the US Library of Congress specializing in astronomical topics, wrote a comprehensive online bibliography of the Hollow Earth idea.[134] The next year, she wrote one for Flat Earth topics, which was expanded by science librarian Sean Bryant in 2018.[135] The reason Bryant expanded Feitag's Flat Earth and not Hollow Earth bibliography after two decades is that in 2015, Flat Earth enjoyed a massive revival courtesy of social media, originally and most prominently on YouTube, where by 2018 it was more popular than the conspiracy mainstay of chemtrails.[136] It is perhaps relevant that of the generational cohorts, Millennials and Gen Z, who grew up with social media, are the most likely to believe not only Flat Earth, but also anti-vaccine, Apollo hoax, and microchip conspiracy

[134] Ruth Freitag, "Hollow-Earth Theories: A List of References," Library of Congress, May, 1997, accessed March 21, 2024, https://www.loc.gov/rr//scitech/SciRefGuides/hollowearth.html.

[135] Sean Bryant, "The Flat Earth and its Advocates: A List of References," Library of Congress, July 28, 2018, accessed March 21, 2024, https://guides.loc.gov/flat-earth.

[136] Mick West, *Escaping the Rabbit Hole: How to Debunk Conspiracy Theories Using Facts, Logic, and Respect* (New York: Skyhorse, 2018), 198–201; Max Fisher, *The Chaos Machine: The Inside Story of How Social Media Rewired Our Minds and Our World* (New York: Little, Brown, and Company, 2022), 220–26; P. W. Singer and Emerson T. Brooking, *LikeWar: The Weaponization of Social Media* (Boston: Houghton Mifflin Harcourt, 2018), 123; Alex Olshansky, Robert M. Peaslee, and Asheley R. Landrum, "Flat-Smacked! Converting to Flat Eartherism," *Journal of Media and Religion*, Vol. 19, No. 2 (Jul., 2020), 46.

theories.[137] This is extensively covered in Kelly Weill's *Off the Edge*, and the fact that social media led to a wave of Flat Earth movement in the 2010s is not of interest, per se, to this work. It is perhaps worth simply noting that if there is one person behind modern social media culture, it is the reality star Tila Tequila (b. 1981, as Nguyễn Thị Thiên Thanh), who led millions to join MySpace to follow her account in 2003.[138] By early 2016, Tequila had become one of the many celebrities endorsing Flat Earth on the social media sphere she helped create, along with various other extremist views, including support for Hitler and white nativist politics (despite the fact that she is of Vietnamese ancestry and an immigrant from Singapore).[139] Tequila's Zetetic development is perhaps a suitable metaphor for the endpoint of social media influencer culture in an age of scientific skepticism and political radicalization.

However, what is notable about the current Western revival is that, in contrast to the past movements—even the Boko Haram and Tunisian movements with which it overlaps but developed independently—there is little anti-imperialism, anti-coloniality, or anti-racism to be seen. Tila Tequila again serves as an example; while she had become an icon for the LGBT community for her 2007–2008 reality show *A Shot at Love*, by the time of her Flat Earth conversion she was siding with the conservative critics who had denounced her show as being immoral and anti-Christian. Another case is Darryle Marble, a Black man from Arkansas who enlisted in the US Army after the 11 September 2001 terrorist attacks out of a desire to fight al-Qaeda, but was instead sent to Iraq in 2003. There, Marble was disillusioned by what he viewed as an illegal war launched by the US government for illegitimate reasons. Returning home, Marble's critique of what he saw as imperialistic US foreign policy drove him first to 9/11 Trutherism, and then Flat Earth, justifying that if the government could lie to manufacture a war, it could lie about other things as well. But in contrast to fellow Black Southerner John Jasper, or the Civil War veterans who turned to Flat Earth, Marble's initially

[137] Lawrence C. Hamilton, "Conspiracy vs. Science: A Survey of U.S. Public Beliefs," *Casey School of Public Policy National Issue Brief*, No. 162 (Spring, 2022), 5.

[138] Taylor Lorenz, *Extremely Online: The Untold Story of Fame, Influence, and Power on the Internet* (New York: Simon & Schuster, 2023), 31.

[139] Weill, *Off the Edge*, 195; West, *Escaping the Rabbit Hole*, 199–200; Donald R. Prothero, *Weird Earth: Debunking Strange Ideas about Our Planet* (Bloomington: Red Lightning Books, 2020), 22.

anti-imperialistic critique took him away from progressive views, soon embracing conspiracy beliefs such as Jewish control of global banking, or that the 2012 mass shooting at an elementary school in Sandy Hook, Connecticut was a false flag engineered by Barack Obama—the first Black US President—to enact gun control measures (in reality, all efforts to pass federal gun control legislation in response to Sandy Hook were defeated in the US Senate).[140]

In contrast to the nineteenth-century Flat Earthers of both the US and UK, therefore, the modern Flat Earth is a movement firmly of the right wing. The majority of its followers are supporters of authoritarianism and militarism, a strain present from the revival's inception and growing as it expanded and became entangled with later conspiracies, like QAnon and covid anti-vaccination. But even with this change, it is worth ending this study with an examination of why Flat Earth no longer promotes the liberation of the oppressed by the hegemony, and of its political trajectory in two case studies: the Flat Earth support of Donald Trump in the United States, and of Jair Bolsonaro in Brazil. In both cases, Flat Earth political aspirations ultimately failed, but not before effecting some gasps of political heterodoxy.

[140] Alan Burdick, "Looking for Life on a Flat Earth," *The New Yorker*, May 30, 2018, accessed July 23, 2024, https://www.newyorker.com/science/elements/looking-for-life-on-a-flat-earth; Ben Kissel, Marcus Parks, and Henry Zebrowski, "Episode 334: Flat Earth," *The Last Podcast on the Left*, September 29, 2018, accessed July 23, 2024, https://www.audible.com/podcast/Episode-334-Flat-Earth/B09RKFCLTB.

CHAPTER 7

Fear of a Flat Planet

In 1950, a psychoanalyst and self-educated amateur historian, Immanuel Velikovsky, published his first and most famous book, *Worlds in Collision*. The thesis was that Venus is not a planet at all, but actually a comet which was ejected from Jupiter several thousand years ago, causing natural disasters on Earth which became the basis for various mythological events around the globe.[1] Velikovsky later expanded this in a number of sequels, but *Worlds in Collision* in particular was a runaway sensation. Its wildly revisionist takes on accepted interpretations of history, mythology, astronomy, and geology led to condemnations by mainstream academics. This in turn led to Velikovsky and his supporters alleging that the academic "establishment" was trying to not only suppress his research, but unfairly slander the man as well, while claiming that his huge popularity with the public highlighted a more populist (and therefore true) element to his work in contrast to out-of-touch academia. The pattern set by the so-called "Velikovsky Affair"—appeal to the public, dismissal by academics whose critiques only set them apart from the public even

[1] Immanuel Velikovsky, *Worlds in Collision* (New York: Macmillan Company, 1950), 3–35.

E. Guimont, *The Power of the Flat Earth Idea*,
Palgrave Studies in the History of Science and Technology,
https://doi.org/10.1007/978-3-032-23981-5_7

more—set the pattern for subsequent fringe writers, such as Erich von Däniken and Graham Hancock.[2]

The Velikovsky Affair and the popularity of his revisionist takes on accepted history would also lead to advocates of other fringe views to attempt to latch onto his theories. One example is Young-Earth Creationism. While the immediate backlash to Velikovsky in the early 1960s led to publishers refusing to print YEC school textbooks, by the 1970s Velikovsky's work was being taken as evidence by advocates of so-called flood geology, and Velikovsky's influence contributed to flood geology and YEC becoming mainstreamed within the US New Christian Right of the 1980s.[3] In the 1990s, Russian mathematician Anatolii Fomenko (b. 1945) would popularize the so-called "New Chronology" conspiracy theory, wherein much of what we know of as the medieval period was actually invented by Western Jesuits as a way of erasing the contributions of the Orthodox Church and Slavic cultures. Fomenko's New Chronology became very popular in Russia after the dissolution of the Soviet Union as a way of restoring a sense of lost historical greatness; one of his main inspirations was the revisionist history of the Russian-born Velikovsky.[4] British academic Martin Bernal (1937–2013), author of the controversial 1987 *Black Athena*, likewise endorsed Velikovsky's claim that the Oedipus myth was a Greek adaptation of the life of the Egyptian pharaoh Akhenaton.[5]

It is perhaps not a surprise to find that Flat Earthers also attempted to latch onto the Velikovsky Affair. As mentioned in chapter three, in 1954,

[2] For details on the Velikovsky Affair, see Donald W. Goldsmith, ed., *Scientists Confront Velikovsky* (Ithaca: Cornell University Press, 1977); Michael D. Gordin, *The Pseudoscience Wars: Immanuel Velikovsky and the Birth of the Modern Fringe* (Chicago: University of Chicago Press, 2012); Patrick Moore, *Can You Speak Venusian? A Guide to the Independent Thinkers* (New York: W. W. Norton & Company, 1972), 57–65; Sharon A. Hill, *Scientifical Americans: The Culture of Amateur Paranormal Researchers* (Jefferson, NC: McFarland & Company, 2017), 126–28.

[3] Ronald L. Numbers, *The Creationists: The Evolution of Scientific Creationism* (New York: Alfred A. Knopf, 1992), 240, 337–38; Gordin, *The Pseudoscience Wars*, 137–48.

[4] Konstantin Sheiko, "Lomonosov's Bastards: Anatolii Fomenko, Pseudo-History, and Russia's Search for a Post-Communist Identity" (PhD diss., University of Wollongong, 2004), 51; Anton Weiss-Wendt, *Putin's Russia and the Falsification of History: Reasserting Control over the Past* (London: Bloomsbury Academic, 2020).

[5] Martin Bernal, *Black Athena Writes Back: Martin Bernal Responds to His Critics* (Durham, NC: Duke University Press, 2001), 335–36.

in the early days of both the civil rights struggle and the popularity of Velikovsky, Isaac James, reverend of Sixth Mount Zion Church, wrote a laudatory biography of his predecessor John Jasper. Included at both the beginning and end of this brief overview was a description of Velikovsky's then-recent book, which has an opening chapter attempting to explain the stillness of the Sun in the Book of Joshua as actually being the perception of Velikovsky's rogue Venus by the Israelites.[6] James asks, "Will Dr. Velikovsky's findings give new stature to the views of John Jasper and perhaps lead the way to new data which will substantiate his theory, or will future students continue to dismiss him as just another unlettered preacher expounding a fundamentalist doctrine?… [Velikovsky's] theories, however, are so far advanced over our usual thinking as to lend hope that some day Jasper's theory that 'the sun moved' may be proved true."[7]

James is wrong in one point: Jasper's claims about the Sun's movements in Joshua were not a theory, but a statement of faith. It also should be noted that unlike YEC or New Chronology, and despite James' efforts, Flat Earth did not seemingly benefit from the continued interest and debate of the Velikovsky Affair. However, James' linking of Jasper to the more popular Velikovsky Affair would prove prescient for developments in the twenty-first century. The revived Flat Earth social media movement of the 2010s would, at the end of the decade, develop a symbiotic relationship with a truly new conspiracy theory byproduct of the late 2010s, one which has come to symbolize the subsequent decade more than any other.

This conspiratorial shift represented by both the re-emergence of Flat Earth and the conspiracies influenced by it in the 2010s finds its origin in a paranormal development a century earlier. In 1910, the Portuguese monarchy was overthrown and a republic established, one which was intensely anticlerical due to the Catholic Church's support for the monarchy and opposition to democracy. In March 1916 the fledgling republic, still politically and culturally divided, entered World War I, leading to periods of dictatorship in the country as a means of preserving unity. On 13 October 1917, tens of thousands of people assembled in the town of Fátima where, since spring of the previous year—when Portugal

[6] Velikovsky, *Worlds in Collision*, 40–46.

[7] Isaac James, *"The Sun Do Move: " The Story of the Life of John Jasper* (Richmond: Hunter's Letter Shop, 1954), 5–6, 41.

entered the war—three children had claimed to see visions of angels and the Virgin Mary speaking of the need for prayer and consecration of Russia to end the war, and promising a miracle on that day to prove the statements. Members of the assembled crowd were widely reported as seeing the Sun either dancing, changing color, dimming, or swooping toward the ground. The reported activities of the Sun in what became known as the Miracle of the Sun were compared to the events of the Book of Joshua.[8]

However, unlike John Jasper's interpretation of the Book of Joshua as a parable of liberation, the Miracle of the Sun emboldened the National Catholic, anti-democratic impulses in the Portuguese Republic, helping contribute to the 1926 military coup and subsequent fascist rule until 1974.[9] Abroad, the Virgin's message to Russia, and the Miracle occurring less than a month before the Bolshevik Revolution, led to the Miracle of the Sun becoming a core aspect of Catholic anti-communism during the Cold War, particularly among American Catholics.[10] What in the 1870s could be interpreted as a sign for progressive liberation had by the 1910s become interpreted as a call for conservative reaction. Due in part to the Cold War and the anti-communism of the West, conspiracy theories and paranormal activities would largely become conservative in nature. Much like NATO and the other political infrastructure of the Cold War, that conservative alignment would not vanish with the Soviet Union in 1991, but undergo reconstruction in the twenty-first century, finding new utility in an age of perpetual War on Terror, financial crises, racial strife, and by the 2010s, domestic political upsets and pandemics.[11]

On 8 November 2016, Donald Trump of the Republican Party was elected President of the United States, an upset given he was widely expected to lose to his Democratic Party rival, the would-be first female President, Hillary Clinton (b. 1947). Against expectations, Republicans

[8] Ian Richard Netton, *Islam, Christianity and the Realms of the Miraculous: A Comparative Exploration* (Edinburgh: Edinburgh University Press, 2020), 158–66.

[9] Netton, Islam, *Christianity and the Realms of the Miraculous*, 166–71; William Thomas Walsh, *Our Lady of Fátima* (New York: The Macmillan Company, 1951), 207–14.

[10] Walsh, *Our Lady of Fátima*, 223–27; Una M. Cadegan, "The Queen of Peace in the Shadow of War: Fatima and U.S. Catholic Anticommunism," *U. S. Catholic Historian*, Vol. 22, No. 4 (Fall, 2004), 1–15.

[11] Roland Imhoff et al., "Conspiracy mentality and political orientation across 26 countries," *Nature Human Behaviour*, Vol. 6 (2022), 392–403.

also held their majority control of both houses of the US Congress, and were able to quickly pack all levels of the US court system with lifelong appointments of conservative judges. This enormous and unexpected political disappointment fueled conspiracy theories among US liberals, with a majority of Democrats convinced—despite no evidence ever emerging to support this—that Russian president Vladimir Putin (b. 1952) directly intervened by altering the vote count from afar to elect Trump and other Republicans, possibly with Trump's collusion.[12] This is perhaps not surprising, given that conspiracy theories serve as a psychological reaction to perceived power and both shape and are shaped by political behavior.[13]

It is also not too surprising, considering that in the first years of the twenty-first century, Americans liberals tended to be more expressive of conspiracy theories and alternative views of science and health, in particular around vaccines and New Age beliefs.[14] An avatar of this was the eponymous 1986–2011 talk show of Oprah Winfrey (b. 1954), the world's first Black female billionaire, which promoted such topics as anti-vaccine fears, Satanic Panic, alien abductions, and psychic phenomena to its largely female, multiracial, middle-class, liberal audience. Among individuals highlighted by Winfrey were alternative medicine advocate Suzanne Somers (1946–2023); anti-vaccination promoter Jenny McCarthy (b. 1972); João Teixeira de Faria (b. 1942), a Brazilian self-proclaimed psychic calling himself John of God who was convicted of multiple counts of rape in 2019; and most prominently surgeon Mehmet

[12] Jacob W. Justice and Brett J. Bricker, "Hacked: Defining the 2016 Presidential Election in the Liberal Media," *Rhetoric & Public Affairs*, Vol. 22, No. 3 (Fall, 2019), 389–420; Anna Merlan, *Republic of Lies: American Conspiracy Theorists and Their Surprising Rise to Power* (New York: Metropolitan Books, 2019), 219–40.

[13] Roland Imhoff and Pia Lamberty, "Conspiracy Theories as Psycho-Political Reactions to Perceived Power," in Michael Butter and Peter Knight, eds., *Routledge Handbook of Conspiracy Theories* (London: Routledge, 2020), 192–205; Hulda Thórisdóttir, Silvia Mari and André Krouwel, "Conspiracy Theories, Political Ideology and Political Behaviour," in Michael Butter and Peter Knight, eds., *Routledge Handbook of Conspiracy Theories* (London: Routledge, 2020), 304–16.

[14] Susan Jacoby, *The Age of American Unreason* (New York: Pantheon Books, 2008), 210–41; Dan Ariely, *Misbelief: What Makes Rational People Believe Irrational Things* (New York: HarperCollins, 2023), 28–30, 138–29; Paul R. Gross and Norman Levitt, *Higher Superstition: The Academic Left and Its Quarrels with Science* (Baltimore: Johns Hopkins University Press, 1994); Alex Berezow and Hank Campbell, *Science Left Behind: Feel-Good Fallacies and the Rise of the Anti-Scientific Left* (New York: PublicAffairs, 2012).

Oz (b. 1960) who went on to become a covid conspiracy advocate, the Republican candidate for the 2022 US Senate election in Pennsylvania, and Administrator of the Centers for Medicare and Medicaid Services in the second Trump administration.[15]

This partisan conspiracy orientation began to change with the 2015 launch of Trump's presidential campaign. Trump had begun his foray into presidential politics in 2012 by stoking the "Birther" conspiracy that Obama (born 1961 in Hawaii) had actually been born in his father's home country of Kenya—again, a conspiracy that had originated with the 2008 presidential campaign of Obama's liberal rival, Hillary Clinton.[16] Perhaps presciently, Obama the next year referred to political opponents as "a meeting of the flat-Earth society."[17] This was not entirely new; as early as the 1996 presidential election, critics of Republicans' plans for a flat tax—already referred to by the paranormal-tinged term "voodoo economics"—compared the belief in its efficacy to belief in a Flat Earth.[18] As Trump's 2015–2016 campaign continued, one of the core groups of support which agglomerated around him was the "alt-right" movement; one of its founders, Iranian-American philosopher Jason Jorjani (b. 1981), was obsessed with such pseudohistorical ideas as the Abrahamic god actually being an alien who used advanced weapons to defend the Israelites, and Atlantis being the origin of the white race.[19] Throughout his campaign, Trump promoted the belief that vaccination causes autism. Anti-vaccination sentiment, along with belief in Flat Earth, UFOs, Bigfoot, and psychic powers, all not only

[15] Janice Peck, *The Age of Oprah: Cultural Icon for the Neoliberal Era* (Boulder, CO: Paradigm Publishers, 2008), 103–34.

[16] Tom Schaller and Paul Waldman, *White Rural Rage: The Threat to American Democracy* (New York: Random House, 2024), 176–77.

[17] Merlan, *Republic of Lies*, 5, 25; Weill, *Off the Edge*, 196; Justin Sink, "Obama mocks skeptics of climate change as 'flat-Earth society'," *The Hill*, June 25, 2013, accessed February 27, 2024, https://thehill.com/blogs/blog-briefing-room/news/154505-obama-mocks-skeptics-of-climate-change-as-flat-earth-society/.

[18] "Flat tax, flat Earth," *The Globe and Mail*, (Feb. 2, 1996), A12.

[19] Harrison Fluss and Landon Frim, "Aliens, Antisemitism, and Academia," *Jacobin*, March 11, 2017, accessed October 27, 2024, https://jacobin.com/2017/03/jason-reza-jorjani-stony-brook-alt-right-arktos-continental-philosophy-modernity-enlightenment/.

surged following Trump's presidential campaign debut, but also underwent partisan realignment.[20] A May 2020 CivicScience poll indicated that not only were Americans who watch conservative Fox News far more likely to believe in Bigfoot and the Loch Ness Monster than viewers of the liberal CNN and MSNBC, but Americans who believe in those cryptids were also less likely to take the threat of covid seriously.[21] 2016 and 2020 Democratic presidential candidate Marianne Williamson (b. 1952) is one of the few modern progressive conspiracists who (at least as of this writing) has not yet followed the path of rightward radicalization common, especially since the covid pandemic, to such figures as Graham Hancock, Robert F. Kennedy Jr. (b. 1954), Naomi Wolf (b. 1962), and the New Age Instagram mommies of the Pastel Q phenomenon.[22]

But once Trump was inaugurated on 20 January 2017, with Republican majorities in the Supreme Court and both houses of Congress, his administration immediately began to flounder, beset by ineptitude and intra-party conflicts. Trump's supporters needed an explanation as to why their president was not enacting the wide-ranging conservative platform he had campaigned on, despite him and his allies holding power in all branches of government. To some, this explanation would come in the form of a new conspiracy movement birthed in the message board website 4chan. 4chan had been the origin point of "Gamergate," a 2014 movement which energized video game players to become active in online conservative political campaigns, a development which led to widespread support for Trump among the formerly apolitical users of 4chan and similar websites.[23] It was therefore not entirely unexpected when, on

[20] Merlan, *Republic of Lies*, 1–34; 1–34; P. W. Singer and Emerson T. Brooking, *LikeWar: The Weaponization of Social Media* (Boston: Houghton Mifflin Harcourt, 2018), 125, 175; John O'Connor, *The Secret History of Bigfoot: Field Notes on a North American Monster* (Naperville, IL: Sourcebooks, 2024), 17–18, 208–09.

[21] Noah Brode, "Bigfoot is Real (For 11% of U.S. Adults)," CivicScience, May 11, 2020, accessed August 4, 2024, https://civicscience.com/bigfoot-is-real-for-11-of-u-s-adults/.

[22] Derek Beres, Matthew Remski, and Julian Walker, *Conspirituality: How New Age Conspiracy Theories Became a Health Threat* (New York: PublicAffairs, 2023), 49–50; Will Sommer, *Trust the Plan: The Rise of QAnon and the Conspiracy That Unhinged America* (New York: HarperCollins, 2023), 153. See also Charles U. Zug and Steven F. Pittz, "Good Conspiracism," in Luke Ritter, ed., *American Conspiracism: An Interdisciplinary Exploration* (London: Routledge, 2024), 159–74.

[23] Angela Nagle, *Kill All Normies: Online Culture Wars from 4chan and Tumblr to Trump and the Alt-Right* (Winchester: Zero Books, 2017); Mike Wendling, *Alt-Right:*

28 October 2017, an individual calling themselves Q posted on 4chan to announce the imminent arrest of Hillary Clinton. From this humble beginning came QAnon, an all-encompassing conspiracy theory whose central premise is a secret war led by Donald Trump and elements of the US military fighting to overthrow a Satan-worshipping "deep state cabal" consisting of most Democratic politicians and celebrities, who engage in pedophilia and consume adrenochrome harvested from children's brains to remain young.[24]

A detailed analysis of QAnon is beyond the scope of this book, although several have been published.[25] Several more are sure to come now that, as of this writing in early 2025, a number of prominent QAnon promoters are now part of the second Trump administration, and others in his Cabinet routinely promote ideas that seemingly originate on QAnon-oriented social medial accounts.[26] Suffice to say that QAnon followers became the most fervent supporters of Trump and participants in the 6 January 2021 effort to overthrow his presidential loss. The identity of Q remains unknown, but most researchers believe there have been several Qs since 2017, and that the most likely candidate for the original Q is a white South African conspiracy theorist named Paul Furber. Furber was a 4chan moderator under the name BaruchtheScribe, a reference to the Old Testament figure Baruch ben Neriah who died in exile in Egypt after the Babylonian conquest of Jerusalem in the sixth century BCE.[27]

From 4Chan to the White House (London: Pluto Press, 2018); Dale Beran, *It Came from Something Awful: How a Toxic Troll Army Accidentally Memed Donald Trump into Office* (New York: All Points Books, 2019).

[24] Schaller and Waldman, *White Rural Rage*, 171–76.

[25] Mike Rothschild, *The Storm Is Upon Us: How QAnon Became a Movement, Cult, and Conspiracy Theory of Everything* (Brooklyn: Melville House Publishing, 2021); Jesselyn Cook, *The Quiet Damage: QAnon and the Destruction of the American Family* (New York: Crown, 2024); Keith Raymond Harris, "The Epistemology of QAnon," in Luke Ritter, ed., *American Conspiracism: An Interdisciplinary Exploration* (London: Routledge, 2024), 19–33; Sommer, *Trust the Plan*.

[26] Anna Merlan, "The Brain Rot Cabinet," *Mother Jones*, May 8, 2025, accessed May 9, 2025, https://www.motherjones.com/politics/2025/05/trump-cabinet-conspiracies-missing-children-epstein/.

[27] Brandy Zadrozny and Ben Collins, "How three conspiracy theorists took 'Q' and sparked Qanon," NBC News, August 14, 2018, accessed April 29, 2019, https://www.nbcnews.com/tech/tech-news/how-three-conspiracy-theorists-took-q-sparked-qanon-n900531; Rothschild, *The Storm Is Upon Us*, 32–39; Sommer, *Trust the Plan*, 18–19, 27, 64–68.

As indicated by the fears of Charles Johnson last chapter, the end of apartheid in South Africa became a motivation for racialized fears among the American right in the twenty-first century, with Furber only one example. Another early QAnon promoter, "Joe M.," was also a white South African who created one of the most prominent QAnon recruitment videos over the course of 2018–2019, called *The Plan to Save the World*.[28] Lara Logan (b. 1971), a white South African journalist who originally began her career in the United States at respectable outlets such as CBS and Fox News, increasingly drifted to the far-right, endorsing QAnon views after the start of the covid pandemic, leading to her firing from Fox in 2022.[29] Endorsed by but existing beyond QAnon, the "Great Replacement"—a conspiracy that Jews are sponsoring non-white migration to the United States to take over the country and eliminate whites, endorsed by mainstream conservative figures as Trump, host Tucker Carlson (b. 1969), and South African-born capitalist Elon Musk (b. 1971)—has its origins in conservative white South African panic, mostly of the Boer population, over alleged land seizures by the Black-dominated government, which have now been integrated into American conservative fears over white 'replacement.' In January 2025, white South African tech billionaire David O. Sacks (b. 1972), who has promoted both cryptocurrency and anti-diversity so-called "race science," became Chair of the President's Council of Advisors on Science and Technology in the second Trump administration. As of writing, this escalation has reached a point where in March 2025 the second Trump administration expelled South African ambassador to the US Ebrahim Rasool (b. 1962) and began resettling white South Africans in the United States under the status of "refugees" from alleged anti-white violence coordinated by the South African government. This reached its apotheosis on 21 May 2025, when Trump met with South African president Cyril Ramaphosa (b. 1952) at the White House, and bombarded him with Facebook memes including those from Paul Hattingh, a white South African Flat Eather. Nor is this

[28] Sommer, *Trust the Plan*, 75–76.

[29] Justin Baragona, "Lara Logan Turns to a Fellow Newsmax Reject to Whine About Her Ban, Double Down on QAnon," *The Daily Beast*, October 27, 2022, accessed February 13, 2024, https://www.thedailybeast.com/lara-logan-turns-to-a-fellow-newsmax-failure-to-whine-about-her-ban-double-down-on-qanon.

only an American phenomenon; Sacha Stone (b. 1966), a white Rhodesian expat in the United Kingdom, has pioneered a number of British 5G- and covid-related conspiracies.[30]

By the time QAnon emerged in late 2017, the social media Flat Earth revival was at its peak. It was only natural that there would be overlap, as there was with other conspiracy ideas like anti-vaccination, Pizzagate, and UFOs. However, Flat Earth believers quickly became one of the key constituents of QAnon. This is not surprising, given that the initial intersection of online Flat Earthers and the most far-right and online wing of Trump's supporters had emerged by the summer of 2016, with one YouTuber suggestively stating that there are "no safe spaces on flat earth."[31] But what began as a symbiotic relationship and example of AIN and PUFT discussed in the last chapter—QAnon spread quickly through the existing Flat Earth social media groups, while Flat Earthers could proselytize among those being newly drawn into conspiracy belief by QAnon—quickly fractured. Stalwart QAnon promoters became frustrated with Flat Earthers, both because their seemingly ridiculous worldview provided easy fodder for those mocking QAnon, but also because they saw the Flat Earthers diluting Q's core message about the cabal.[32]

Things came to a head on 18 December 2018, when Q hosted an AMA ("Ask Me Anything") event where followers could ask him questions directly. As a sign of just how deeply integrated various conspiracies

[30] Michael Feola, *The Rage of Replacement: Far Right Politics and Demographic Fear* (Minneapolis: University of Minnesota Press, 2024); Derek Beres, Matthew Remski, and Julian Walker, *Conspirituality: How New Age Conspiracy Theories Became a Health Threat* (New York: PublicAffairs, 2023), '89; David Frum, "The Dangerous Myths of South African Land Seizures," *The Atlantic*, August 30, 2018, accessed May 20, 2019, https://www.theatlantic.com/ideas/archive/2018/08/the-danger-of-south-african-land-seizure/568960/; Michael Bueckert, "Flight of the Boers," *Africa is a Country*, May 29, 2018, accessed July 1, 2020, https://africasacountry.com/2018/05/flight-of-the-boers; Alex Park, "The blankest spot on Trump's world map," *Africa is a Country*, June 26, 2020, accessed July 1, 2020, https://africasacountry.com/2020/06/the-blankest-spot-on-trumps-world-map; J. D. Wolf, "Exclusive: Trump Shared Racist, Flat-Earth Facebook Account With South African President," Meidas+, May 22, 2025, accessed May 22, 2025, https://www.meidasplus.com/p/exclusive-trump-shared-racist-flat.

[31] Aric Toler, ""No Safe Spaces on the Flat Earth"—Emergent Alt-Right Inspired Flat Earth Online Communities," Bellingcat, June 7, 2017, accessed December 10, 2019, https://www.bellingcat.com/resources/articles/2017/06/07/flat-earth-online-communities/.

[32] Weill, *Off the Edge*, 160–62; Sommer, *Trust the Plan*, 77; Julian Feeld, Travis View, Jake Rockatansky, and Kelly Weill, "Episode 17: Our Thicc Flat Earth feat. Kelly Weill," *QAnon Anonymous*, November 28, 2018, accessed August 21, 2019, https://www.stitcher.com/podcast/qanon-anonymous/e/57443731.

had become with QAnon, four of the sixteen questions were about Q's stances on different conspiracy beliefs. One of those questions was from an irate believer who asked, "Just to shut the Flat Earthers up Q, Is the Earth flat?" Q's response was a simple, "No." However, what was intended to shore up continuity of message in the QAnon movement failed, as Flat Earthers and other heterodox conspiracists—such as those who believe that John F. Kennedy, Jr. (1960–1999) faked his death, which Q also rejected in the AMA—instead used Q's own motto, "disinformation is necessary," against him. Flat Earth—and JFK Jr. revivalism—therefore continued to comfortably exist within the QAnon broad tent despite Q's explicit rejection, and alongside rival but incompatible conspiracies, such as the Earth indeed being a globe regularly visited by alien inhabitants of other globes. An example is the popularity in the Flat Earth community of the phrase "enjoy the show," which had earlier become one of the catchphrases of QAnon believers anticipating the coming mass arrests (Fig. 7.1).[33]

Especially after the start of the covid pandemic, core tenets of QAnon became mainstreamed in broader US conservatism, especially among those

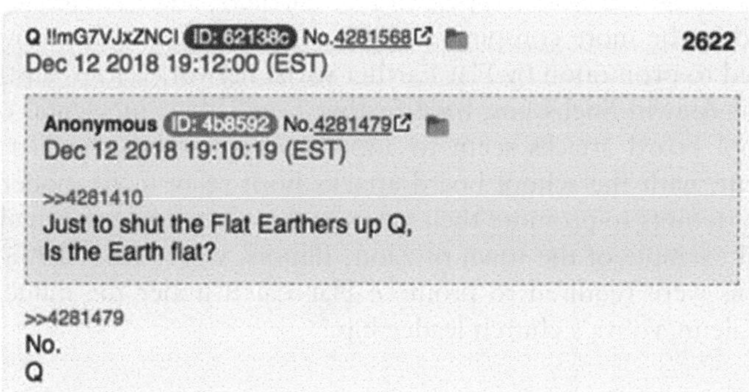

Fig. 7.1 Q rejects Flat Earth[34]

[33] Weill, *Off the Edge*, 130; Sommer, *Trust the Plan*, 45–47; Dappergander, "Explainer: Why did QAnon Factionalize?," dappergander, April 1, 2021, accessed August 24, 2021, https://www.patreon.com/posts/49472926; Q, "Dec 12 2018," Qmap, December 12, 2018, accessed August 24, 2021, https://qanon.pub/#2622.

[34] Q, "Dec 12 2018."

who rejected Trump's initial re-election defeat on 3 November 2020. It is not impossible that the Flat Earth movement has provided some conservative elements of its own. In the aftermath of Trump's 2021 departure from office, the US Republicans have increasingly advocated anti-transgender policies; these views were very prominent in US Flat Earthers prior to them becoming as widespread among mainstream conservatives, paired with the deliberate appropriation of the LGBT phrase "coming out of the closet" to signify their publicly endorsing Flat Earth. Also used is the phrase "Once You Go Flat, You Can't Go Back," connotating interracial sexual relations as equivalent to the addictive pleasure of endorsing Flat Earth.[35] This is in line with other Flat Earth appropriations of traditionally liberal terminology. An example is Mark Sargent describing the 2019 Third Flat Earth International Conference as a "safe space" for believers who feel ostracized in the real world for their views.[36]

Other broad aspects of post-first Trump US conservatism are a fixation on so-called "globalists" as a nebulous international enemy, and both metaphorical and literal attacks on local school boards for allegedly teaching critical race theory (a term loosely applied to any acknowledgement of past or present racism). While globalism is a widely-used term, it is easy to see how it would appeal to Flat Earthers, and that, like QAnon, its spread in the more conspiratorial wing of the conservative movement was linked to promotion by Flat Earther social networks. Journalists Kelly Weill and Asawin Suebsaeng have further noted that, incidental or not, the school board attacks seem to adopt concepts from the Flat Earth movement, with the school board attacks both relevant to modern Flat Earther attempts to promote their views in the United States, but also the historical example of the town of Zion, Illinois, where from 1905–1935 its schools were required to promote Flat Earth under the influence of Wilbur Glenn Voliva's church leadership.[37]

[35] Weill, *Off the Edge*, 127, 142–43; Olshansky, Peaslee, and Landrum, "Flat-Smacked!," 50.

[36] Rob Picheta, "The flat-Earth conspiracy is spreading around the globe. Does it hide a darker core?," CNN, November 18, 2019, accessed August 3, 2020, https://edition.cnn.com/2019/11/16/us/flat-earth-conference-conspiracy-theories-scli-intl/index.html. See also Amanda Montell, *Cultish: The Language of Fanaticism* (New York: HarperCollins, 2021).

[37] Asawin Suebsaeng and Kelly Weill, "Phat Earth, Feat. Dylan Gelula," *Fever Dreams*, February 23, 2022, accessed February 23, 2022, https://podcasts.apple.com/us/podcast/phat-earth-feat-dylan-gelula/id1558716930?i=1000551957397.

However, it was the effort to extend Trump's time in office that resulted in the most direct integration of Flat Earth into the Republican mainstream. Lauren Witzke (b. 1988), the 2020 Republican candidate to take Joe Biden's former US Senate seat in Delaware, was both an avowed Flat Earther and QAnon supporter; she failed, but still received over 186,000 votes, nearly 40% of those cast in the race.[38] Successfully elected from Colorado in the 2020 election was Lauren Boebert (b. 1986), who was notable for being one of the two QAnon supporters to win election to the US House of Representatives in 2020. On 13 November 2024, a week after both she and Trump were re-elected, Boebert was part of a House panel questioning alleged whistleblowers of a supposed US military UFO retrieval program. During her period of questioning, Boebert not only endorsed the idea that the US military was keeping UFOs and using them to create alien-human hybrids, but also briefly stated her views that the Earth is flat, the Apollo Moon landings were fake, and that Biden lost the 2020 election to Trump.[39]

The legal effort to "de-certify" the 2020 presidential election results in the courts to keep Trump in office was led by Lin Wood (b. 1952), a lawyer for several high-profile Republican politicians who had gradually embraced QAnon over the span of 2020, and in 2022 made public his belief in Flat Earth.[40] At least two prominent Flat Earthers participated in the 6 January 2021 attack on the US Capitol, the culmination of the Trump election-denial movement that Wood pushed.[41] By this point, perhaps sensing the political winds and fearing (ultimately without cause)

[38] "Republican nominee for US Senate in Delaware has repeatedly promoted QAnon conspiracy theory, once called self 'flat earther'," *Chicago Tribune*, September 17, 2020, accessed August 24, 2021, https://www.chicagotribune.com/2020/09/17/republican-nominee-for-us-senate-in-delaware-has-repeatedly-promoted-qanon-conspiracy-theory-once-called-self-flat-earther/; Weill, *Off the Edge*, 196. The Republican candidate for that same seat in the 2010 election, Christine O'Donnell (b. 1969), saw her race flounder when it was revealed that she had once been a practicing witch.

[39] Kase Wickman, "Lauren Boebert Told Congress She Is Worried About Space Aliens Operating From Underwater "Bases"," *Vanity Fair*, November 14, 2024, accessed November 23, 2024, https://www.vanityfair.com/style/story/rep-lauren-boebert-secret-underwater-alien-base-congress-ufo-hearing.

[40] Benjamin Hart, "Who Are You Going to Believe, Lin Wood or NASA?," *Intelligencer*, June 13, 2022, accessed March 1, 2023, https://nymag.com/intelligencer/2022/06/lin-wood-pro-trump-attorney-is-a-flat-earther.html; Sommer, *Trust the Plan*, 99–110.

[41] Weill, *Off the Edge*, 191.

reprisals from the incoming Biden administration Q had already vanished, making his final post on 8 December 2020.

It is worth noting that there is an additional, if tangential, Flat Earth connection to 6 January, beyond the participants of actual Flat Earthers in the mob. One of the few legislators who urged on the rioters assembled outside the Capitol on 6 January, and then defended their actions (and Trump's election claims) on the Senate floor after the attack several hours later, was Josh Hawley (b. 1979), a Republican who served as junior Senator from Missouri since 2019. Hawley is a strong admirer of Dutch politician and theologian Abraham Kuyper (1837–1920), Prime Minister of the Netherlands from 1901–1905, including the final months of the Second Boer War in which he supported Paul Kruger and his partisans. A strict Calvinist, Kuyper led a schism in the Dutch Reformed Church inspired by the theology of Kruger's Dopper Church in the Transvaal. Hawley has cited Kuyper to justify his belief that Christians must take over the US government and end secularism. Several months earlier, Trump's education secretary Betsy DeVos (b. 1958), had also cited Kuyper in a speech denouncing public, secular education.[42] The same radical Calvinism which led to Kruger's belief in a Flat Earth helped motivate the 2020 attacks on the US government, rhetorical and actual, which drew in Kruger's cosmological descendants.

Although having vanished by the time Hawley urged on the Flat Earthers eager to overthrow the government on 6 January, Q would return to make thirteen posts in mid-2022. But by that point the QAnon movement had already morphed to the point that even a returned Q could not reclaim the central leadership of the movement. The evolution of QAnon tied in with the changing Flat Earth movement, which had faded somewhat with the start of the covid pandemic in 2020. Most Flat Earthers became more interested in covid conspiracies, social media platforms cracked down on Flat Earth groups as part of their attempts to combat covid misinformation, and several prominent Flat Earthers died from the disease they refused to believe in. As an example, the Flat Earthers and anti-vaccine activist Rob Skiba, who in 2018 had debated the Traditionalist Catholic geocentric Robert Sungenis, would contract

[42] James Bratt, "Using Kuyper," *Reformed Journal*, May 10, 2021, accessed March 5, 2024, https://reformedjournal.com/using-kuyper/; Gerrit Schutte, "Abraham Kuyper and his South African Brethren," *Rozenberg Quarterly*, June 22, 2011, accessed March 5, 2024, https://rozenbergquarterly.com/abraham-kuyper-and-his-south-african-brethren/.

covid at a "biblical flat earth conference" in 2021 and die from it shortly after.[43]

Early on, some observers optimistically hoped that the covid pandemic might kill off Flat Earth belief entirely, along with the nebulous threat of political populism, but this proved to be a forlorn hope.[44] An illustrative example comes from journalist Will Sommer's interviews with QAnon supporters and their families. One of the figures he covered was "Nathan," a teenager who lived with his parents and who already believed in Bigfoot and UFOs. One month into the covid pandemic lockdown, "Nathan" was drawn into QAnon. Disillusioned by Trump's electoral failure, "Nathan" had left QAnon by November 2021, only to be subsequently re-converted—by fellow members of a Flat Earth online group.[45] Much like the coronavirus, Flat Earth groups had not been wiped out by the pandemic, but served as a reserve incubator for conspiracy ideas to mutate to new strains and re-emerge.

While "Nathan" is only a general QAnon advocate, the post-pandemic role of Flat Earth can be seen in its highest echelons as well. Dylan Wheeler, more commonly known by his online handle "Education4Libs" (and similar derivatives adopted to avoid successive social media bans) was one of the largest names in QAnon, having been one of the first promoters of the movement and a major backer of Witzke's Senate campaign. In August 2020, Wheeler left QAnon, saying that his Christian faith would not allow him to follow Q, who he now viewed as a false messiah. In February 2021, Wheeler announced that those same religious convictions had led him to become a Flat Earther, seemingly the first big-name shift from QAnon to explicit Flat Earth.[46] Conversely, Jacob Chansley (b. 1988), a prominent figure in the QAnon movement in 2020 and well known for his role as the so-called "Q Shaman" who took

[43] Kelly Weill, "Anti-Vax Flat Earth Preacher Dies of COVID-19," *The Daily Beast*, October 14, 2021, accessed March 16, 2024, https://www.thedailybeast.com/flat-earth-preacher-rob-skiba-dies-of-covid-19.

[44] Frank Jacobs, "Will COVID-19 kill off flat Earthism once and for all?," *Big Think*, March 20, 2020, accessed April 1, 2020, https://bigthink.com/strange-maps/end-of-flat-earth/; Tim Binnall, "Flat Earther Busted for Coronavirus Conspiracy Graffiti in Texas," *Coast to Coast AM*, July 28, 2020, accessed July 28, 2020, https://www.coasttocoastam.com/article/flat-earther-busted-for-coronavirus-conspiracy-graffiti-in-texas/.

[45] Sommer, *Trust the Plan*, 147–58.

[46] Alex Kaplan, Twitter, February 2, 2021, accessed February 18, 2024, https://twitter.com/alkapdc/status/1356612532404969476.

center stage in the 6 January 2021 attack on the US Capitol, denounced Flat Earth after being released from prison in 2023. Perhaps significantly, Chansley is one of the few QAnon followers who is not an evangelical Protestant.[47] In contrast, one of the figures who emerged to try and take control of QAnon in 2021 after Q's departure was Robert Smart, known online as "GhostEzra." Smart explicitly revived QAnon Flat Earth belief—in contrast to Q's earlier condemnation—as part of his bid to replace Q while also escalating the far-right connections of both movements. An example is a 4 May 2021 post on the social media platform Telegram where Smart asked, "The two most taboo topics are the shape of the earth and the Holocaust. I wonder why this is? We should find out."[48]

Nor was the changing Flat Earth relation to American far-right politics limited to QAnon. Kandiss Taylor, a Georgia Republican who had run in the party primaries for both the 2020 US Senate and 2022 governor candidacies, and chaired the state's first congressional district local Republican Party, came out as a Flat Earther in May 2023; in May 2025, she announced that she is running for the Republican nomination for US representative of her district.[49] In 2022, an Alabama pastor, Dean Odle, unsuccessfully ran in the Republican primary for governor of Alabama; Odle claimed that God revealed the truth of the Earth's

[47] Julian Feeld, Travis View, Jake Rockatansky, and Jacob Chansley, "Episode 234: Jacob Chansley, America's Shaman," *QAnon Anonymous*, June 10, 2023, accessed June 11, 2023, https://podcasts.apple.com/us/podcast/episode-234-jacob-chansley-americas-shaman/id1428209307?i=1000616446338. Of note is that, despite major media sources stating that Chansley had abandoned QAnon in prison after being convicted for his role in 6 January, this is not the case.

[48] Ewan Palmer, "QAnon Telegram Account Is World's 'Largest Antisemitic Online Channel'," *Newsweek*, May 26, 2021, accessed February 18, 2024, https://www.newsweek.com/qanon-ghostezra-holocaust-denial-telegram-1594907.

[49] Nikki McCann Ramirez, "Georgia GOP Chair Goes Full Flat-Earth, Says Globes Are Part of a Conspiracy," *Rolling Stone*, May 24, 2023, accessed June 4, 2023, https://www.rollingstone.com/politics/politics-news/kandiss-taylor-globes-anti-flat-earth-brainwashing-1234741082/; Terry Collins, "Flat Earth on the ballot? Debunked claim pops up in US politics," *USA Today*, May 17, 2025, accessed May 19, 2025, https://www.usatoday.com/story/news/2025/05/17/flat-earth-us-politicians/83492658007/.

flatness to him in 2015, and he attended the 2017 First Flat Earth International Conference. In 2025, Odle announced he will run for Alabama's lieutenant-governor race in 2026.[50]

In December 2023, shortly after an interview with conspiracy theorist Alex Jones (b. 1974) and after years of promoting UFOs (and several months after, like Logan, being fired from Fox News for increasingly extreme views), Tucker Carlson—whose podcast is, as of writing, the second most popular on Spotify, behind conspiracy theorist Joe Rogan's (b. 1967)—announced his critical endorsement of Flat Earth, claiming that supposed coverups by the liberal media made it plausible.[51] Perhaps related, this came several months after what Carlson later said was a turn to evangelical Christianity and literalist Bible study stemming from an alleged physical assault by a demon in Carlson's bed.[52] In January 2025, it was reported that Carlson was so concerned over UFOs that he had lobbied Elon Musk and Donald Trump to be placed in charge of UFO investigations in the latter's second presidential administration.[53]

During the 2023–2024 primaries that would solidify Trump's return to office, the right-wing video platform Rumble (which received a large cash infusion in 2021 from future Trump vice president JD Vance) served as the official live-streaming platform of the Republican Party candidate debates. On 24 February 2024, Rumble promoted a pro-Flat Earth video as one of its "editor's picks."[54] In July 2024, prominent Black supporter of Trump Candace Owens (b. 1989) likewise made a qualified nod to Flat Earth as a conservative cultural signifier, stating that she read Flat Earth

[50] Collins, "Flat Earth on the ballot?".

[51] Daniel Kreps, "Tucker Carlson Says He's 'Open' to the Flat Earth Theory," *Rolling Stone*, December 16, 2023, accessed December 16, 2023, https://www.rollingstone.com/politics/politics-news/tucker-carlson-flat-earth-theory-1234931330/.

[52] Zachary Folk, "Tucker Carlson Says a Demon Attack Left Him Bleeding in Bed," *The Daily Beast*, October 31, 2024, accessed November 23, 2024, https://www.thedailybeast.com/tucker-carlson-says-he-was-attacked-by-a-demon-in-new-doc/.

[53] "Alien Nut Tucker Carlson 'Set to Beg Donald Trump to Let Him Get to Bottom of America's UFO Mysteries'," Radar Online, January 2, 2025, accessed January 7, 2025, https://radaronline.com/p/tucker-carlson-beg-donald-trump-get-to-bottom-ufo-mysteries-america/.

[54] Alex Kaplan, "Rumble promotes the flat Earth conspiracy theory among its "editor picks"," Media Matters for America, February 28, 2024, accessed March 16, 2025, https://www.mediamatters.org/rumble/rumble-promotes-flat-earth-conspiracy-theory-among-its-editor-picks.

materials sent to her on the conservative social media platform Minnect, and when questioned why by her husband, "I said to him: 'Listen. I'm not a flat-earther. I'm not a round-earther.' Actually, what I am is, I am somebody who has left the cult of science. I have left the megachurch of science because what I have now realized is that science—what it is actually, if you think about it—is a pagan faith."[55] On 29 March 2025, Army veteran Bret Bussman was elected chair of the Republican Party of Minnesota's 7th Congressional District—the most heavily-Republican district in the state—defeating the incumbent chair who was accused of not being sufficiently right-wing; among many conspiracy theories—including chemtrails, 9/11 Trutherism, Apollo denial, anti-vaccination, and support for the 6 January 2021 attack on the US Capitol—Bussman is an avowed Flat Earther.[56]

Although Flat Earth is not as popular as it was prior to the covid pandemic, it seems as though it may be seeing a resurrection as a sort of cultural signifier for far-right politics in the United States, a sign of one's political affiliation more than genuine belief as it had been before.[57] The Polar, Environment and Science Survey, conducted in summer and early fall 2021 by University of New Hampshire sociologist by Lawrence Hamilton, indicated that in the immediate aftermath of both covid and the 2020 election, not only did a full 10% of the US population believe the Earth is flat, but Flat Earth was an extremely high signifier of support for Donald Trump; 21% of Trump supporters believed the Earth was flat, in contrast to 14% of Trump critics.[58] Likewise, a 2022 survey by Fairleigh Dickinson University found that while 11% of the general US

[55] Ryan Smith, "Candace Owens Describes Science as 'Pagan Faith'," *Newsweek*, July 3, 2024, accessed July 5, 2024, https://www.newsweek.com/candace-owens-science-pagan-faith-christianity-podcast-1920510. Of note is that Owen's husband, George Farmer, is son of British House of Lords member Michael Farmer, Baron Farmer, former treasurer of the UK Conservative Party.

[56] Christopher Ingraham, "Minnesota Republicans elect a flat-earther to a party leadership post," *Minnesota Reformer*, April 16, 2025, accessed April 17, 2025, https://minnesotareformer.com/2025/04/16/minnesota-republicans-elect-a-flat-earther-to-a-party-leadership-post/; Collins, "Flat Earth on the ballot?".

[57] Tim Binnall and Edward Guimont, "Binnall of America: The Revival - E18 | Dr. Edward Guimont," *Binnall of America*, February 21, 2024 (accessed February 21, 2024), https://www.youtube.com/watch?v=thUdGTIX6q8.

[58] Lawrence C. Hamilton, "Trumpism, Climate and COVID: Social Bases of the New Science Rejection," University of New Hampshire Scholars Repository, August 6, 2022 (accessed May 19, 2025), https://scholars.unh.edu/faculty_pubs/1520/, 5, 7; Lawrence

population thought the Earth was either "somewhat" or "very" likely to be flat, that number went up to 13% of Republicans and 17% of those who rejected Trump's defeat in the 2020 election, in contrast to 9% of non-Republicans and 2020 election accepters. This was speculated as being due to Flat Earth being spread through the same social media vectors as not only pro-Trump conspiracy theories, but Trump support in general.[59] Outside of the United States, there is also evidence for a revival. In early 2020, Italy was one of the first and hardest-hit countries in the covid pandemic; by the end of 2021, not only did an estimated 6% of Italians believe covid did not exist, but the same percentage expressed Flat Earth belief, and 10% doubted the Apollo landings.[60]

The contemporary Flat Earth movement in the United States has clearly drifted toward support for Republicans broadly, and Trump specifically. This is in line with other conspiracy movements, particularly anti-vaccine and UFO belief, as mentioned earlier. But the efficacy of mobilizing Flat Earthers to tap into an unrecognized political consistency is somewhat questionable. For one, unlike the time of John Jasper, modern Flat Earthers tend to be overwhelmingly white and evangelical Christian, a group already incredibly supportive of Trump. Even the few Black Flat Earthers in the United States are the subject of racist invectives from some of the white leaders of the movement, cutting down on the likelihood they will be mobilized for Republicans.[61] The other major issue is that, even at its height, Flat Earth support in the United States is far lower than that of anti-vaccination or UFO belief. In 2018, Chapman University polling indicated 35% of Americans believed in contemporary UFOs and 41% believed in ancient aliens—although this was a decline from the 52% of Americans who believed in UFOs in 1975.[62] The same

C. Hamilton, "Conspiracy vs. Science: A Survey of U.S. Public Beliefs," *Casey School of Public Policy National Issue Brief*, No. 162 (Spring, 2022), 3.

[59] Dan Cassino, "2020 Election Conspiracy Believers More Likely to Embrace Bigfoot, Flat Earth," FDU Poll, May 5, 2022 (accessed May 19, 2025), https://portal.fdu.edu/fdupoll-archive/220505/.

[60] "Conspiracy theories take root in Italy, survey shows," *Reuters*, March 16, 2019, accessed December 3, 2021, https://www.reuters.com/lifestyle/oddly-enough/conspiracy-theories-take-root-italy-survey-shows-2021-12-03/.

[61] Weill, *Off the Edge*, 188.

[62] Colin Dickey, *The Unidentified: Mythical Monsters, Alien Encounters, and Our Obsession with the Unexplained* (New York: Viking, 2020), 3; Erich von Däniken, trans. Michael

year, a YouGov poll indicated that only 2% of the US public were fully Flat Earthers and an additional 3% were uncertain about the Earth's shape (although as noted above, this percentage may have increased heavily after 2020).[63] However, during the late 2010s, there was a country undergoing a political realignment which (at least at the time) had an even larger percentage of Flat Earthers in its population: Brazil.

As mentioned in the previous chapter, Nigeria has the largest Black population on Earth. However, thanks to the trans-Atlantic slave trade discussed in earlier chapters, it is Brazil that has the second-largest Black population in the world. It also has the largest Arab population outside of the Middle East. Because of this, the social and political hierarchies and allegiances of race are much more nuanced than in the United States, though the Reconstruction period in the South was perhaps closer to it.[64] Brazil also has a large Indigenous population, with astronomical knowledge dating back thousands of years and often incorporating the fact that they inhabit a region that spans both sides of the spherical world's equator. One group in the central north of Brazil, the Apinajé, has a ritual celebrating the passage of the Sun from one hemisphere to the other.[65]

The history of Brazil likewise complicates traditional Western metropole-colony and racial category distinctions. The first Portuguese colonial settlements in what is now Brazil were established in 1500, less than a decade after Columbus reached the Caribbean. The enormous number of slaves meant a large number of not only slave insurrections, but autonomous communities of escaped slaves. These quilombos were legally recognized to some degree by the colonial administration, most notably the quilombo of Palmares which existed from 1605–1694, was governed by royalty taken from the Kingdom of Kongo in modern Angola, and

Heron, *According to the Evidence: My Proof of Man's Extraterrestrial Origins* (London: Souvenir Press, 1977), 291.

63 Olshansky, Peaslee, and Landrum, "Flat-Smacked!," 47.

64 Kevin Ochieng Okoth, *Red Africa: Reclaiming Revolutionary Black Politics* (London: Verso, 2023), 62–64; Carl N. Degler, *Neither Black Nor White: Slavery and Race Relations in Brazil and the United States* (Madison, WI: The University of Wisconsin Press, 1971); Anthony W. Marx, *Making Race and Nation: A Comparison of South Africa, the United States, and Brazil* (Cambridge: Cambridge University Press, 1998).

65 Rodolfo Langhi, "Some Brief Reflexions About the Astronomy Education in Brazil," *Academia Letters*, No. 1088 (Jun., 2021), 1; Germano Bruno Afonso, "The Brazilian Indigenous Planetary-Observatory," *Boletim da Sociedade Astronômica Brasileira*, Vol. 21, No. 1 (Aug., 2003), 69.

which incorporated not only Africans, but also Indigenous people and white Portuguese settlers, particularly Jews and others escaping the Inquisition. In 1808, the Portuguese royal family relocated to Brazil to escape Napoleon's armies, and remained there until 1821, years after Napoleon's defeat and exile. For over a decade, Portugal was in the unique position of being governed from its own colony. In 1822, Brazil became independent, the son of the Portuguese king being crowned emperor of Brazil. In 1888, Brazil became the last nation to abolish slavery, nearly a generation after the end of the Civil War in the United States. The end of slavery led to a policy of so-called "racial democracy" arguing that Brazil after emancipation avoided the hardline racial issues that dominated other post-slavery societies, particularly the legal racial segregation of the US South.[66]

Unrest by former slaveowners in Brazil led to the abolition of the monarchy in 1889, with a republic being established. Brazil's republic underwent various periods of oligarchic, democratic, and dictatorial periods. In particular, the era of 1945–1964 was a notably liberal period of democracy in which socialist and communist parties were legalized. In 1964, a military dictatorship seized power in Brazil from the elected socialist president with the assistance of the United States, arguing that it was an anti-communist Cold War necessity. The dictatorship period was marked with widespread imprisonment, torture, and murder of political dissidents. Democracy was restored to Brazil in 1985, but the presidency remained dominated by conservative parties that had supported the dictatorship.[67] This changed in 2002 with the election of Luiz Inácio da Silva (known simply as Lula, b. 1945) of the Workers' Party (in Portuguese, the Partido dos Trabalhadores or PT), the first socialist to hold the presidency since the previous one was overthrown in 1964. Lula held office from 2003–2010, during which time he steadily improved the standard of living in Brazil—particularly among Afro-Brazilians, who were 10% of the population and an important voting bloc for his party—and worked

[66] Leslie Bethell, *Brazil: Essays on History and Politics* (London: University of London Press, 2018), 113–44; Edward E. Telles, *Race in Another America: The Significance of Skin Color in Brazil* (Princeton: Princeton University Press, 2006), 24–77; Darcy Ribeiro, trans. Gregory Ribassa, *The Brazilian People: The Formation and Meaning of Brazil* (Gainesville, FL: University of Florida Press, 2000 [1995]), 49–112.

[67] Bethell, *Brazil*, 147–74; Ribeiro, *The Brazilian People*, 115–86.

to expand ties with African countries and leftist Latin American governments as part of a policy to build up Global South ties independent of US and European influence.[68]

On leaving office, Lula's vice-president, Dilma Rousseff (b. 1947), was elected to succeed him; she was re-elected in 2014. However, the PT administrations of Lula and Rousseff, with their economic nationalization and foreign policy opposed to the United States, had made powerful enemies. The US administration of Barack Obama collaborated with conservative members of the Brazilian judiciary and law enforcement in a vast, allegedly anti-corruption probe called Operation Car Wash. Car Wash resulted in the impeachment of Rousseff in 2016 and imprisonment of Lula the next year, in both cases on charges that were widely considered more political than criminal. With the leaders of the PT barred from office, the 2018 presidential election had been opened to a very different political figure: Jair Bolsonaro (b. 1955), a far-right member of parliament who had previously served in the army during the dictatorship. In the runoff, Bolsonaro received 55% of the vote in contrast to the 45% received by the PT's candidate, the Lebanese-Brazilian mayor of São Paulo and former education minister Fernando Haddad (b. 1963).[69]

In contrast to his PT predecessors, Bolsonaro's constituency was almost entirely white, and both before and after his 2018 election he opposed the culture and political aims of Afro-Brazilians, Indigenous groups, and non-Christians, being a Catholic married to an evangelical Baptist. Domestically, in addition to cutting the social services and wealth redistribution programs of the PT, Bolsonaro championed an attack on not only funding, but political independence of universities and other education institutions, predicting the path that US Republicans would take after Trump's first term ended. In foreign relations, Bolsonaro abandoned the south-south orientation of the PT governments in favor of increased ties to the US and Israel mirroring Trump's foreign policy, albeit

[68] Yvonne Captain, "Brazil's Africa Policy under Lula," *The Global South*, Vol. 4, No. 1 (Spring, 2010), 183–98; Telles, *Race in Another America*, 75–77.

[69] Bethell, *Brazil*, 195–222; Brian Winter, "Messiah Complex: How Brazil Made Bolsonaro," *Foreign Affairs*, Vol. 99, No. 5 (Sep./Oct., 2020), 120–23; Gideon Rachman, *The Age of the Strongman: How the Cult of the Leader Threatens Democracy Around the World* (New York: Other Press, 2022), 159–72.

while retaining close connections with its fellow BRICS allies (Russia, India, China, and South Africa).[70]

After the United States, Brazil is YouTube's second-largest market. Far-right YouTube videos exploded in Brazil beginning with the 2016 crisis surrounding Rousseff's impeachment, reaching their peak with the 2018 election. As with the 2016 election in the United States, this coincided with an expansion of racist and conspiratorial notions across Brazilian society, along with a political realignment of them toward the right—including Flat Earth.[71] In Brazilian Portuguese, the term for "Flat Earther" is "Terraplanista," and the discussion and promotion of it by right-wing media channels became so widespread that by the late 2010s Terraplanista became a catch-all term for all far-right conspiracy theorizing and historical revisionism.[72] However general the term has become, it does not mask that there are indeed proponents of literal Terraplanismo in Brazil. One of its major evangelists is a construction business owner turned paranormal promoter named Urandir Fernandes de Oliveira, who goes by his apt initials, "UFO."[73] In 1997, UFO founded the Dakila Pesquisas Research Institute, which promotes a wide range of alternative views, including Flat Earth and ancient aliens. In addition to beginning a planned city called Zigurats in the Mato Grosso state of Brazil in 2019 (with funding from both the state and local governments), in 2023 Dakila signed an agreement with the private Abu Dhabi University in the UAE to promote their views through the guise of a joint technology development program.[74] The government support of UFO and Dakila perhaps

[70] Adam Bledsoe, "Racial Antagonism and the 2018 Brazilian Presidential Election," *Journal of Latin American Geography*, Vol. 18, No. 2 (Jun., 2019), 165–70; Cyril Prinsloo, "Bolsonaro and the BRICS: Bull in a china shop?," *South African Institute of International Affairs*, Policy Briefing 180 (Mar., 2019), 6–7; Ana Garcia, "Brazil under Bolsonaro: Social base, agenda and perspectives," *Journal of Global Faultlines*, Vol. 6, No. 1 (Aug.-Oct., 2019), 62–69; James N. Green, "Brazil's Far-Right President, University Autonomy, and Academic Freedom," *Academe*, Vol. 105, No. 4 (Fall, 2019), 18–21.

[71] Fisher, *The Chaos Machine*, 266–95; Weill, *Off the Edge*, 106–07.

[72] Discussion with Andrew Fisher, 7 December 2024.

[73] UFO is of course the acronym for the English term "unidentified flying object." In Portuguese, it is OVNI, for "objeto voador não identificado." That Oliveira uses the English term may indicate a vector for the origin of his interests.

[74] "About the Dakila Institute," Dakila Association, accessed February 21, 2024, https://dakila.si/en/about-the-institute/; "Abu Dhabi University and Dakila Research Social Sciences UAE collaborate to advance technologies," Abu Dhabi University News &

makes more sense given that a 2019 Datafolha Institute survey indicated that 7% of Brazil's population were Terraplanistas, in contrast with 2% of Flat Earthers in the United States.[75] On 10 November 2019, at the end of Bolsonaro's first year in office, the first Brazilian Flat Earth conference was held in São Paulo.[76]

It also is perhaps not a surprise that Bolsonaro courted the Terraplanistas early in his run, via arguably the most prominent Brazilian Terraplanista advocate, even moreso than UFO. This was the right-wing Traditionalist Catholic philosopher Olavo de Carvalho (1947–2022), who was described as the "chief ideologue" of the Bolsonaro coalition and one of Bolsonaro's primary political advisors, playing a similar role as Steve Bannon (b. 1953) did to Trump and the so-called "alt-right" in the United States. Carvalho had a stint as an astrologer beginning in the late 1970s, before making a name for himself as a public intellectual writing for major publications in Brazil. Starting in 2002, Carvalho abandoned mainstream written media, first for his own website, then starting in 2004 the social media platform Orkut (enormous in Brazil from its 2004 launch until it was shut down in 2014, dominating Brazilian internet usage from 2005–2007), then starting one of the earliest podcasts in 2006, and finally moving to YouTube in 2009, which became his primary platform. The pioneering switch to online platforms allowed Carvalho to relocate to the United States in 2005, moving to John Jasper's former home of Richmond, Virginia. Carvalho became an embodiment of the AIN/PUFT confluence, mixing political invectives opposing the PT, leftist intellectuals, globalism, Marxism, and LGBT rights with the promotion of Car Wash, astrology, alternative medicine, and questions of the Earth's shape. Carvalho's followers, called Olavistas, became a core

Articles, August 28, 2023, accessed February 21, 2024, https://www.adu.ac.ae/news-and-events/news/news/detail/2023/08/28/abu-dhabi-university-and-dakila-research-social-sciences-uae-collaborate-to-advance-technologies.

[75] Rafael Garcia, "7% of Brazilians Believe that Earth Is Flat," *Folha de S.Paulo*, July 15, 2019, accessed February 24, 2022, https://www1.folha.uol.com.br/internacional/en/scienceandhealth/2019/07/7-of-brazilians-believe-that-earth-is-flat.shtml.

[76] Cristiano Mattos, Felipe Sanches Lopez, José Luis Ortega, and André Rodrigues, "The Public Discussion on Flat Earth Movement: An Analysis Based on the Esperantist-Epideictic Discourse," *Science & Education*, Vol. 31 (Jan., 2022), 1349.

component of the Brazilian New Right which emerged in opposition to Lula's administration and formed the central constituency of Bolsonaro.[77]

So influential were the Olavistas that Bolsonaro often placated them at the expense of other segments of his coalition, including the free marketeers and the military bloc that he himself came from.[78] Carvalho directly nominated two of Bolsonaro's key cabinet ministers in his initial government: Foreign Minister Ernesto Araújo (b. 1967) and Education Minister Ricardo Vélez Rodríguez (b. 1943). Araújo claims to believe in the spherical Earth but has voiced support for Terraplanistas in questioning science, while Rodríguez is an outright Flat Earther. Under the influence of Carvalho, therefore, the figures in Bolsonaro's government involved in education and in traveling the globe were both Terraplanistas. Nor was it only high-ranking cabinet ministers who shared this view; Dante Mantovani, appointed head of the National Arts Foundation in 2019, was a Terraplanista. It should be pointed out that, unlike in the US, Brazilian presidential politics are far more shifting coalition blocs whose individual components rise and fall, and to use an astrological term, the Olavista star was in the descendent. Rodríguez was fired within several months (although his replacement, Abraham Weintraub, b. 1971, was also an Olavista); Araújo retired under pressure in 2021; and Mantovani's two-year term also ended in 2021. In January 2022, Carvalho himself died from covid, the disease he had denied existing.[79]

[77] Camila Rocha, "From Orkut to Brasília: the origins of the New Brazilian Right," in Katerina Hatzikidi and Eduardo Dullo, eds., *A Horizon of (Im)possibilities: A Chronicle of Brazil's Conservative Turn* (London: University of London Press, 2021), 87–90; Benjamin R. Teitelbaum, *War for Eternity: Inside Bannon's Far-Right Circle of Global Power Brokers* (New York: Dey Street, 2020), 125–40; Marko Monteiro, "Science is a war zone: some comments on Brazil," *Tapuya: Latin American Science, Technology and Society*, Vol. 3, No. 1 (Jan., 2020), 4–8; Daniel Edler Duarte, Pedro Benetti, and Marcos César Alvarez, "A "war on science?" Far-right movements and the disputes over epistemic authority in Brazil," *Tapuya: Latin American Science, Technology and Society*, Vol. 7, No. 1 (Apr., 2024), 1–14; Rafael Antunes Almeida, ""Eles nos dizem tudo o que estão fazendo": ciência e conspiração nos discursos terraplanistas," *Anuário Antropológico*, Vol. 48, No. 2 (2023), 102–22; Rachman, *The Age of the Strongman*, 215; Brian Winter, "Jair Bolsonaro's Guru," *America's Quarterly*, December 17, 2018, accessed February 21, 2024, https://www.americasquarterly.org/article/jair-bolsonaros-guru/.

[78] Winter, "Messiah Complex," 128; Garcia, "Brazil under Bolsonaro," 66.

[79] Green, "Brazil's Far-Right President," 20; Monteiro, "Science is a war zone," 5; Teitelbaum, *War for Eternity*, 249–60; Michael Royster, "Brazil's Bolsonaro: How's He Doing After 100 Days? Part I: Foreign Affairs," *The Rio Times*, April 10, 2019,

By this time, Bolsonaro's re-election campaign was in full swing, his opponent none other than Lula. On taking office in 2019, Bolsonaro appointed Sergio Moro (b. 1972), the judge involved with Car Wash who had convicted Lula, to be his justice minister, a move which drew widespread condemnation. In the ensuring scrutiny of Moro's prosecution of Lula, leaked transcripts revealed the case had been politically motivated and coordinated with the Bolsonaro movement with the express purpose of removing the popular Lula's potential for running against Bolsonaro. Lula was released from prison in 2019, Moro resigned the next year, and in 2021 all charges against Lula were annulled. Running again as the PT's presidential candidate, in the October 2022 presidential election he defeated Bolsonaro by 51% to 49%, and succeeded Bolsonaro as President of Brazil on 1 January 2023. As with Trump and the 6 January 2021 attack on the US Congress, Bolsonaro refused to accept the legitimacy of the returns, and on 8 January 2023, his followers attacked the Brazilian Congress, Presidential Palace, and Supreme Court in an unsuccessful effort to overthrow Lula and return Bolsonaro to power. On 30 June 2023, the Superior Electoral Court banned Bolsonaro from running for office again until 2030 due to his inciting his followers to violence in rejecting democratic outcomes. In a final act of indignity, one of the judges in the trial barring him from office compared Bolsonaro's attack on the electoral process as president with a teacher promoting Terraplanismo to his students.[80] This echoed Virginia representative

accessed April 19, 2019, https://riotimesonline.com/brazil-news/opinion-editorial/brazils-bolsonaro-hows-he-doing-after-100-days-part-i-foreign-affairs/; Sam Cowie, "Brazil's flat Earthers to get their day in the sun," *The Guardian*, November 9, 2019, accessed June 7, 2023, https://www.theguardian.com/world/2019/nov/06/brazil-flat-earth-conference-terra-plana; Tom Phillips, "Covid denialist and Bolsonaro ally Olavo de Carvalho died of virus, says daughter," *The Guardian*, January 25, 2022, accessed June 7, 2023, https://riotimesonline.com/brazil-news/opinion-editorial/brazils-bolsonaro-hows-he-doing-after-100-days-part-i-foreign-affairs/; Tom Phillips and Dom Phillips, "'Unqualified, dangerous': the oddball officials running Bolsonaro's Brazil," *The Guardian*, January 2, 2020, accessed February 21, 2024, https://www.theguardian.com/world/2020/jan/02/bolsonaro-brazil-government-oddball-officials; Gustavo Freire Barbosa, "Além da Terra plana: o terraplanismo como método do governo Bolsonaro," *CartaCapital*, January 15, 2020, accessed July 5, 2020, https://www.cartacapital.com.br/opiniao/alem-da-terra-plana-o-terraplanismo-como-metodo-do-governo-bolsonaro/.

[80] "Voto contra Bolsonaro no TSE tem menção a 'Terra plana'," *CartaCapital*, June 29, 2023, accessed July 5, 2023, https://www.cartacapital.com.br/politica/voto-contra-bolsonaro-no-tse-tem-mencao-a-terra-plana/.

Denver Ruggleman (b. 1970), one of the few elected Republicans to oppose Trump's 2020 election denial claims, who did so by comparing them to Flat Earth belief.[81]

Given the narrow margin by which Bolsonaro was elected in 2018, and the even narrower margin by which he lost re-election in 2022 after parting ways with the key Terraplanistas and the death of Carvalho, it is not inconceivable that Terraplanistas formed one of the (even if not a primary) political constituencies that won him the office and whose lack of support cost him re-election. Given Bolsonaro and his movement's opposition to the Afro-Brazilian community, Terraplanismo (as well as the alternative medicine espoused by Olavistas) may have provided one of the few, if clearly tenuous, means of garnering support from that community, though this would require anthropological research beyond the scope of this work to verify. But Terraplanismo may have a deeper relevance to the Bolsonaro administration than merely its use as an electoral strategy. Writing in January 2020, Brazilian lawyer Gustavo Freire Barbosa argued that Terraplanismo was, beyond its literal connections to the government, symbolic of a method of governance under Bolsonaro, a metaphor for not just anti-intellectualism but wider flights from practical reality that the government was unable to extricate itself from even while it crippled effective governance. Barbosa noted both the fact that Bolsonaro's democratic election had relied on the undemocratic Car Wash, but also that the rejection of empirical science was present in the supposedly grounded "technical" wing of the administration outside of the Olavistas. A key example was finance minister Paulo Guedes (b. 1949), who received his PhD in economics at the University of Chicago. Despite not being an Olavista, and in fact coming into conflict with them in the government, Guedes still relied on astrology to guide his economic policies for the entire country.[82]

January 2020 was an apt date for Barbosa's article. Within two months, the covid pandemic would emerge worldwide, with the United States and Brazil among the highest casualty rates. This was partly due to the fact that both Trump and Bolsonaro rejected medical science in favor of the alternative medicines promoted by allied ideologues and Flat Earthers alike. Their subsequent mishandling of the pandemic in both

[81] Weill, *Off the Edge*, 199.

[82] Winter, "Messiah Complex," 129–30; Barbosa, "Além da Terra plana."

resistance to government mandates and promotion of quack cures like MMS contributed to the re-election defeats of both presidents.[83] As early as April 2020, one Brazilian journalist referred to Bolsonaro's covid (non-)policy as "epidemiologia terraplanista" (Flat Earth epidemiology).[84] Studies have indicated that Brazilian Terraplanistas are broadly evangelical Protestants, but not nearly as monolithic as US Flat Earthers, though evangelicalism was similarly the framework by which Terraplanismo was integrated with New Right ideological current.[85] This is particularly interesting given that Carvalho, the key to what might be called political Terraplanismo, was a Traditionalist Catholic, and Bolsonaro a more mainstream conservative Catholic. As mentioned in the previous chapter, Traditionalist Catholicism is more generally associated with "regular" geocentrism instead of Flat Earth. Further, a number of evangelical Terraplanistas denounce Catholicism, seeing it and science as being ideologically identical, given the Catholic Church's opposition to Luther's desire for all believers to read and interpret scripture.[86] The embrace of a traditionally evangelical concept like Terraplanismo by the Catholic Bolsonaro may have played political utility in a society where Catholicism and evangelicalism represent highly contested electoral blocs. Indeed, polling around the 2018 election indicated only 51% of Catholics planned to vote for Bolsonaro, in contrast to 69% of evangelicals, an unheard-of political unity among that religious set.[87] While Terraplanismo is certainly not the sole cause, it may serve as a reflection of that realignment.

It should also be noted that while Flat Earthers and Terraplanistas both widely lined up behind Trump and Bolsonaro, only the latter directly

[83] Rachman, *The Age of the Strongman*, 230–31; Schaller and Waldman, *White Rural Rage*, 56–59.

[84] Naomar de Almeida Filho, "Previsões terraplanistas sobre a Covid-19 no Brasil têm validade?" *Abrasco*, April 27, 2020, accessed May 17, 2024, https://abrasco.org.br/previsoes-terraplanistas-sobre-a-covid-19-no-brasil-tem-validade-artigo-de-naomar-de-almeida-filho/.

[85] Mattos, Lopez, Ortega, and Rodrigues, "The Public Discussion on Flat Earth Movement," 1341, 1344.

[86] Mattos, Lopez, Ortega, and Rodrigues, "The Public Discussion on Flat Earth Movement," 1353.

[87] David Lehmann, "Ritual, text and politics: the evangelical mindset and political polarisation," in Katerina Hatzikidi and Eduardo Dullo, eds., *A Horizon of (Im)possibilities: A Chronicle of Brazil's Conservative Turn* (London: University of London Press, 2021), 103–19.

appointed them to the highest positions in his government. But if Terraplanismo in Brazil was potentially in decline in the Brazilian New Right after the death of Carvalho and loss of Bolsonaro, the actions of figures like Wood, Carlson, and Owens indicate that it was on the rise in the Trump movement into his ultimately successful 2024 re-election to the presidency. Indeed, at a rally at the start of his Republican primary campaign in January 2024, Trump made the unorthodox claim that water could destroy magnets, paralleling a longstanding fascination and misunderstanding of magnetism in the Flat Earth movement.[88] It remains to be seen if Trump's re-election will herald a broader resurgence of Flat Earth, but even if Flat Earthers were not a major constituency in his 2024 coalition, it is hard not to see its promotion by key cultural figures in his movement as fitting comfortably with the other wide-ranging pseudoscientific agglomeration around Trump, particularly but not exclusively embodied by the likes of Elon Musk, Joe Rogan, and Robert F. Kennedy, Jr.—a mirror of Kipling's fictional "Village that Voted the Earth was Flat" from over a century earlier. A Flat Earth resurgence would certainly benefit from the fact that, following Twitter's 2022 purchase by Elon Musk, in January 2025 Mark Zuckerberg announced that Meta platforms (Facebook, Instagram, and WhatsApp) would also cease content moderation and fact-checking of posts as part of his rapprochement with Trump, shortly before Google (owner of YouTube) made similar policy changes for similar reasons.[89] The social media ecosystem appears primed to return to the era when Flat Earth began its latest return a decade prior.[90]

[88] Kelly McClure, "Trump presents interesting theory on how magnets work during rally in Iowa," *Salon*, January 6, 2024, accessed March 11, 2024, https://www.salon.com/2024/01/06/presents-interesting-theory-on-how-magnets-work-during-rally-in-iowa/; Elizabeth Blount and Albert Smith, *Zetetic Astronomy: or The Sun's Motions North and South; with the Moon's Motions; Fancied and Real. Showing the Uselessness of the Gravitation Theory,* etc. (Kingston Hill, Surrey: Lady Blount, 1904), 14–19.

[89] Drew Gillis, "Apparently, fact-checking is now too political for Facebook," The A. V. Club, January 7, 2025, accessed January 7, 2025, https://www.avclub.com/facebook-no-longer-fact-checking; Emma Keates, "Google becomes latest tech giant to send Trump money," The A. V. Club, January 9, 2025, accessed January 12, 2025, https://www.avclub.com/google-trump-donation.

[90] That being said, there is evidence that Facebook's fact-checking efforts, at least before the covid-era revamp aimed at conspiracy theories, had limited impact: R. Kelly Garrett and Shannon Poulsen, "Flagging Facebook Falsehoods: Self-Identified Humor Warnings Outperform Fact Checker and Peer Warnings," *Journal of Computer-Mediated Communication*, Vol. 24, No. 5 (Sp., 2019), 240–58.

Nor is the political utility of Flat Earth limited to conservatives. A month after Trump's magnet claim, he spoke at the annual Conservative Political Action Conference (CPAC). Also speaking there was former Conservative UK Prime Minister Liz Truss (b. 1975), whose fifty-day tenure as Prime Minister in 2022 made her the shortest holder of that office. At CPAC, Truss tapped into QAnon terminology to blame her short period in office on "sabotage" by the "deep state." This resulted in Labour Party leader Keir Starmer (b. 1962) referring to the Conservative Party as the "political wing of the Flat Earth Society."[91] Another speaker at the 2024 CPAC was Javier Milei (b. 1970), the far-right President of Argentina elected the prior year. In that election, as with the US election of 1996, Milei's libertarian support for a flat tax caused his election opponent to term him "un terraplanismo en economía."[92] On taking office, his government quickly began exhibiting the same scientific views as Bolsonaro's did in Brazil.[93] More concretely, in May 2024 Milei appointed his campaign makeup artist and former girlfriend, YouTube influencer Lilia Lemoine (b. 1980), to be head of the Science and Technology Committee of the lower house of the legislature, to which she had been elected in 2023. This was particularly controversial given Lemoine has asked, "Why do the governments of the world want to hide from humanity that the Earth is flat and that there is a large wall of ice that surrounds it?," answering her own question with the claim that it was a plot by scientists to undermine the authority of the Church. Lemoine has also rejected the truth of the Apollo Moon landings, and given support to the venerable conspiracy claim that in 1947, US Navy admiral Richard Byrd (1888–1957) carried out a secret expedition to the Antarctic where he was taken into the center of the Hollow Earth by flying saucers.[94]

[91] Holly Patrick, "Tories are 'political wing of Flat Earth Society', taunts Starmer," *The Independent*, February 28, 2024, accessed March 27, 2024, https://www.independent.co.uk/tv/news/sunak-starmer-liz-truzz-deep-state-b2503941.html.

[92] Martín Sivak, "Axel Kicillof: "Milei quiere resolver los problemas de Argentina con ideas ridículas del siglo XIX"," *El País*, October 5, 2023, accessed May 19, 2024, https://elpais.com/argentina/2023-10-05/axel-kircillof-milei-quiere-resolver-los-problemas-de-argentina-con-ideas-ridiculas-del-siglo-xix.html.

[93] Hebe Vessuri, "Milei charges against Argentine science," *Tapuya: Latin American Science, Technology and Society*, Vol. 7, No. 1 (Apr., 2024), 1–7.

[94] "Lilia Lemoine asumió como secretaria de la Comisión de Ciencias: el video que se viralizó para criticar su nombramiento," *La Nacion*, May 8, 2024, accessed May

The actions of Milei and Lemoine have a curious bit of synchronicity with another government. In 2023, several months before Milei and Lemoine's elections, the ostensibly social democratic Robert Fico (b. 1964) began his third term as Prime Minister of Slovakia, as the head of a coalition government including the far-right Slovak Nationalist Party, and began a program of bringing the media under greater government control. As with Lemoine, in May 2024 it was revealed that the Nationalist Party official who would be in charge of promoting media education to schoolchildren, Lukáš Machala, was a Flat Earther, in addition to holding various other conspiracy views on vaccines and the Illuminati.[95] This comes several months after, in nearby Serbia, the far-right politician Branimir Nestorović (b. 1954) entered that country's National Assembly; while Nestorović's political rise began with his endorsement of covid-related conspiracies in 2020, he has also expressed many other conspiracies, including Flat Earth.[96] Nor is Flat Earth the only alternate cosmology gaining political representation in Eastern Europe. In 2023, Krzysztof Bosak (b. 1982), a far-right member of parliament in Poland, was made deputy leader of the Polish parliament's lower house. In addition to his white supremacist advocacy and Holocaust denialism, Bosak—a Traditionalist Catholic—also rejects heliocentrism and has publicly stated several times his belief that the Sun orbits the Earth.[97] Notably, Bosak also has ties to Jack Posobiec (b. 1984), a far-right Polish-American activist

19, 2024, https://www.lanacion.com.ar/politica/lilia-lemoine-asumio-como-secretaria-de-la-comision-de-ciencias-el-video-que-se-viralizo-para-nid08052024/.

[95] "Is the Earth round or flat? Culture Ministry chief official says proof is still needed," *The Slovak Spectator*, May 3, 2024, accessed May 15, 2024, https://spectator.sme.sk/c/23325477/is-the-earth-round-or-flat-culture-ministry-chief-official-says-proof-is-still-needed.html?ref=njctse; Peter Dlhopolec, "News digest: Flat earther might teach pupils how to use media and think critically," *The Slovak Spectator*, May 14, 2024, accessed May 15, 2024, https://spectator.sme.sk/c/23330114/news-digest-flat-earther-might-teach-pupils-how-to-use-media-and-think-critically.html.

[96] Mirko Dautović, "Serbia's Elections – Defeat for Far Right, as well as Moderate Opposition," *Balkan Insight*, January 5, 2024, accessed November 11, 2024, https://balkaninsight.com/2024/01/05/serbias-elections-defeat-for-far-right-as-well-as-moderate-opposition/.

[97] "Krzysztof Bosak i jego teoria heliocentryzmu. "Nie ma fizycznych dowodów, na to które krąży wokół którego"," *Dziennik Gazeta Prawna*, March 26, 2018, accessed March 16, 2025, https://wiadomosci.dziennik.pl/wydarzenia/artykuly/571442,krzysztof-bosak-o-ziemi-i-sloncu.html.

who popularized Pizzagate (the precursor to QAnon) and played a major role in Trump's three presidential campaigns.[98]

In Argentina, Slovakia, and Serbia, at least, Flat Earth is certainly represented within right-wing populist governments. It should also be noted that Trump and Bolsonaro's usage of Flat Earth is not the only way that the concept has been used by right-wing authoritarian movements in modern democracies. In 2014, Narendra Modi (b. 1950) became Prime Minister of India, the world's largest democracy, when his Bharatiya Janata Party (Indian People's Party; BJP) won that year's elections. The BJP is a right-wing Hindu nationalist party which has attempted to turn back India's secular multicultural status and promote a Hindu-only society, largely by attacking Muslims, metaphorically as well as physically. Prior to becoming Prime Minister, Modi had served as chief minister of the state of Gujarat from 2001–2014, during which time his BJP administration encouraged a massive pogrom against Muslims in 2002. In foreign relations, Modi's government has funded a number of both pro-India and pro-Hindu politicians and activists, particularly in North America.[99] The consolidation of Modi's rule has seen both the BJP government and its grassroots supporters endorse a number of Hindutva—revisionist Hindu supremacist—views of history that had been promoted since the 1980s by author Purushottam Nagesh Oak (1917–2007) and which have exploded on Facebook. Most famously is the attempt to claim that the Taj Mahal was not a tomb to a Muslim ruler, but actually a Hindu temple taken over by Muslims, but also included are incendiary claims that the Kaaba—the center of Islam in Mecca—was a former Hindu temple, the Catholic Church originated as a Vedic priesthood, and that Roman ruins are actually remains of an ancient Indian ATAC that had colonized Europe.[100]

[98] "OANN's Posobiec Met With Polish Neo-Fascists and Amplified Their Messages Online," Southern Poverty Law Center, July 22, 2020, accessed March 16, 2025, https://www.splcenter.org/resources/hate-watch/oanns-posobiec-met-polish-neo-fascists-and-amplified-their-messages-online/.

[99] Siddhartha Deb, *Twilight Prisoners: The Rise of the Hindu Right and the Fall of India* (Chicago: Haymarket Books, 2024), 13–43; Mishra, *Age of Anger*, 252–72; Rachman, *The Age of the Strongman*, 73–88.

[100] Cynthia Ann Humes, "Hindutva, Mythistory, and Pseudoarchaeology," *Numen*, Vol. 59, No. 2/3 (2012), 178–201; Kathinka Frøystad, "Affective digital images: Shiva in the Kaaba and the smartphone revolution," in Paul Rollier, Kathinka Frøystad, and Arild Engelsen Ruud, eds., *Outrage: The Rise of Religious Offence in Contemporary*

It is not only history that has become susceptible to Hindutva in Modi's India, but science as well, with its literalist interpretation of Vedic scriptures promoting concepts similar to ancient alien supporters in the West. This included Modi in 2014 stating that ancient Hindus had developed genetic engineering and plastic surgery, and Hindutva attendees of the 102nd Indian Science Congress in 2015 claiming that Indians had developed interplanetary spacecraft and anti-gravity technology 7000 years ago.[101] The BJP government of both India and the states it controls has not only promoted these Hindutva views of science in subsidized private schools, but increasingly ostensibly-secular public schools as well.[102] Along with severely underfunding public education in general, a 2023 revision to high school textbooks by Modi's government completely removed such key science topics as the periodic table of elements, evolution, and the Pythagorean theorem.[103]

As mentioned in earlier chapters, similar literalist interpretations of certain Hindu texts would give support for a Flat Earth cosmology. Some of the figures in the Indian diaspora, including those lobbied by Modi and the BJP, have endorsed those views. A notable example was Makhar Singh "Mak" Parhar (1973–2021), an Indian-Canadian owner of a Bikram Yoga studio in British Columbia. After initially supporting alternative medical views, he adopted Flat Earth in January 2017—incidentally, the month of Trump's (first) inauguration—and from there veered to covid conspiracies in March 2020. In October 2021, Parhar broke Canadian quarantine guidelines to attend a Flat Earth convention in South Carolina, where

South Asia (London: UCL Press, 2019), 123–48; Manimugdha S. Sharma, "Faking history starts online," *The Times of India*, November 19, 2017, accessed February 25, 2024, https://timesofindia.indiatimes.com/home/sunday-times/faking-history-starts-online/articleshow/61705453.cms; Pankaj Mishra, *From the Ruins of Empire: The Intellectuals Who Remade Asia* (New York: Farrar, Straus and Giroux, 2012), 95–96; Benjamin B. Olshin, *Lost Knowledge: The Concept of Vanished Technologies and Other Human Histories* (Leiden: Brill, 2019), 150; Deb, *Twilight Prisoners*, 139–54.

[101] Nand Teltumbde, "Saffron Science," *Economic and Political Weekly*, Vol. 50, No. 6 (7 Feb., 2015), 10–11; Deb, *Twilight Prisoners*, 139–41.

[102] Snigdhendu Bhattacharya, "How RSS textbooks are reshaping Indian history and science under Modi," *Al Jazeera*, February 19, 2024, accessed December 12, 2024, https://www.aljazeera.com/news/2024/2/19/how-rss-textbooks-are-reshaping-indian-history-and-science-under-modi.

[103] Robert F. Worth, "Narendra Modi's Populist Facade Is Cracking," *The Atlantic*, January 2, 2025, accessed January 2, 2025, https://www.theatlantic.com/magazine/archive/2025/02/modi-nationalism-demagoguery-limitations/681094/.

he got covid. In early November, Parhar accidentally killed himself by consuming a mixture of fentanyl, ethanol, and cocaine he seemingly believed would cure him of the virus he publicly claimed not to have contracted.[104]

However, in contrast to Hindutva interpretations of Vedic science and Hindutva supporters in the diaspora, the Hindu supremacists within Modi's India have developed a different political usage of Flat Earth. There, BJP members have cited the claims by Abd al-Aziz ibn Baz and other recent Salafist clerics to argue that Islam is inherently in support of Flat Earth, as a way to reject the compatibility of Islam with modern thought, denigrate it in contrast to traditional Hindu science—an inversion of al-Biruni's views of a thousand years prior—and justify the BJP's attempts to remove Muslims and Islam from the public sphere in India. In framing Islam in such a manner, the BJP has drawn from the online claims of Western "Ex-Muslim" organizations that are linked with the right-wing New Atheist movement, which are ostensibly just as incompatible with the Hindu literalism of Hindutva but are willing to form alliances with anti-Muslim religious figures.[105]

The BJP use of *rejecting* Flat Earth shows that the concept has right-wing political viability beyond its endorsement. Consider that in 1998, David Duke (b. 1950), former leader of the Ku Klux Klan, published his memoirs. The foreword was written by geneticist Glayde Whitney (1939–2002), in which he wrote that "racial egalitarianism is the scientific equivalent of the flat-Earth theory."[106] Clearly things have changed for the far-right in the United States, as well as United Kingdom and Brazil (if not India) in between 1998 and 2015. A 2019 study of Flat Earth videos concluded that while those which mentioned religion were universally positive toward it, only a very small number made explicit references to "conservative" political views, leading the study author

[104] Beres, Remski, and Walker, *Conspirituality*, 87–89.

[105] Aditi Joshi, "Islam's Flat Earth Controversy: Unveiling The Debate Over Earth's Shape And The Role Of Fatwas In Islamic Thought," *The Jaipur Dialogues*, August 15, 2023, accessed August 15, 2023, https://www.thejaipurdialogues.com/islam/islams-flat-earth-controversy-unveiling-the-debate-over-earths-shape-and-the-role-of-fatwas-in-islamic-thought/; Taner Edis, "Finding an Enemy: Islam and the New Atheism," in Ruqayya Yasmine Khan, ed., *Muhammad in the Digital Age* (Austin: University of Texas Press, 2015), 174–92.

[106] Oksana Yakushko, *Scientific Pollyannaism: From Inquisition to Positive Psychology* (London: Palgrave Macmillan, 2019), 173.

to state "that (at least explicitly) political views were not a particularly popular component of the flat-Earth discourse in the YouTube videos sampled."[107] As demonstrated above, even by 2019 the idea that there was almost no endorsement of right-wing politics by online Flat Earth promotion is very difficult to sustain. Instead, it points to a similar issue as brought up by Barbosa on Terraplanismo as a method of government under Bolsonaro. By the early twenty-first century, as with other types of paranormal and conspiratorial views, Flat Earth had become both a signifier and a methodology of the far-right. Flat Earthers did not need to explicitly support Trump, Bolsonaro, or other figures. By simply stating their view of the Earth's shape, their views toward science, government, experts, regulations, and religion would be clear—and not merely to other Flat Earthers, but to other paranormal believers and far-right activists alike, in line with the AIN/PUFT models.

As mentioned in chapter one, it is possible that the very first instance of Flat Earth politics was the speculative involvement by Lactantius in a 326 CE coup attempt against Emperor Constantine. It is perhaps apt that modern Flat Earth politics were associated with coup attempts in 2021 and 2023. In each of those cases, Flat Earth political aspirations ultimately failed, but not before effecting some final gasps of political heterodoxy, which has had a major impact on the wider political culture. If not able to fully throw the hated establishment off the edge, then Flat Earth politics has, to turn a phrase, been able to reshape at least a section of the world beyond its own horizons.

In 2023, Mark Sargent stated that "Flat Earth is taking over."[108] This was a bold claim, not least of which given the decline of outward Flat Earthers. However, it seems to indicate a realization that Sargent himself might not have been aware of: the methodology of Flat Earth is indeed on the ascent, Terraplanismo remaining even when Bolsonaro's cabinet ministers have long since departed. To those who view themselves as an oppressed minority, whether with justification (Black Pan-Africanists in the nineteenth century) or not (white Christian Republicans in the twenty-first century), this perhaps gives further context to the appeal

[107] Shaheed N. Mohammed, "Conspiracy Theories and Flat Earth Videos on YouTube," *The Journal of Social Media in Society*, Vol. 8, No. 2 (Fall, 2019), 94–95.

[108] SciManDan, "Flat Earther Thinks They are Taking Over the World," YouTube, December 1, 2023, accessed December 2, 2023, https://www.youtube.com/watch?v=SVFl9DKf18o.

for Flat Earth. If they are a minority on a spherical Earth, perhaps a Flat Earth would be more equal. In 1931, Polish-American philosopher Alfred Korzybski (1879–1950) wrote his most famous phrase, "The map is not the territory." An addendum might be added: that Flat Earth is not merely the physical shape of the world. Dr. James Lattin, curator of the Museum of Imaginative Knowledge, expounded on this concept by stating that "Flatness is more than just a physical thing, it is a means of dissolving hierarchies of knowledge."[109] This is perhaps the most apt summary of the thesis of this book on the appeal of Flat Earth to oppressed people, especially in modern times, and sets up the concluding topic of the book.

[109] Kay Burns, *Museum of the Flat Earth Strategic Plan 2016–2020* (Fogo Island, NL: Museum of the Flat Earth, 2016), 4.

Conclusion: From Afronauts to Mad Mike

In 1960, the Virginian journalist Burke Davis (1913–2006) published *Our Incredible Civil War*, a book that aimed to present interesting details from the US Civil War to a young audience. The last chapter of the book was on folklore of the war, with the final two pages of both the chapter and book dedicated to the story of the "Confederate rocket." This was an alleged two-stage liquid-fuel rocket built at the Tredegar Iron Works in Richmond with contributions from Lord Kelvin, Austrian physicist Ernst Mach (1838–1916), and Matthew Fontaine Maury (1806–1873), a Confederate naval officer who had earlier served as Superintendent of the United States Naval Observatory. The Confederate Army's Torpedo Bureau supposedly launched the rocket from Richmond in early March 1865 toward Washington, DC, with Confederate President Jefferson Davis (1808–1889) in attendance. However, the rocket never hit the US Capitol—because it entered orbit, where it remained as of 1960. Davis claimed that the story originated in 1958 by "C. R. Johnson," supposedly the Vienna correspondent of a "Southern newspaper" and referred to by Davis as a "folklorist or prankster."[1]

The Confederate rocket story remained in circulation for decades, especially once the Internet emerged, and it gained a new audience when

[1] Burke Davis, *The Civil War: Strange & Fascinating Facts* (Avenel, NJ: Wings Books, 1994 [1960]), 245–46.

E. Guimont, *The Power of the Flat Earth Idea*,
Palgrave Studies in the History of Science and Technology,
https://doi.org/10.1007/978-3-032-23981-5

it was the subject of a 2005 episode of the popular US science show *MythBusters* (2003–2016).[2] The 1958 origin of the story is particularly telling. Not only was this the second year of Virginia's so-called "massive resistance" campaign against school desegregation, it was also the year after the Soviet Union beat the United States into space with the launch of Sputnik I. By making Confederate Virginia the location of the first space launch ninety years earlier would justify Virginia's resistance against the Federal government and show that the Confederacy, rather than the United States, had been able to lead the world in technological development and exploration of the final frontier. The Johnson/Davis folktale aside, the Confederacy did manage to achieve a rocket—of a sort. In 1968, the US Air Force donated a now-obsolete Titan I missile to the Rotary Club of Cordele, Georgia for placement in a parking lot next to a restaurant. The display was officially dedicated on 17 July 1969, and given the title of "Confederate Air Force Pad One." Thus, the day after the launch of Apollo 11 from the next state down and three days before the first Moon landing, the Confederacy unveiled its first nuclear missile.[3]

1957 was not only the year of Sputnik and of the first civil rights act in the United States since 1875. It was also the year when the splintering of the British Empire in Africa began, when the colony of the Gold Coast became the independent nation of Ghana. The Empire itself is traditionally seen as ending only eleven years later, with the independence of Swaziland (modern Eswatini) and Mauritius in Africa, and the British Commonwealth Office merging into the Foreign Office. Two of the major imperial holdouts after 1968 were the unresolved statuses of the white settler regimes in South Africa and Southern Rhodesia. But while Southern Rhodesia would remain a diplomatic ulcer until it became Zimbabwe in 1980, Northern Rhodesia, with a much smaller white population, had led a more successful transition to independence as Zambia in 1964.

[2] Erik A. Zavrel, "How the Discovery Channel Television Show "Mythbusters" Accurately Depicts Science and Engineering Culture," *Journal of Science Education and Technology*, Vol. 20, No. 2 (Apr., 2011), 205.

[3] Lenny Flank, "Photo Diary: A Titan I Nuclear Missile in Georgia," *Daily Kos*, October 22, 2022, accessed March 3, 2024, https://www.dailykos.com/stories/2022/10/22/2128343/-Photo-Diary-A-Titan-I-Nuclear-Missile-in-Georgia.

One of the delegates to the convention that drafted Zambia's constitution that year was Edward Mukuka Nkoloso (1919–1989). A World War II veteran who had been educated in a British mission school, he was also a member of the nationalist movement who had been imprisoned by the British administration from 1956–1957 and served as both a security officer for the pro-independence United National Independence Party and a schoolteacher.[4] But Nkoloso was most known for something outside of traditional politics: his founding of the so-called Zambia National Academy of Science, Space Research and Philosophy upon his nation achieving its independence. Nkoloso's goal was the creation of a Zambian space program whose aims were to land an "Afronaut," in his case a teenaged girl named Matha Mwambwa, on the Moon prior to the US or USSR and to make peaceful contact with what he believed to be the primitive, but intelligent, native Martians whom he warned missionaries to avoid forcing Christianity on. In line with this was Nkoloso's support for traditional witch doctors, lobbying the government to give them the same support it did Western physicians.[5]

Although widely mocked by Western media, Nkoloso's Zambian space program was a symbol for the progress that he hoped the newly independent nation could achieve—not only technically, but in developing a different path from either the former colonizers or the two superpowers, using the exploration of space as a metaphor for the synthesis of Western modernity and pre-colonial traditions.[6] After all, the Space Race, especially in the 1960s, was perhaps the most visible aspect of the Cold War, which in the United States internationalized domestic issues of science

[4] Walima T. Kalusa, "Death, Christianity, and African Miners: Contesting Indirect Rule in the Zambian Copperbelt, 1935—1962," *The International Journal of African Historical Studies*, Vol. 44, No. 1 (2011), 89, 109–110.

[5] Joan Ruman Perkal, "Witchcraft Involves Millions in Africa," *Western Folklore*, Vol. 29, No. 4 (Oct., 1970), 284.

[6] Namwali Serpell, "The Afronaut Archives: Reports from a Future Zambia," *Public Books*, March 28, 2019, accessed January 14, 2020, https://www.publicbooks.org/the-afronaut-archives-reports-from-a-future-zambia/.

and race.[7] In this way, the Zambian space program of 1964 serves an almost completely opposite goal than the Confederate missile program of 1958.

2 February 2020 saw the death of a very different figure from Rhodesian history than Nkoloso: Thomas Michael Hoare (b. 1919), known as "Mad Mike." Hoare identified as British but was born in colonial India to Irish parents, at the time when Ireland's partially-successful war for independence had just begun. Hoare served in the British army in Burma during World War II, before moving to South Africa. After attempting to locate the so-called Lost City of the Kalahari Desert—a myth invented by Canadian adventurer William Leonard Hunt (1838–1929) in the 1880s—Hoare became a mercenary, serving South African and Rhodesian interests in the Congo and Seychelles. Despite this adventuring, he managed to die at the age of 100 at his home in Durban. Mad Mike's career demonstrates the fluidity of identity offered by imperial structures, as well as the blending of pseudohistory and adventurism.

Exactly twenty days after Mad Mike Hoare, another Mad Mike died halfway around the world—a daredevil named Michael "Mad Mike" Hughes (b. 1956). A stuntman, in the 2010s Hughes became interested in building a homemade steam-powered rocket that would enable him to ascend high enough to determine whether the Earth was flat or not. In a 2019 documentary about him, Hughes is shown to be cagey about his personal beliefs regarding Flat Earth; depending on his audience he either claims to be a true believer, an agnostic, or a Flat Earth skeptic just using the conspiracy in an ultimately unsuccessful attempt to raise funds for his rocket passion project. However, Kelly Weill, who interviewed Hughes extensively, was convinced of his genuine belief in Flat Earth, after initially doubting it.[8]

[7] Helen Tilley, *Africa as a Living Laboratory: Empire, Development, and the Problem of Scientific Knowledge, 1870–1950* (Chicago: The University of Chicago Press, 2011), 69–114; Audra J. Wolfe, *Freedom's Laboratory: The Cold War Struggle for the Soul of Science* (Baltimore: Johns Hopkins University Press, 2018), 113–34; Thomas Borstelmann, *The Cold War and the Color Line: American Race Relations in the Global Arena* (Cambridge, MA: Harvard University Press, 2001), 135–71; Naomi Oreskes, "Science in the Origins of the Cold War," in Naomi Oreskes and John Krige, eds., *Science and Technology in the Global Cold War* (Cambridge, MA: The MIT Press, 2014), 11–30.

[8] Kelly Weill, *Off the Edge: Flat Earthers, Conspiracy Culture, and Why People Will Believe Anything* (Chapel Hill: Algonquin Books, 2022), 1–5; *Rocketman: Mad Mike's*

Hughes certainly held views in line with many Flat Earthers. In the documentary, reflecting the PUFT concept, Hughes is shown to have had a cat named Shasta, presumably after the New Age/militia hotspot discussed in chapter six; he believed that vaccines cause cancer and that UFOs were a false-flag meant to terrify the world into joining a one-world government and religion; identified NASA, SpaceX (and its CEO Elon Musk specifically), Blue Horizon, and Virgin Galactic as part of the Flat Earth coverup; professed disbelief in gravity, evolution, advanced math, and "science" in general; and stated that Flat Earth was what brought him back into being a Christian after years as an atheist. He also was a sovereign citizen, shown to make his own license plates and attempt to make citizen's arrests when his car was subsequently impounded. Weill noted that Hughes was anti-Trump because his sovereign citizen beliefs placed him so far to the right of Trump that the president was unacceptably leftist. This did not prevent Hughes from attempting to run for governor of California in the November 2018 election, or for using the court system to engage in frivolous lawsuits against celebrities' names to establish them as "legal entities," causing him to be charged with attempted extortion.[9]

On 24 March 2018, after several delays, Hughes flew a short distance in a test launch of a prototype rocket, meant to gain exposure for fundraising purposes to build his larger rocket. Inadvertently connecting with earlier Flat Earth interests, such as those of "Professor" John Wise from chapter two, Hughes' final rocket was to be launched from a balloon high in the atmosphere to provide maximum lift. Hughes chose as his test launch site the town of Amboy, California—reputed to be a haunted ghost town, again demonstrative of the PUFT principle.[10] Showing his Flat Earth heterodoxy, the water used as reaction mass in his initial, aborted February test launch contained holy water blessed by Pope Francis.[11] The test launch was successful and gained Hughes publicity, if not a large amount of funding. Hughes' promotional campaign included him being

Mission to Prove the Flat Earth, directed by Toby Brusseau and Michael Linn (New York: Flix Premiere, 2019).

[9] Weill, *Off the Edge*, 137–38, 149; *Rocketman*.

[10] Weill, *Off the Edge*, 152; Alan Burdick, "Looking for Life on a Flat Earth," *The New Yorker*, May 30, 2018, accessed July 23, 2024, https://www.newyorker.com/science/elements/looking-for-life-on-a-flat-earth.

[11] *Rocketman*.

featured on the show *Homemade Astronauts*, whose crew were on location to film his second test flight. In that 22 February 2020 launch, his rocket's parachutes accidentally detached on takeoff, causing it to crash into the desert, killing Hughes. Although the footage of the crash was not used, other footage of him was shown when *Homemade Astronauts* finally aired in 2021. Weill has called this the first death of the Flat Earth movement, although as demonstrated in chapter six, this is not the case (Boko Haram, and Muslim Flat Earthers more broadly, are absent from Weill's book, one of her several historical blind spots).[12]

In comparing the two figures of Edward Mukuka Nkoloso and "Mad" Mike Hughes, it is inarguable that Hughes—or rather, the colleagues who donated their time and knowledge to help him—had more technical expertise than Nkoloso. But Nkoloso perhaps had a firmer grasp of science. His scientific views were mocked by Western media, and in some cases were at odds with Western science, but that does not mean they were unscientific, merely reflecting a different framework. Even in the twenty-first century, South African science educator Sibusiso Biyela has spoken of the need to explain the concept of "the Sun" to an isiZulu speaker, who did not "know" about the Sun before, just as there is no isiZulu word for "planet," despite living on one. As part of his efforts to develop a scientific lexicon in the language, Biyela adopted the isiZulu word for "wanderer," *umhambi*, given that is the meaning of "planet" in Greek.[13]

In terms of nationalism, Hughes rejected the national impact and pride of space exploration that came from NASA—and *Homemade Astronauts* makes it clear that Hughes' other non-Flat Earth rivals also viewed themselves as populist upstarts against both NASA and billionaires like

[12] Weill, *Off the Edge*, 170.

[13] Sibusiso Biyela, "Decolonize Science Writing in South Africa." Lecture, British Society for the History of Science Global Digital History of Science Festival, 6 July, 2020; Sibusiso Biyela, "Decolonizing Science Writing in South Africa," *The Open Notebook*, February 12, 2019, accessed August 18, 2022, https://www.theopennotebook.com/2019/02/12/decolonizing-science-writing-in-south-africa/. See also Marissa Rollnick, "The Growth of Science Education Organisations in the Old and New South Africa," 27–41; Paul Webb, "Science Education in South Africa: Issues of Language and Scientific Literacy," 119–31, in Femi S. Otulaja and Meshach B. Ogunniyi, eds., *The World of Science Education: Handbook of Research in Science Education in Sub-Saharan Africa* (Rotterdam: Sense Publishers, 2017); and Jörg Matthias Determann, *Diversity, Equity, and Inclusion in Astronomy: A Modern History* (Cham: Springer Nature, 2023), 165–70.

Musk—while Nkoloso embraced the zeitgeist of the Space Race era as a reflection of building a new nation free of the errors of the past, paradoxically closer to the goal of the Confederate rocket urban legend. But even so, there is a piece of mutual DNA that can link the aspiring Afronaut and the Flat Earther daredevil. According to Weill, Hughes' route to Flat Earth belief began with learning about the 1890 massacre of three hundred Lakota Native Americans by the US Army at Wounded Knee Creek, South Dakota, and deciding to never trust the government again.[14] Even in the individualistic, sovereign citizen Hughes, there was perhaps a hint of the anti-imperial.

Ryan Nefdt, professor of philosophy at the University of the Western Cape, has noted that in line with French philosopher Edouard Glissant (1928–2011)'s famous dictum that "the West is not the West: it is a project not a place," that "Perhaps… we can appreciate an African philosophical project not bound by geography or history but not ignorant of them either."[15] In line with that, Nefdt may have appreciated the comment that Nkoloso made at the press conference announcing the foundation of his Zambian space academy: "Most Westerners don't even know whereabouts in Africa we are."[16] The Western scientific project of mapping the colonized space of Africa proved as incomplete and inaccurate as the Boer imagination of Africa as an outpost of King Solomon along the edge of the Flat Earth.

As mentioned earlier, the most common conception of the Flat Earth is the world as a disk—the azimuthal equidistant projection of al-Biruni—with the North Pole at the center and Antarctica as the outer rim. In this cosmology, Antarctica is actually an ice wall that keeps our world bottled up. Flat Earthers are not the only conspiracists to place a premium in Antarctica, whose remoteness, harshness, uninhabited nature, and relatively recent discovery (1820) make it ideal as a final refuge for all notions of forbidden marvels. As noted in chapter two, Edgar Allan Poe's 1837 novel *The Narrative of Arthur Gordon Pym* drew influence from Hollow Earth theories in its concluding establishment of vanished ATACs and

[14] Weill, *Off the Edge*, 149–50.

[15] Ryan M. Nefdt, "Decolonizing philosophy," *Africa is a Country*, June 30, 2017, accessed March 19, 2019, https://africasacountry.com/2017/06/decolonizing-philosophy.

[16] Serpell, "The Afronaut Archives."

strange animals in Antarctica. In 1931, H. P. Lovecraft wrote a spiritual sequel to Poe's *Pym*, *At the Mountains of Madness*. Lovecraft's novella established the connection of the continent with both ancient alien settlements and flying saucer bases that have become so prominent in arguably the most emblematic conspiracy of the current zeitgeist, UFO mythology.[17]

But just as Antarctica has taken on different meanings for various conspiracy beliefs, it has no static role among Flat Earthers, for whom there is no consensus view on what lies beyond the Antarctic ice wall. Those following a more Biblical literalist evangelical view will say that there is only void beyond, the domed firmament above, and the ice wall keeps the oceans from spilling out into the abyss. But more speculative Flat Earthers imagine that our world is on an infinite plane, and if one were to find an opening in the ice wall, an expedition across the interplanetary ocean to other archipelagic worlds would be possible. In 2022, Argentine Flat Earther Claudio Nocelli wrote the book *The Navigator Who Crossed the Ice Walls: Worlds Beyond the Antarctica* [*sic*], an allegedly nonfiction account of a man named William Morris. A captain in the Continental Navy during the American Revolutionary War, in 1830 (45 years after the dissolution of the Continental Navy) Morris sails to Antarctica, finding a hole in the ice wall and navigating to other worlds beyond our own on the Flat Earth. Nocelli's book has been taken as fact by Flat Earthers on social media, particularly Facebook and TikTok, where believers have drawn maps of the 178 worlds that Nocelli claims exist upon the immense plane ocean beneath the dome of the firmament.[18] (Fig. A1).

Nocelli was beaten to his idea by British science fiction author Charles Stross (b. 1964), who in 2005 published the novella *Missile Gap*. In it, a copy of the Earth on the eve of the 1962 Cuban Missile Crisis is transported to an Alderson disk, an enormous construct around a star that provides fifty million times the living space than the spherical Earth, beyond even the Ringworld of chapter two. The spherical Earth

[17] John M. Navroth, "Lovecraft and the Polar Myth," *Lovecraft Annual*, No. 3 (2009), 190–98; Allegra Rosenberg, "The Truth About Antarctica," *The Long Now Foundation*, February 21, 2023, accessed February 21, 2023, https://longnow.org/ideas/the-truth-about-antarctica/.

[18] Claudio Nocelli, *The Navigator Who Crossed the Ice Walls: Worlds Beyond the Antarctica* (Buenos Aires: Claudio Nocelli, 2022), 3–9.

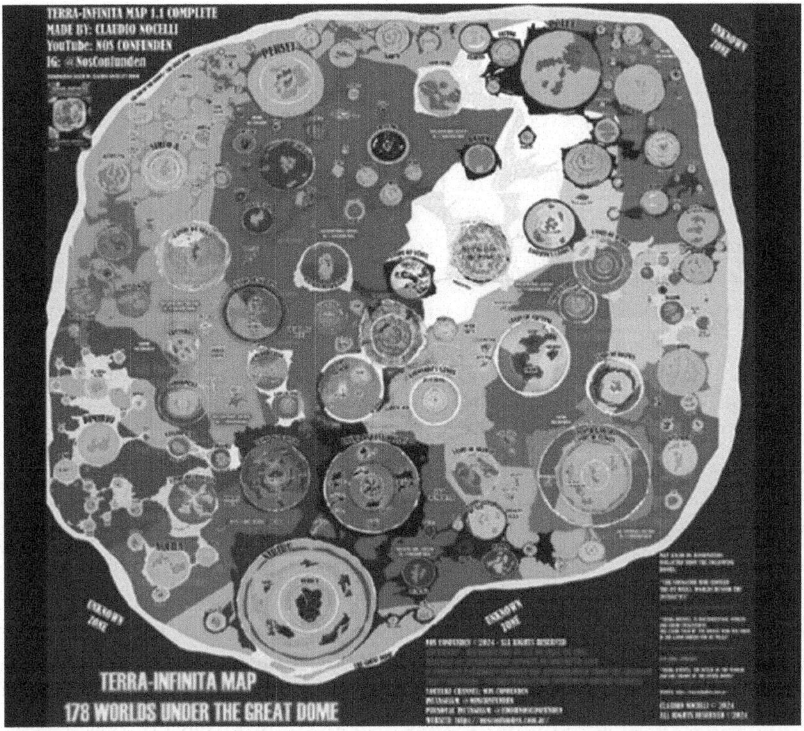

Fig. A1 Claudio Nocelli's map of "Terra Infinita:" Earth and 177 other worlds, imagined as flattened-projection archipelagos, existing on the immense Flat Earth beneath the dome of the celestial firmament and linked by a navigable ocean[19]

is "unpeeled" into the flat shape of the azimuthal equidistant projection map; as in Nocelli, the ocean that extends beyond Antarctica leads to other archipelagos of worlds millions of kilometers away. The story is set in 1976, and one of the main characters is Yuri Gagarin (1934–1968), the Soviet cosmonaut who was the first person in space in 1961. The historical Gagarin died earlier, but in this setting he carries out expeditions to these nearby worlds on a giant nuclear-powered ground-effect vehicle—rockets

[19] Nocelli, *The Navigator Who Crossed the Ice Walls*, 2.

not being effective on an Alderson disk due to the immense gravitational shear of a near-endless disk.[20]

Gagarin, in *Missile Gap* as in reality a living symbol of Soviet scientific achievement and superpower status, transforms the prestige of spaceflight into the prestige of exploration beyond the boundaries of the Flat Earth, bridging the Flat Error beliefs of Columbus (as in Philip José Farmer's "Sail On! Sail On!" from chapter two), the nation- and future-building dream of Nkoloso, the rocket-building project of Mad Mike, and the attempts of Mike's contemporary Flat Earthers to fund ship or aerial expeditions to Antarctica to cross the ice wall, which were being planned in 2019 before the following year's covid pandemic halted those efforts.[21] Just as Homer viewed the Earth as flat in the *Odyssey*, and *2001: A Space Odyssey* helped fuel the Apollo hoax narratives which fed into Flat Earth beliefs, the fictional Gagarin—and his successor, Captain Morris—embarked on Zetetic odysseys of their own (Fig. A2).

Appropriately, 2020 also saw the debut of the comic *The Department of Truth*, about a titular US government agency which investigates a plot by right-wing billionaire brothers—based on the real-life brothers Charles and David Koch—who finance various conspiracy theories, in the hope of instantiating them into reality using the notion of the tulpa, a Theosophical concept loosely adapted from Buddhism and popularized by media such as *The X-Files* and *Twin Peaks*, whose basis is that something believed in by enough people will become physically real.[22] The debut issue of the comic features an FBI researcher who, while studying "The Flat Earth Society Annual Gathering" (and noticing a vendor selling a book titled *Flat Earth and the Protocols of the Elders of Zion*), is recruited by the Koch stand-ins, and flown on private jet to see the Antarctic ice

[20] Charles Stross, "Missile Gap," in Gardner Dozois, ed., *One Million A.D.* (Garden City, NY: Science Fiction Book Club, 2005), 283–347.

[21] Weill, *Off the Edge*, 153–57; Jim Dobson, "Flat Earth Supporters Now Plan An Antarctica Expedition To The Edge Of The World," *Forbes*, March 16, 2019, accessed March 20, 2024, https://www.forbes.com/sites/jimdobson/2019/03/16/flat-earth-supporters-now-plan-an-antarctica-expedition-to-the-edge-of-the-world/?sh=335d9bd59165.

[22] Natasha L. Mikles and Joseph P. Laycock, "Tracking the Tulpa: Exploring the "Tibetan" Origins of a Contemporary Paranormal Idea," *Nova Religio*, Vol. 19, No. 1 (Aug., 2015), 87–97.

Fig. A2 "Objevitelé vesmíru/Člověk a kosmos" ("Discoverers of the Universe/ Man and Cosmos"), 1975 sculpture by Vàclav Uruba (1928–1983) in Ostrava, modern Czech Republic. The sculpture commemorates the 1543 publication of *On the Revolutions of the Heavenly Spheres* by Nicolaus Copernicus, and the 1961 flight of *Vostok 1* by Yuri Gagarin[23]

wall. Through their funding of Flat Earthers, the Flat Earth itself had managed to manifest as a tulpa.[24]

As recounted about, *The Department of Truth* was slightly misguided in expecting Flat Earth advocacy to continue to grow after 2020,

[23] "Václav Uruba: Objevitelé vesmíru / Člověk a kosmos," Ostravské sochy, n. d., accessed February 18, 2025, https://ostravskesochy.cz/dilo/752-Objevitele-vesmiru-Clovek-a-kosmos.

[24] James Tynion IV and Martin Simmonds, "The Department of Truth #1," Image Comics, September 30, 2020, accessed January 23, 2025, https://imagecomics.com/comics/releases/the-department-of-truth-1.

although perhaps less so in relation to the movement's connections to funders of conservative politics. Nevertheless, it did anticipate the fruition of the Flat Earther plans to actually go to Antarctica to investigate. This mission finally took place in December 2024 as "The Final Experiment," an expedition led by Colorado pastor Will Duffy. The expedition included both Flat Earth believers and skeptics, allowing them to observe firsthand both the land of the ice wall and the twenty-four hour sun, an undertaking which as of publication has had mixed reception from the Flat Earth community.[25]

In his 2024 study of Bigfoot in US culture, American journalist John O'Connor proposed that Bigfoot hunting has become so popular in the twenty-first century because it offers an alternative to religion. Bigfoot represents confirmation that modern science and understanding of the world is not complete, and marvels exist beyond what official sources state. While the promise of religion can never be empirically proved, at least in this life, Bigfoot hypothetically could be, at least by those who believe he is still to be found wandering the Pacific Northwest. Thus the search for Bigfoot serves as a stand-in for the search for religious truth.[26]

Flat Earthers are similar in that, to believers, it is something that can be scientifically confirmed—via the classic Bedford Level test or more modern ARIG experiments with homemade rockets, chartered flights over Antarctica, sailing up to the ice wall, or testing the Earth's (lack of) motion with gyroscopes. But unlike Bigfoot, the scientific confirmation of the Flat Earth *would* also be the scientific confirmation of the veracity of religion, proving that God did indeed build the firmament over our small island disc in the center of creation, only six thousand years ago. In this way, the Flat Earth would not only be the synthesis of religion and science, tradition and modernity, but a revelation on the level of every religious reform and scientific revolution combined—the impact of Luther, Columbus, Copernicus and Darwin in one. This would be

[25] Laura Simmons, "Flat-Earthers Travel To Antarctica To Test Theories, But Are Quickly Humbled," IFLScience, December 17, 2024, accessed December 19, 2024, https://www.iflscience.com/flat-earthers-travel-to-antarctica-to-test-theories-but-are-quickly-humbled-77254.

[26] John O'Connor, *The Secret History of Bigfoot: Field Notes on a North American Monster* (Naperville, IL: Sourcebooks, 2024), 228–61. For Bigfoot hunting as a (failed) populist attempt to understand and apply the scientific method, see Sharon A. Hill, *Scientifical Americans: The Culture of Amateur Paranormal Researchers* (Jefferson, NC: McFarland & Company, 2017), 56–68.

an ultimate millenarian development, to be eclipsed only by the Second Coming, an apocalyptic transformation such as the Storm promised by QAnon, but without the apocalypse. The utopia would arrive, and all strife—the racial oppression of John Jasper's parishioners, the colonialist and statist oppression of Nigerians and Tunisians, the economic pitfalls of Mad Mike's compatriots—would all be swept away.

On the Flat Earth, after all, everyone is on an equal level.

Bibliography

Misbah Abdulbaqi, "Higher Education in Afghanistan," *Policy Perspectives*, Vol. 6, No. 2 (Jul.–Dec., 2009), 99–117.

Charlie Abizaid, "Re: The Enlightment of the World," The Flat Earth Society, January 3, 2009, accessed February 8, 2024, https://www.theflatearthsociety.org/forum/index.php?topic=25630.0.

John G. Abizaid, *The Enlightenment of the World, Second Edition* (Boston: John G. Abizaid, 1912).

John G. Abizaid, "New correct map of the flat surface, stationary earth [I]," Library of Congress, c. 1920, accessed January 6, 2024, https://www.loc.gov/item/2013585077/.

John G. Abizaid, "New correct map of the flat surface, stationary earth [II]," Library of Congress, c. 1920, accessed January 6, 2024, https://www.loc.gov/resource/g9930.ct003852v/?r=-0.832,-0.034,2.664,1.241,0.

"About the Dakila Institute," Dakila Association, accessed February 21, 2024, https://dakila.si/en/about-the-institute/.

"Abu Dhabi University and Dakila Research Social Sciences UAE collaborate to advance technologies," Abu Dhabi University News & Articles, August 28, 2023, accessed February 21, 2024, https://www.adu.ac.ae/news-and-events/news/news/detail/2023/08/28/abu-dhabi-university-and-dakila-research-social-sciences-uae-collaborate-to-advance-technologies.

H. H. D'Arcy Adams, "He Only Saw the Plane," *The Earth*, Vol. III, Nos. 33 & 34 (Apr./May, 1903), 178–80.

Hunter Havelin Adams III, "African Observers of the Universe: The Sirius Question," in Ivan Van Sertima, ed., *Blacks in Science: Ancient and Modern* (New Brunswick, NJ: Transaction Books, 1983), 27–46.

E. Guimont, *The Power of the Flat Earth Idea*,
Palgrave Studies in the History of Science and Technology,
https://doi.org/10.1007/978-3-032-23981-5

"Advertisements," *The Journal of the Manchester Geographical Society*, Vol. 18, Nos. 10–12 (Oct.–Dec., 1902), 246.

The Aeronauts. Directed by Tom Harper. Burbank, CA: Mandeville Films, 2019.

Germano Bruno Afonso, "The Brazilian Indigenous Planetary-Observatory," *Boletim da Sociedade Astronômica Brasileira*, Vol. 21, No. 1 (Aug., 2003), 69.

Agora. Directed by Alejandro Amenábar. Madrid: Telecinco Cinema, 2009.

Grace Oluremi Akanbi and Alice Arinlade Jekayinfa, "Education and emancipation, educational policies and «de-emancipation»: A history of the Nigerian education system from 1914 to 2014," *Tiempo y Educación*, Vol. 6, No. 2 (Jul.–Dec., 2019), 177–96.

Fred Femi Akinfala, Gabriel Aunde Akinbode, and Ibi-Agiobu Kemmer, "Boko Haram and Terrorism in Northern Nigeria: A Psychological Analysis," *British Journal of Arts and Social Sciences*, Vol. 17, No. 1 (2014), 115–36.

Gerardo Aldana, *The Apotheosis of Janaab' Pakal: Science, History, and Religion at Classic Maya Palenque* (Boulder: University Press of Colorado, 2007).

Buzz Aldrin and John Barnes, *Encounter with Tiber* (London: Hodder & Stoughton, 1996).

Amir Alexander, *Proof!: How the World Became Geometrical* (New York: Farrar, Straus and Giroux, 2019).

"Al-Hoda Newspapers, ," Khayrallah Center for Lebanese Diaspora Studies, n. d., accessed January 7, 2024, https://lebanesestudies.omeka.chass.ncsu.edu/collections/show/58.

"Alien Nut Tucker Carlson 'Set to Beg Donald Trump to Let Him Get to Bottom of America's UFO Mysteries'," Radar Online, January 2, 2025, accessed January 7, 2025, https://radaronline.com/p/tucker-carlson-beg-donald-trump-get-to-bottom-ufo-mysteries-america/.

James J. Allegro, "The bottom of the universe: Flat earth science in the Age of Encounter," *History of Science*, Vol. 55, No. 1 (2017), 61–85.

Rafael Antunes Almeida, ""Eles nos dizem tudo o que estão fazendo": ciência e conspiração nos discursos terraplanistas," *Anuário Antropológico*, Vol. 48, No. 2 (2023), 102–22;

Kyle Almond, "How ISIS changed Iraqi schools," CNN, July, 2017, accessed February 19, 2021, https://www.cnn.com/interactive/2017/07/world/iraq-schools-cnnphotos/.

David D. Anderson, *Ignatius Donnelly* (Boston: Twayne Publishers, 1980).

David S. Anderson, "Black Olmecs and White Egyptians: A Parable for Professional Archaeological Responses to Pseudoarchaeology," in Jeb J. Card and David S. Anderson, eds., *Lost City, Found Pyramid: Understanding Alternative Archaeologies and Pseudoscientific Practices* (Tuscaloosa: University of Alabama Press, 2016), 68–80.

Eric D. Anderson, "Black responses to Darwinism, 1859–1915," in Ronald L. Numbers and John Stenhouse, eds., *Disseminating Darwinism: The Role of Place, Race, Religion, and Gender* (Cambridge: Cambridge University Press, 1999), 247–66.

A. J. Angulo, "William Barton Rogers and the Southern Sieve: Revisiting Science, Slavery, and Higher Learning in the Old South," *History of Education Quarterly*, Vol. 45, No. 1 (Spring, 2005), 18–37.

Zayde Antrim, *Routes and Realms: The Power of Place in the Early Islamic World* (Oxford: Oxford University Press, 2012).

The Appeal, 23 November 1911, 3.

Dan Ariely, *Misbelief: What Makes Rational People Believe Irrational Things* (New York: HarperCollins, 2023).

Olivier Arvisais, Marie-Hélène Bruyère, Chirine Chamsine, and Mohamed Amine Mahhou, "The educational intentions of the Islamic State through its textbooks," *International Journal of Educational Development*, Vol. 87 (Nov., 2021), 1–9.

Olivier Arvisais and Mathieu Guidère, "Education in conflict: how Islamic State established its curriculum," *Journal of Curriculum Studies*, Vol. 52, No. 4 (2020), 498–515.

Olivier Arvisais and Mathieu Guidère, "The Integration of Religious Elements into ISIS Textbooks," *Religion & Education*, Vol. 47, No. 2 (2020), 188–203.

Mick Ashworth, *Why North is Up: Map Conventions and Where They Came From* (Oxford: Bodleian Library Publishing, 2019).

St. Augustine, trans. Henry Bettenson, *Concerning the City of God against the Pagans* (London: Penguin Books, 1984).

Ariella Aïsha Azoulay, *Potential History: Unlearning Imperialism* (New York: Verso Books, 2019).

Christopher D. Bader, Joseph O. Baker, L. Edward Day, and Ann Gordon, *Fear Itself: The Causes and Consequences of Fear in America* (New York: NYU Press, 2020).

Balaam's Ass, "Faith and Science," *The Earth not a Globe Review*, Vol. III, No. 2 (May, 1896), 16–19.

Chris Ballard, "Strange Alliance: Pygmies in the Colonial Imaginary," *World Archaeology*, Vol. 38, No. 1, Race, Racism and Archaeology (Mar., 2006), 131–51.

Justin Baragona, "Lara Logan Turns to a Fellow Newsmax Reject to Whine About Her Ban, Double Down on QAnon," *The Daily Beast*, October 27, 2022, accessed February 13, 2024, https://www.thedailybeast.com/lara-logan-turns-to-a-fellow-newsmax-failure-to-whine-about-her-ban-double-down-on-qanon.

Gustavo Freire Barbosa, "Além da Terra plana: o terraplanismo como método do governo Bolsonaro," *CartaCapital*, January 15, 2020, accessed July 5, 2020, https://www.cartacapital.com.br/opiniao/alem-da-terra-plana-o-terraplanismo-como-metodo-do-governo-bolsonaro/.

Michael Barkun, *Religion and the Racist Right: The Origins of the Christian Identity Movement* (Chapel Hill: University of North Carolina Press, 1997).

Geremie R. Barmé, "China's Flat Earth: History and 8 August 2008," *The China Quarterly*, Vol. 197 (Mar., 2009), 64–86.

Timothy D. Barnes, *Constantine: Dynasty, Religion and Power in the Later Roman Empire* (Chichester: Wiley-Blackwell, 2011).

Timothy D. Barnes, "Lactantius and Constantine," *The Journal of Roman Studies*, Vol. 63 (1973), 29–46.

Ralph Bauer, *The Alchemy of Conquest: Science, Religion, and the Secrets of the New World* (Charlottesville: University of Virginia Press, 2019).

Mia Bay, "The Historical Origins of Afrocentrism," *Amerikastudien / American Studies*, Vol. 45, No. 4, Time and the African-American Experience (2000), 501–12.

Walter E. A. van Beek, "Dogon Restudied: A Field Evaluation of the Work of Marcel Griaule," *Current Anthropology*, Vol. 32, No. 2 (Apr., 1991), 139–67.

Gillian Beer, "'The Death of the Sun': Victorian Solar Physics and Solar Myth," in *The Sun is God: Painting, Literature and Mythology in the Nineteenth Century*, ed. J. D. Bullen (Oxford: Clarendon Press, 1989), 159–80.

Nawal Morcos Bell, "The Age of Mansa Musa of Mali: Problems in Succession and Chronology," *The International Journal of African Historical Studies*, Vol. 5, No. 5 (1972), 221–34.

Dale Beran, *It Came from Something Awful: How a Toxic Troll Army Accidentally Memed Donald Trump into Office* (New York: All Points Books, 2019).

Derek Beres, Matthew Remski, and Julian Walker, *Conspirituality: How New Age Conspiracy Theories Became a Health Threat* (New York: PublicAffairs, 2023).

Alex Berezow and Hank Campbell, *Science Left Behind: Feel-Good Fallacies and the Rise of the Anti-Scientific Left* (New York: PublicAffairs, 2012).

Jerry Bergman, "Colonel James Irwin: Creationist Astronaut," Institute for Creation Research, October 31, 2013, accessed January 24, 2024, https://www.icr.org/article/7724/.

Anna Berkes, "Who is the liar now?," *Thomas Jefferson's Monticello*, November 14, 2008, accessed September 1, 2022, https://www.monticello.org/research-education/blog/who-is-the-liar-now/.

J. D. Bernal, *Science in History, Volume 1: The Emergence of Science* (Cambridge, MA: The MIT Press, 1971 [1954]).

Martin Bernal, *Black Athena Writes Back: Martin Bernal Responds to His Critics* (Durham, NC: Duke University Press, 2001).

Andrés Bernáldez, *Historia de los reyes católicos D. Fernando y Doña Isabel* (Seville: Impr. que fué de J. M. Geofrin, 1870).

Leslie Bethell, *Brazil: Essays on History and Politics* (London: University of London Press, 2018).

Francis S. Betten, "The Knowledge of the Sphericity of the Earth during the Earlier Middle Ages," *The Catholic Historical Review*, Vol. 9, No. 1 (Apr., 1923), 74–90.

Snigdhendu Bhattacharya, "How RSS textbooks are reshaping Indian history and science under Modi," *Al Jazeera*, February 19, 2024, accessed December 12, 2024, https://www.aljazeera.com/news/2024/2/19/how-rss-textbooks-are-reshaping-indian-history-and-science-under-modi.

Michael J. Bielawa, "American Frankensteins: George Porter and George Poe, and Their Attempts to Reanimate the Dead in New England," in *Lovecraftian Proceedings No. 4*, eds. Dennis P. Quinn and Elena Tchougounova-Paulson (New York: Hippocampus Press, 2022), 149–60.

Michael J. Bielawa, *Wicked Bridgeport* (Charleston: The History Press, 2012).

Tim Binnall, "Flat Earther Busted for Coronavirus Conspiracy Graffiti in Texas," *Coast to Coast AM*, July 28, 2020, accessed July 28, 2020, https://www.coasttocoastam.com/article/flat-earther-busted-for-coronavirus-conspiracy-graffiti-in-texas/.

Tim Binnall and Edward Guimont, "Binnall of America: The Revival - E18 | Dr. Edward Guimont," *Binnall of America*, February 21, 2024 (accessed February 21, 2024), https://www.youtube.com/watch?v=thUdGTIX6q8.

Tim Binnall and Edward Guimont, "Binnall of America Summer of Strangeness with Dr. Edward Guimont," *Binnall of America*, August 14, 2020 (accessed August 17, 2020), http://www.binnallofamerica.com/boaa081720.html.

Tim Binnall and Mark Sargent, "BoA Holiday Season 2019: Mark Sargent," *Binnall of America*, November 11, 2019 (accessed November 20, 2019), http://www.binnallofamerica.com/boaa111119.html.

Sibusiso Biyela, "Decolonize Science Writing in South Africa." Lecture, British Society for the History of Science Global Digital History of Science Festival, 6 July, 2020;.

Sibusiso Biyela, "Decolonizing Science Writing in South Africa," *The Open Notebook*, February 12, 2019, accessed August 18, 2022, https://www.theopennotebook.com/2019/02/12/decolonizing-science-writing-in-south-africa/.

Stephen P. Blake, *Astronomy and Astrology in the Islamic World* (Edinburgh: Edinburgh University Press, 2016).

Adam Bledsoe, "Racial Antagonism and the 2018 Brazilian Presidential Election," *Journal of Latin American Geography*, Vol. 18, No. 2 (Jun., 2019), 165–70.

Elizabeth Blount, *Adrian Galilio; or, a Song Writer's Story* (Malvern: The Faith Press, 1898).

Elizabeth Blount, "Ancient and Modern Astronomy," *The Earth*, Vol. VI, Nos. 72 & 73 (Jul./Aug., 1906), 97–101.

Elizabeth Blount, "Answers to Correspondents," *The Earth not a Globe Review*, No. 1, New Series (Sep., 1894), 22–23.

Elizabeth Blount, "Answers to Correspondents," *The Earth not a Globe Review*, No. 4, New Series (Jul., 1895), 95–96.

Elizabeth Blount, "Does the Earth Rotate?," *The Earth*, Vol. III, Nos. 37 & 38 (Aug./Sep., 1903), 260–63.

Elizabeth Blount and E. V. Mulgrave, *Mr. Blatchford's Great Mistake, or, Clarion's "Science" Versus God's Truth* (Kingston Hill: Lady Blount, 1905).

Elizabeth Blount and Albert Smith, *Flat or Spherical?* (Kingston Hill: Lady Blount, 1905).

Elizabeth Blount and Albert Smith, Zetetic Astronomy: or The Sun's Motions North and South; with the Moon's Motions; Fancied and Real. Showing the Uselessness of the Gravitation Theory, etc. (Kingston Hill, Surrey: Lady Blount, 1904).

W. Jeffrey Bolster, *Black Jacks: African American Seamen in the Age of Sail* (Cambridge, MA: Harvard University Press, 1997).

Sarah L. Bosha, Michelle Adeniyi, Jenna Ivan, Roya Ghiaseddin, Fabakary Minteh, Lamin F. Barrow, and Rex Kuye, "The Impact of the Presidential Alternative Treatment Program on People Living with HIV and the Gambian HIV Response," *Health and Human Rights Journal*, Vol. 21, No. 1 (Jun., 2019), 239–52.

Gerardus D. Bouw, "Flat Earth Mythology and Fact," *The Biblical Astronomer*, Vol. 10, No. 94 (Fall, 2000), 23–30.

Gerardus D. Bouw, *Geocentricity: Christianity in the Woodshed* (Miamitown, OH: Daystar Publishing, 2013).

Gerardus D. Bouw, "Panorama," *The Biblical Astronomer*, Vol. 19, No. 127 (Winter, 2009), 9–19.

Joe Boyle, "Nigeria's 'Taliban' enigma," BBC News, July 31, 2009, accessed August 18, 2020, http://news.bbc.co.uk/2/hi/8172270.stm.

Edward Wilmot Blyden, *Christianity, Islam and the Negro Race* (Baltimore, MD: Black Classic Press, 1994 [1888]).

Arna Bontemps, "Introduction," in Langston Hughes and Arna Bontemps, eds., *The Book of Negro Folklore* (New York: Dodd, Mead & Company, 1958), vii-xv.

Daniel Boorstin, *The Discoverers: A History of Man's Search to Know His World and Himself* (New York: Random House, 1983).

Thomas Borstelmann, *The Cold War and the Color Line: American Race Relations in the Global Arena* (Cambridge, MA: Harvard University Press, 2001).

Earl Bousquet, "From Emancipation Through Recognition, Reconciliation, Respect," *The Voice*, August 20, 2015. Accessed July 21, 2020. https://www.thevoiceslu.com/2015/08/from-emancipation-through-recognition-reconciliation-respect/.

John R. Bovee, "Doctor Huguet: Donnelly on Being Black," *Minnesota History*, Vol. 41, No. 6 (Summer, 1969), 286–94.

H. C. Bowker, "Letter to the Editor," *The Earth not a Globe Review*, No. 5 (Jan., 1894), 112.

John P. Boyd, "Meteorology and Oceanography on a Flat Earth," *arXiv* (Mar., 2020), 1–43.

John Bradley and A. Hottentot, "Letters to the Editor," *The Earth not a Globe Review*, Vol. IV (Jul./Sep., 1894), 78.

James Bratt, "Using Kuyper," *Reformed Journal*, May 10, 2021, accessed March 5, 2024, https://reformedjournal.com/using-kuyper/.

Michael T. Bravo, "Precision and Curiosity in Scientific Travel: James Rennell and the Orientalist Geography of the New Imperial Age (1760–1830)," in Jás Elsner and Joan-Pau Rubiés, eds., *Voyages and Visions: Towards a Cultural History of Travel* (London: Reaktion Books, 1999), 162–83.

Ebenezer Breach, *Fifty Scientific Facts For Surrender to Nature's Fixed Truths* (London: John Williams, 1896).

Jeutonne P. Brewer, *The Federal Writers' Project: A Bibliography* (Metuchen, NJ: The Scarecrow Press, 1994).

Noah Brode, "Bigfoot is Real (For 11% of U.S. Adults)," CivicScience, May 11, 2020, accessed August 4, 2024, https://civicscience.com/bigfoot-is-real-for-11-of-u-s-adults/.

Edward Brooke-Hitching, *The Phantom Atlas: The Greatest Myths, Lies and Blunders on Maps* (San Francisco: Chronicle Books, 2018).

B. Charles Brough, "History of Cosmography," *The Zetetic: A Monthly Journal of Cosmographical Science*, No. 1 (Jul. 1872), 7.

C. Mackenzie Brown, *Hindu Perspectives on Evolution: Darwin, Dharma, and Design* (Abingdon: Routledge, 2012).

William Wells Brown, *The Black Man, His Antecedents, His Genius, and His Achievements* (New York: Thomas Hamilton, 1863).

William Wells Brown, *My Southern Home: Or, the South and Its People* (Boston: A. G. Brown & Co., Publishers, 1880).

Dickson D. Bruce, *Black American Writing from the Nadir: The Evolution of a Literary Tradition, 1877–1915* (Baton Rouge: Louisiana State University Press, 1989).

Bernd Brunner, *Moon: A Brief History* (New Haven: Yale University Press, 2010).

Martin Brückner, "Maps, Movement, and American Literature," Mapping Movement in American History and Culture, 2016. Accessed May 24,

2017. https://mappingmovement.newberry.org/essay/maps-movement-and-american-literature.

Martin Brückner, "Popular Map Genres in American Literature," in *Literature and Cartography: Theories, Histories, Genres, ed. Anders Engberg-Pedersen* (Cambridge, MA.: The MIT Press, 2017), 325–60.

Sean Bryant, "The Flat Earth and its Advocates: A List of References," Library of Congress, July 28, 2018, accessed March 21, 2024, https://guides.loc.gov/flat-earth.

E. A. Wallis Budge, *The Gods of the Egyptians, or, Studies in Egyptian Mythology, Vol. I* (London: Methuen & Co., 1904).

Michael Bueckert, "Flight of the Boers," *Africa is a Country*, May 29, 2018, accessed July 1, 2020, https://africasacountry.com/2018/05/flight-of-the-boers.

Michael Buehler, "Extraordinarily rare 1892 flat Earth map by Alexander Gleason," *Boston Rare Maps*, n.d., accessed August 4, 2022, https://bostonraremaps.com/inventory/alexander-gleason-flat-earth-1892/.

Joshua Blu Buhs, *Think to New Worlds: The Cultural History of Charles Fort and His Followers* (Chicago: University of Chicago Press, 2024).

Alan Burdick, "Looking for Life on a Flat Earth," *The New Yorker*, May 30, 2018, accessed July 23, 2024, https://www.newyorker.com/science/elements/looking-for-life-on-a-flat-earth.

Kay Burns, *Museum of the Flat Earth Strategic Plan 2016–2020* (Fogo Island, NL: Museum of the Flat Earth, 2016).

Guy Burrows, *The Land of the Pygmies* (London: C. Arthur Pearson, 1898).

Richard F. Burton, *Two Trips to Gorilla Land and the Cataracts of the Congo, Vol. I* (London: Sampson Low, Son, & Co., 1876).

Una M. Cadegan, "The Queen of Peace in the Shadow of War: Fatima and U.S. Catholic Anticommunism," *U. S. Catholic Historian*, Vol. 22, No. 4 (Fall, 2004), 1–15.

Allen Dwight Callahan, "Perspectives for a Study of African American Religion: From the Valley of Dry Bones," *Nova Religio: The Journal of Alternative and Emergent Religions*, Vol. 7, No. 1 (Jul., 2003), 44–59.

Allen Dwight Callahan, *The Talking Book: African Americans and the Bible* (New Haven: Yale University Press, 2006).

L. Sprague de Camp and Willy Ley, *Lands Beyond: A Fascinating Expedition Into Unknown Lands* (New York: Rinehart & Company, 1952).

Yvonne Captain, "Brazil's Africa Policy under Lula," *The Global South*, Vol. 4, No. 1 (Spring, 2010), 183–98.

Jeb J. Card, "*America Before* as a Paranormal Charter," *The SAA Archaeological Record*, Vol. 19, No. 5 (Nov., 2019), 26–30.

John Carey, "Ireland and the Antipodes: The Heterodoxy of Virgil of Salzburg," *Speculum*, Vol. 64, No. 1 (Jan., 1989), 1–10.

William Carpenter, *One Hundred Proofs that the Earth is Not a Globe, Fifth Edition* (Baltimore: William Carpenter, 1886).

William B. Carpenter, "Nature and Law," *The Modern Review*, Vol. 1, No. 4 (Oct., 1880), 748–70.

Giancarlo Casale, "Did Alexander the Great Discover America? Debating Space and Time in Renaissance Istanbul," *Renaissance Quarterly*, Vol. 72, No. 3 (Fall, 2019), 863–909.

Bartolomé de las Casas, "Journal of the First Voyage of Columbus," Early Americas Digital Archive, 2003, accessed December 10, 2023, http://eada.lib.umd.edu/text-entries/journal/.

Vincent H. Cassidy, "Other Fortunate Islands and Some that were Lost," *Terrae Incognitae: The Journal of the Society for the History of Discoveries*, Vol. 1, No. 1 (1969), 35–39.

Dan Cassino, "2020 Election Conspiracy Believers More Likely to Embrace Bigfoot, Flat Earth," FDU Poll, May 5, 2022 (accessed May 19, 2025), https://portal.fdu.edu/fdupoll-archive/220505/.

Charles A. Cerami, *Benjamin Banneker: Surveyor, Astronomer, Publisher, Patriot* (New York: John Wiley & Sons, 2002).

Allen J. Christenson, trans., *Popol Vuh: The Sacred Book of the Maya* (Norman, OK: University of Oklahoma Press, 2003).

Joseph Steer Christopher, *Natal, Cape of Good Hope* (London: Effingham Wilson, 1850).

Mark Chu-Carroll, "Zombie Math in the Vortex," *Good Math/Bad Math*, January 2, 2018, accessed July 23, 2024, http://www.goodmath.org/blog/2018/01/02/zombie-math-in-the-vortex/.

Stuart Cloete, *Against These Three: A Biography of Paul Kruger, Cecil Rhodes, and Lobengula, Last King of the Matabele* (Cambridge, MA: Riverside Press, 1945).

Jason Colavito, "Did Columbus Find an Ancient Mosque in Cuba?," JasonColavito.com, June 22, 2013, accessed December 10, 2023, https://www.jasoncolavito.com/blog/did-columbus-find-an-ancient-mosque-in-cuba.

Elizabeth Dabney Coleman, "Black Boanerges: John Jasper's Conviction that "De Sun Do Move" Brought Him Worldwide Fame," *Virginia Cavalcade*, Vol. 4, No. 3 (Winter, 1954), 18–22.

John William Colenso, *The Pentateuch and Book of Joshua Critically Examined, Volume 1* (London: Longman, Green, Longman, Roberts, & Green, 1862).

Terry Collins, "Flat Earth on the ballot? Debunked claim pops up in US politics," *USA Today*, May 17, 2025, accessed May 19, 2025, https://www.usatoday.com/story/news/2025/05/17/flat-earth-us-politicians/83492658007/.

Victoria J. Collis-Buthelezi, "Caribbean Regionalism, South Africa, and Mapping New World Studies," *Small Axe*, Vol. 19, No. 1 (Mar., 2015), 37–54.

Congressional Record: Proceedings and Debates of the 89th Congress Volume 111—Part 5 (Washington, D.C.: United States Government Printing Office, 1965).

"Conspiracy theories take root in Italy, survey shows," *Reuters*, March 16, 2019, accessed December 3, 2021, https://www.reuters.com/lifestyle/oddly-enough/conspiracy-theories-take-root-italy-survey-shows-2021-12-03/.

Jesselyn Cook, *The Quiet Damage: QAnon and the Destruction of the American Family* (New York: Crown, 2024).

William Cooper, *Behold a Pale Horse* (Flagstaff: Light Technology Publishing, 1991).

Nicolaus Copernicus, trans. A. M Duncan, *Copernicus: On the Revolutions of The Heavenly Spheres: A New Translation from the Latin by A. M. Duncan* (Newton Abbot: David & Charles, 1976).

Miranda Corcoran, "Science and the Supernatural: Spiritualism, Psychical Research and Pseudo-Science in the Nineteenth Century," in Clive Bloom, ed., *The Palgrave Handbook of Steam Age Gothic* (Houndsmills: Palgrave Macmillan, 2021), 509–26.

Cosmas, trans. and ed. J. W. McCrindle, *The Christian Topography* (Edinburgh: The Hakluyt Society, 1897).

Sam Cowie, "Brazil's flat Earthers to get their day in the sun," *The Guardian*, November 9, 2019, accessed June 7, 2023, https://www.theguardian.com/world/2019/nov/06/brazil-flat-earth-conference-terra-plana.

John Wesley Cromwell, *The Negro in American History* (Washington, D.C.: The American Negro Academy, 1914).

Gerald Roe Crone, "The Origin of the Name Antillia," *The Geographic Journal*, Vol. 91, No. 3 (Mar., 1938), 260–62.

Michael J. Crowe, *The Extraterrestrial Life Debate 1750–1900: The Idea of a Plurality of Worlds from Kant to Lowell* (Cambridge: Cambridge University Press, 1986).

Christopher Cullen, "A Chinese Eratosthenes of the Flat Earth: A Study of a Fragment of Cosmology in Huai Nan tzu," *Bulletin of the School of Oriental and African Studies, University of London*, Vol. 39, No. 1 (1976), 106–27.

Christopher Cullen and Hiraoka Ryuji, "The Geneva Sphere: An Astronomical Model from Seventeenth-Century Japan," *Technology and Culture*, Vol. 60, No. 1 (Jan., 2019), 219–51.

"Current Items," *American Sociological Review*, Vol. 8, No. 5 (Oct., 1943): 602–04.

"D. J. Faour & Bros. Importers Exporters & Commission Merchants," James Cummins, Bookseller, n. d., accessed January 7, 2024, https://www.jamescumminsbookseller.com/pages/books/352074/arab-american/d-j-faour-bros-importers-exporters-commission-merchants-in-notions-dry-goods-cutlery-jewelry-c.

Patrick Gautier Dalché, "The Reception of Ptolemy's *Geography* (End of the Fourteenth to Beginning of the Sixteenth Century)," in David Woodward, ed., *The History of Cartography, Volume 3: Cartography in the European Renaissance* (Chicago: The University of Chicago Press, 2018), 285–364.

Ahmad S. Dallal, "Early Islam," in John Hedley Brooke and Ronald L. Numbers, eds., *Science and Religion Around the World* (Oxford: Oxford University Press, 2011), 120–47.

Erich von Däniken, trans. Michael Heron, *According to the Evidence: My Proof of Man's Extraterrestrial Origins* (London: Souvenir Press, 1977).

Erich von Däniken, *The Gold of the Gods* (London: Souvenir Press, 1973).

Robert Dankoff, *An Ottoman Mentality: The World of Evliya Çelebi* (Leiden: Brill, 2006).

Dappergander, "Explainer: Why did QAnon Factionalize?," dappergander, April 1, 2021, accessed August 24, 2021, https://www.patreon.com/posts/49472926.

Jill Darrell and Diana Clements, "Fossil Collector," in Oxford University Museum of Natural History, eds., *Strata: William Smith's Geological Maps* (Chicago: The University of Chicago Press, 2020), 184–99.

Mirko Dautović, "Serbia's Elections – Defeat for Far Right, as well as Moderate Opposition," *Balkan Insight*, January 5, 2024, accessed November 11, 2024, https://balkaninsight.com/2024/01/05/serbias-elections-defeat-for-far-right-as-well-as-moderate-opposition/.

Burke Davis, *The Civil War: Strange & Fascinating Facts* (Avenel, NJ: Wings Books, 1994 [1960]).

Craig Davis, ""A" Is for Allah, "J" Is for Jihad," *World Policy Journal*, Vol. 19, No. 1 (Spring, 2002), 90–94.

Lacy Davis, ed., John Holman, *On the Wings of Ezekiel: The Story of Reverend Burrell Cannon's Remarkable Airship* (Pittsburg, TX: Pittsburg/Camp County Museum Association, 2002).

Thulani Davis, *The Emancipation Circuit: Black Activism Forging a Culture of Freedom* (Raleigh, NC: Duke University Press, 2022).

Kevin Dawson, *Undercurrents of Power: Aquatic Culture in the African Diaspora* (Philadelphia: University of Pennsylvania Press, 2018).

Beth Day, *Joshua Slocum, Sailor: A Biography* (Cambridge, MA: Riverside Press, 1953).

Michael A. B. Deakin, *Hypatia of Alexandria: Mathematician and Martyr* (Amherst, NY: Prometheus Books, 2007).

Siddhartha Deb, *Twilight Prisoners: The Rise of the Hindu Right and the Fall of India* (Chicago: Haymarket Books, 2024).

Carl N. Degler, *Neither Black Nor White: Slavery and Race Relations in Brazil and the United States* (Madison, WI: The University of Wisconsin Press, 1971).

Natana J. DeLong-Bas, *Wahhabi Islam: From Revival and Reform to Global Jihad* (Oxford: Oxford University Press, 2016).

David Deming, *Science and Technology in World History, Volume 1: The Ancient World and Classical Civilization* (Jefferson, NC: McFarland & Company, 2010).

Donald H. DeMeules, "Ignatius Donnelly: A Don Quixote in the World of Science," *Minnesota History*, Vol. 37, No. 6 (Jun., 1961), 229–34.

Jörg Matthias Determann, *Diversity, Equity, and Inclusion in Astronomy: A Modern History* (Cham: Springer Nature, 2023).

Jörg Matthias Determann, *Space Science and the Arab World: Astronauts, Observatories and Nationalism in the Middle East* (London: I. B. Tauris, 2018).

Peter Dlhopolec, "News digest: Flat earther might teach pupils how to use media and think critically," *The Slovak Spectator*, May 14, 2024, accessed May 15, 2024, https://spectator.sme.sk/c/23330114/news-digest-flat-earther-might-teach-pupils-how-to-use-media-and-think-critically.html.

Colin Dickey, *Under the Eye of Power: How Fear of Secret Societies Shapes American Democracy* (New York: Viking, 2023).

Colin Dickey, *The Unidentified: Mythical Monsters, Alien Encounters, and Our Obsession with the Unexplained* (New York: Viking, 2020).

Alexander C. Diener and Joshua Hagen, "The power of place in place attachment," in Alexander C. Diener and Joshua Hagen, eds., *The Power of Place in Place Attachment* (Abingdon: Routledge, 2023), 1–5.

Father Divine, "A Transparent Airplane That Can Take on Any Color," in Langston Hughes and Arna Bontemps, eds., *The Book of Negro Folklore* (New York: Dodd, Mead & Company, 1958), 275–76.

Ernst von Dobschütz, "The 'Decretum Gelasianum de Libris Recipiendis et non Recipiendis'," *The Tertullian Project*, December 2, 2000, accessed June 22, 2023, https://www.tertullian.org/decretum_eng.htm.

Jim Dobson, "Flat Earth Supporters Now Plan An Antarctica Expedition To The Edge Of The World," *Forbes*, March 16, 2019, accessed March 20, 2024, https://www.forbes.com/sites/jimdobson/2019/03/16/flat-earth-supporters-now-plan-an-antarctica-expedition-to-the-edge-of-the-world/?sh=335d9bd59165.

Peter Doerrie, "Insiders Insight: The serious side of the Buhari clone story," *African Arguments*, December 4, 2018, accessed February 7, 2024, https://africanarguments.org/2018/12/serious-side-buhari-clone-story/.

Jim Downs, *Maladies of Empire: How Colonialism, Slavery, and War Transformed Medicine* (Cambridge, MA: Harvard University Press, 2021).

Jacob S. Dorman, *Chosen People: The Rise of American Black Israelite Religions* (Oxford: Oxford University Press, 2013).

Arthur Conan Doyle, *The Coming of the Fairies* (New York: George H. Doran Company, 1922).
Arthur Conan Doyle, *A Study in Scarlet: A Detective Story* (London: Ward, Lock, Bowden, and Co., 1892).
Arthur Conan Doyle, *The Valley of Fear: A Sherlock Holmes Novel* (New York: George H. Doran, 1914).
H. A. Drake, *Constantine and the Bishops: The Politics of Intolerance* (Baltimore: The Johns Hopkins University Press, 2000).
Richard Drayton, *Nature's Government: Science, Imperial Britain and the 'Improvement' of the World* (New Haven: Yale University Press, 2000).
W. E. B. Du Bois, *Black Reconstruction in America, 1860–1880* (New York: Touchstone, 1995 [1935]).
W. E. B. Du Bois, *The World and Africa: An Inquiry into the Part which Africa has Played in World History, New Enlarged Edition* (New York: International Publishers, 1965).
Daniel Edler Duarte, Pedro Benetti, and Marcos César Alvarez, "A "war on science?" Far-right movements and the disputes over epistemic authority in Brazil," *Tapuya: Latin American Science, Technology and Society*, Vol. 7, No. 1 (Apr., 2024), 1–14.
John R. Durant, "Scientific Naturalism and Social Reform in the Thought of Alfred Russel Wallace," *The British Journal for the History of Science*, Vol. 12, No. 1 (Mar., 1979), 31–58.
Pieter Duvenage, "Afrikaner intellectual history: An interpretation," in Joel Netshitenzhe, ed., *Whiteness Afrikaans Afrikaners: Addressing Post-Apartheid Legacies, Privileges and Burdens* (Johannesburg: Mapungubwe Institute for Strategic Reflection, 2018), 91–114.
Katharine L. Dvorak, *An African-American Exodus: The Segregation of the Southern Churches* (Brooklyn: Carlson Publishing, 1991).
Taner Edis, "Finding an Enemy: Islam and the New Atheism," in Ruqayya Yasmine Khan, ed., *Muhammad in the Digital Age* (Austin: University of Texas Press, 2015), 174–92.
Evelyn Edson, *The World Map, 1300–1492: The Persistence of Tradition and Transformation* (Baltimore: Johns Hopkins University Press, 2007).
Ron Eglash, "The African Heritage of Benjamin Banneker," *Social Studies of Science*, Vol. 27, No. 2 (Apr., 1997), 307–15.
Muhammed Jamaludddin El-Fandy, trans. Ismail Cashmiry, revised by Shawki Sukkary, *On Cosmic Verses in the Quran* (Cairo: Supreme Council for Islamic Affairs, 1967).
Paul Ellenberger, David J. Mossman, Alexander D. Mossman and Martin G. Lockley, "Bushmen Cave Paintings of Ornithopod Dinosaurs: Paleolithic Trackers Interpret Early Jurassic Footprints," *Ichnos: An International Journal for Plant and Animal Traces*, Vol. 12, No. 3 (2005), 223–26.

Arthur Jackson Ellis, *The Divining Rod: A History of Water Witching* (Washington, D.C.: Government Printing Office, 1917).

Marwa Elshakry, "When Science Became Western: Historiographical Reflections," *Isis*, Vol. 101, No. 1 (Mar., 2010), 98–109.

Birgitta Elweskiöld. "John Philoponus against Cosmas Indicopleustes: A Christian Controversy on the Structure of the World in Sixth-Century Alexandria." PhD diss., Lund University, 2005.

Jim Endersby, *Imperial Nature: Joseph Hooker and the Practices of Victorian Science* (Chicago: The University of Chicago Press, 2008).

Philip C. England, Peter Moinar, and Frank M. Richter, "Kelvin, Perry and the Age of the Earth," *American Scientist*, Vol. 95, No. 4 (Jul., 2007), 342–49.

Charlie English, *The Storied City: The Quest for Timbuktu and the Fantastic Mission to Save Its Past* (New York: Riverhead Books, 2017).

Charles Watkyns de Lacy Evans, *How to Prolong Life: An Inquiry into the Cause of "Old Age" and "Natural Death"* (London: Central Publishing Company, 1897).

Robert Evans and Jamie Loftus, "Part One: Bill Cooper: The Man Who Killed Truth," *Behind the Bastards*, July 14, 2020 (accessed July 20, 2020), https://podcasts.apple.com/us/podcast/part-one-bill-cooper-the-man-who-killed-truth/id1373812661?i=1000484862608.

"The Ezekiel Airship," *Scientific American*, Vol. 85, No. 15 (Oct. 12, 1901), 228.

A. Babs Fafunwa, *History of Education in Nigeria* (London: George Allen and Unwin, 1974).

Seb Falk, *The Light Ages: The Surprising Story of Medieval Science* (New York: W. W. Norton & Company, 2020).

Richard Fallon, *Contesting Earth's History in Transatlantic Literary Culture, 1860–1935: Believers and Visionaries on the Borderlines of Geology and Palaeontology* (Oxford: Oxford University Press, 2025).

Richard Fallon and Edward Guimont, "Introduction: Conceptualising heterodox palaeoscience," *Interdisciplinary Science Reviews*, Vol. 49, No. 3–4 (Jul., 2024), 307–15.

Philip José Farmer, "Sail On! Sail On!," *Startling Stories*, Vol. 28, No. 2 (Dec., 1952), 55–61.

Danny Faulkner, "Second Flat Earth International Conference," Answers in Genesis, December 14, 2018, accessed March 16, 2024, https://answersingenesis.org/blogs/danny-faulkner/2018/12/14/second-flat-earth-international-conference/.

Danny Faulkner, "Testing a Flat-Earth Prediction: Is the Moon's Light Cooling?," *Answers Research Journal*, Vol. 12 (2019), 1–11.

Arthur Huff Fauset, *Black Gods of the Metropolis: Negro Religious Cults of the Urban North* (Philadelphia: University of Pennsylvania Press, 1971 [1944]).

François-Xavier Fauvelle, trans. Troy Tice, *The Golden Rhinoceros: Histories of the African Middle Ages* (Princeton: Princeton University Press, 2018).
Kenneth L. Feder, "Answering Pseudoarchaeology," in Jeb J. Card and David S. Anderson, eds., *Lost City, Found Pyramid: Understanding Alternative Archaeologies and Pseudoscientific Practices* (Tuscaloosa: University of Alabama Press, 2016), 199–210.
Kenneth L. Feder, *Archaeological Oddities: A Field Guide to Forty Claims of Lost Civilizations, Ancient Visitors, and Other Strange Sites in North America* (London: Rowman & Littlefield, 2019).
Kenneth L. Feder, *Encyclopedia of Dubious Archaeology: From Atlantis to the Walam Olum* (Santa Barbara, CA: ABC-CLIO, 2010).
Federal Writers' Project, *The Negro in Virginia* (New York: Hastings House Publishers, 1940).
Julian Feeld, Travis View, and Jake Rockatansky, "Episode 77: Drink Bleach To Own The Libs," *QAnon Anonymous*, February 3, 2020, accessed February 4, 2020, https://soundcloud.com/qanonanonymous/episode-77-drink-bleach-to-own-the-libs.
Julian Feeld, Travis View, Jake Rockatansky, and Jacob Chansley, "Episode 234: Jacob Chansley, America's Shaman," *QAnon Anonymous*, June 10, 2023, accessed June 11, 2023, https://podcasts.apple.com/us/podcast/episode-234-jacob-chansley-americas-shaman/id1428209307?i=1000616446338.
Julian Feeld, Travis View, Jake Rockatansky, and Arif Hasan, "Episode 263: Aaron Rodgers feat Arif Hasan," *QAnon Anonymous*, January 19, 2024, accessed January 21, 2024, https://soundcloud.com/qanonanonymous/episode-263-aaron-rodgers-feat-arif-hasan.
Julian Feeld, Travis View, Jake Rockatansky, and Kelly Weill, "Episode 17: Our Thicc Flat Earth feat. Kelly Weill," *QAnon Anonymous*, November 28, 2018, accessed August 21, 2019, https://www.stitcher.com/podcast/qanon-anonymous/e/57443731.
William N. Fenton, "This Island, the World on the Turtle's Back," *The Journal of American Folklore*, Vol. 75, No. 298 (Oct.–Dec., 1962), 283–300.
Felipe Fernández-Armesto, *Before Columbus: Exploration and Colonization from the Mediterranean to the Atlantic, 1229–1492* (Philadelphia: University of Pennsylvania Press, 1987).
Felipe Fernández-Armesto, *Columbus and the Conquest of the Impossible* (New York: Saturday Review Press, 1974).
J. Carlos Viana Ferreira, "Orwell on Kipling: an imperialist, a gentleman and a great artist," *Anglo Saxonica*, Ser. III, No. 13 (2017), 61–71.
Steven Feierman and John M. Janzen, "African Religions," in John Hedley Brooke and Ronald L. Numbers, eds., *Science and Religion Around the World* (Oxford: Oxford University Press, 2011), 229–51.

Jacqueline Feke, *Ptolemy's Philosophy: Mathematics as a Way of Life* (Princeton: Princeton University Press, 2018).

Michael Feola, *The Rage of Replacement: Far Right Politics and Demographic Fear* (Minneapolis: University of Minnesota Press, 2024).

Leo Ferrari, "Feminism and education in a Flat Earth perspective," *McGill Journal of Education*, Vol. 10, No. 1 (Spring, 1975), 77–81.

Robert Fikes Jr., "Careers of African Americans in Academic Astronomy," *The Journal of Blacks in Higher Education*, No. 29 (Autumn, 2000), 132–34.

Naomar de Almeida Filho, "Previsões terraplanistas sobre a Covid-19 no Brasil têm validade?" *Abrasco*, April 27, 2020, accessed May 17, 2024, https://abrasco.org.br/previsoes-terraplanistas-sobre-a-covid-19-no-brasil-tem-validade-artigo-de-naomar-de-almeida-filho/.

Max Fisher, *The Chaos Machine: The Inside Story of How Social Media Rewired Our Minds and Our World* (New York: Little, Brown, and Company, 2022).

Lenny Flank, "Photo Diary: A Titan I Nuclear Missile in Georgia," *Daily Kos*, October 22, 2022, accessed March 3, 2024, https://www.dailykos.com/stories/2022/10/22/2128343/-Photo-Diary-A-Titan-I-Nuclear-Missile-in-Georgia.

"Flat Earth and Anti-Evolution," *The Science News-Letter*, Vol. 11, No. 305 (Feb. 12, 1927), 95–96.

"Flat Earth? The Hidden Side of a Hoax," Society of Saint Pius X, July 11, 2023, accessed July 11, 2023, https://sspx.org/en/news-events/news/flat-earth-hidden-side-hoax-84054.

Flat Earth Brooklyn Live, "How Could The Earth be Flat | The Earth Is A Plane John Edward Quinlan," YouTube, November 22, 2017, accessed July 22, 2020, https://www.youtube.com/watch?v=cZ4VUJ94J1s.

"Flat tax, flat Earth," *The Globe and Mail*, (Feb. 2, 1996), A12.

Harrison Fluss and Landon Frim, "Aliens, Antisemitism, and Academia," *Jacobin*, March 11, 2017, accessed October 27, 2024, https://jacobin.com/2017/03/jason-reza-jorjani-stony-brook-alt-right-arktos-continental-philosophy-modernity-enlightenment/.

Holly Folk, "Raymond W. Bernard, Hollow Earth, and UFO s," in Ben Zeller, ed., *Handbook of UFO Religions* (Leiden: Brill, 2021), 312–25.

Zachary Folk, "Tucker Carlson Says a Demon Attack Left Him Bleeding in Bed," *The Daily Beast*, October 31, 2024, accessed November 23, 2024, https://www.thedailybeast.com/tucker-carlson-says-he-was-attacked-by-a-demon-in-new-doc/.

Eric Foner, *Reconstruction: America's Unfinished Revolution, 1863–1877* (New York: Harper & Row, 1988).

"Footballer arrested to hide faked moonwalk," *The Independent*, July 21, 1994, accessed February 18, 2024, https://www.independent.co.uk/news/world/footballer-arrested-to-hide-faked-moonwalk-1415442.html.

Charles Fort, *New Lands* (New York: Boni & Liveright, 1923).

B. Fraser, "Comets in Bushman Paintings," *African Skies/Cieux Africains*, No. 11 (Jul., 2007), 16.

John Freely, *Aladdin's Lamp: How Greek Science Came to Europe Through the Islamic World* (New York: Alfred A. Knopf, 2009).

John Freely, *Light from the East: How the Science of Medieval Islam Helped to Shape the Western World* (London: I. B. Tauris & Co., 2011).

Charles Freeman, *The Reopening of the Western Mind: The Resurgence of Intellectual Life from the End of Antiquity to the Dawn of the Enlightenment* (New York: Alfred A. Knopf, 2023).

Colin Freeman, "Geography teachers targeted by Boko Haram," *The Daily Telegraph* (12 April, 2016), 12.

Ruth Freitag, "Hollow-Earth Theories: A List of References," Library of Congress, May, 1997, accessed March 21, 2024, https://www.loc.gov/rr//scitech/SciRefGuides/hollowearth.html.

Thomas L. Friedman, "A Plan to Get America Back to Work," *The New York Times*, March 22, 2020, accessed March 23, 2020, https://www.nytimes.com/2020/03/22/opinion/coronavirus-economy.html.

Thomas L. Friedman, *The World Is Flat: A Brief History of the Twenty-first Century* (New York: Farrar, Straus and Giroux, 2005).

Ronald Fritze, *Invented Knowledge: False History, Fake Science, and Pseudo-Religions* (London: Reaktion Books, 2009).

Kathinka Frøystad, "Affective digital images: Shiva in the Kaaba and the smartphone revolution," in Paul Rollier, Kathinka Frøystad, and Arild Engelsen Ruud, eds., *Outrage: The Rise of Religious Offence in Contemporary South Asia* (London: UCL Press, 2019), 123–48.

David Frum, "The Dangerous Myths of South African Land Seizures," *The Atlantic*, August 30, 2018, accessed May 20, 2019, https://www.theatlantic.com/ideas/archive/2018/08/the-danger-of-south-african-land-seizure/568960/.

Gladys-Marie Fry, *Night Riders in Black Folk History* (Knoxville: University of Tennessee Press, 1975).

Dexter J. Gabriel, "The British Emigration Scheme and the African American Emigration Movement to the Caribbean," in Ronald Angelo Johnson and Ousmane K. Power-Greene, eds., *In Search of Liberty: African American Internationalism in the Nineteenth-Century Atlantic World* (Atlanta: The University of Georgia Press, 2021), 165–96.

Dexter J. Gabriel, *Jubilee's Experiment: The British West Indies and American Abolitionism* (Cambridge: Cambridge University Press, 2023).

Karine Gadré, "Le Lever du Soleil sur la Terre d'Egypte," *African Skies/Cieux Africains*, No. 6 (Dec., 2001), 6–9.

Diego Gambetta and Steffen Hertog, *Engineers of Jihad: The Curious Connection between Violent Extremism and Education* (Princeton: Princeton University Press, 2016).

Michael D. Gambone, *Modern Conspiracies in America: Separating Fact from Fiction* (Lanham, MD: Rowman & Littlefield, 2022).

Ana Garcia, "Brazil under Bolsonaro: Social base, agenda and perspectives," *Journal of Global Faultlines*, Vol. 6, No. 1 (Aug.–Oct., 2019), 62–69.

Rafael Garcia, "7% of Brazilians Believe that Earth Is Flat," *Folha de S.Paulo*, July 15, 2019, accessed February 24, 2022, https://www1.folha.uol.com.br/internacional/en/scienceandhealth/2019/07/7-of-brazilians-believe-that-earth-is-flat.shtml.

Zen Garcia, *The Firmament: Vaulted Dome of the Earth* (Augusta, GA: Sacred Word Publishing, 2016).

Timothy F. Garrard, *Akan Weights and the Gold Trade* (London: Longman Group, 1980).

R. Kelly Garrett and Shannon Poulsen, "Flagging Facebook Falsehoods: Self-Identified Humor Warnings Outperform Fact Checker and Peer Warnings," *Journal of Computer-Mediated Communication*, Vol. 24, No. 5 (Sp., 2019), 240–58.

Giles Gasper, "A Grand Tour of the Medieval Cosmos" (lecture, British Society for the History of Science Global Digital History of Science Festival, 10 July, 2020).

Christine Garwood, *Flat Earth: The History of an Infamous Idea* (New York: Thomas Dunne Books, 2007).

Henry Louis Gates, Jr., *The Black Church: This Is Our Story, This Is Our Song* (New York: Penguin Press, 2021).

Eugene D. Genovese, *Roll, Jordan, Roll: The World the Slaves Made* (New York: Vintage Books, 1972.

"George Augustus Rogers Balloonist," New England Aviation History, April 3, 2015, accessed February 10, 2024, https://newenglandaviationhistory.com/tag/george-augustus-rogers-balloonist/.

Pankaj Ghemawat, "Why the World Isn't Flat," *Foreign Policy*, October 14, 2009, accessed June 3, 2023, https://foreignpolicy.com/2009/10/14/why-the-world-isnt-flat/.

Jerry H. Gill, *Native American Worldviews: An Introduction* (Amherst, NY: Humanity Books, 2002).

Drew Gillis, "Apparently, fact-checking is now too political for Facebook," The A. V. Club, January 7, 2025, accessed January 7, 2025, https://www.avclub.com/facebook-no-longer-fact-checking.

Owen Gingerich, *The Eye of Heaven: Ptolemy, Copernicus, Kepler* (New York: American Institute of Physics, 1993).

Carlo Ginzburg, trans. John and Anne Tedeschi, *The Cheese and the Worms: The Cosmos of a Sixteenth-Century Miller* (Baltimore: Johns Hopkins University Press, 1992 [1976]).

Kathering Giraldo, "Boston's Syriatown: Preserving Cultural Heritage Through Religious Centers," *Medium*, June 5, 2019, accessed January 7, 2024, https://medium.com/@kgiraldo/bostons-syriatown-94cca93626cc.

Martin Ira Glassner, "The Foreign Relations of Jamaica and Trinidad and Tobago, 1960–1965," *Caribbean Studies*, Vol. 10, No. 3 (Oct., 1970), 116–53.

Zachary J. Goldberg and Sean Richey, "Anti-Vaccination Beliefs and Unrelated Conspiracy Theories," *World Affairs*, Vol. 183, No. 2 (Summer, 2020), 105–24.

Stephen A. Goldman, *One More War to Fight: Union Veterans' Battle for Equality through Reconstruction, Jim Crow, and the Lost Cause* (Lanham, MD: Rowman and Littlefield, 2023).

Donald W. Goldsmith, ed., *Scientists Confront Velikovsky* (Ithaca: Cornell University Press, 1977).

Matthew Goodman, *The Sun and the Moon: The Remarkable True Account of Hoaxers, Showmen, Dueling Journalists, and Lunar Man-Bats in Nineteenth-Century New York* (New York: Basic Books, 2008).

Nicolás Wey Gómez, *The Tropics of Empire: Why Columbus Sailed South to the Indies* (Cambridge, MA: The MIT Press, 2008).

Wendy Gonaver, *The Peculiar Institution and the Making of Modern Psychiatry, 1840–1880* (Chapel Hill: University of North Carolina Press, 2019).

Bryce Gordon, "Flat-Earth Revival: A Distortion of Capitalism," *Socialist Revolution*, No. 21 (Mar., 2020), 12–13.

Daniel W. Graham, *Explaining the Cosmos: The Ionian Tradition of Scientific Philosophy* (Princeton: Princeton University Press, 2006).

Caesar Grant, "All God's Chillen Had Wings," in Langston Hughes and Arna Bontemps, eds., *The Book of Negro Folklore* (New York: Dodd, Mead & Company, 1958), 62–65.

Stan Grayson, *A Man for All Oceans: Captain Joshua Slocum and the First Solo Voyage Around the World* (Thomaston, ME: Tilbury House Publishers, 2017).

Erica L. Green, "Baltimore educators recognize official who started black schools," *The Baltimore Sun*, November 12, 2010. Accessed July 31, 2020. https://www.baltimoresun.com/news/bs-xpm-2010-11-12-bs-ci-first-superintendent-recognitio20101112-story.html.

James N. Green, "Brazil's Far-Right President, University Autonomy, and Academic Freedom," *Academe*, Vol. 105, No. 4 (Fall, 2019), 18–21.

John Wyatt Greenlee and Anna Fore Waymack, "Thinking Globally: Mandeville, Memory, and Mappaemundi," *The Medieval Globe*, Vol. 4, No. 2 (2018), 69–106.

Marcel Griaule and Germaine Dieterlen, "The Dogon of the French Sudan," in Daryll Forde, ed., *African Worlds: Studies in the Cosmological Ideas and Social Values of African Peoples* (Oxford: James Currey Publishers, 1999 [1954]), 83–110.

Michael D. Gordin, *The Pseudoscience Wars: Immanuel Velikovsky and the Birth of the Modern Fringe* (Chicago: University of Chicago Press, 2012).

Paul R. Gross and Norman Levitt, *Higher Superstition: The Academic Left and Its Quarrels with Science* (Baltimore: Johns Hopkins University Press, 1994).

Nidhal Guessoum, *Islam's Quantum Question: Reconciling Muslim Tradition and Modern Science* (London: I. B. Tauris, 2011 [2009]).

Nidhal Guessoum, "Lessons from the 'Earth does not rotate' debate," *Gulf News Opinion*, February 23, 2015, accessed February 5, 2024, https://gulfnews.com/opinion/op-eds/lessons-from-the-earth-does-not-rotate-debate-1.1461490.

Nidhal Guessoum, "PhD thesis: The earth is flat," *Gulf News Opinion*, April 10, 2017, accessed March 12, 2019, https://gulfnews.com/opinion/op-eds/phd-thesis-the-earth-is-flat-1.2009202.

Edward Guimont, "As Above, So Below: The Role of Astronomical Evolution, Imperialism, and Religion in the Long Nineteenth Century," in *Evolutionary Theories and Religious Traditions: National, Transnational, and Global Perspectives, 1800–1920*, eds. Bernard V. Lightman and Sarah A. Qidwai (Pittsburgh: University of Pittsburgh Press, 2023), 77–94.

Edward Guimont. "From King Solomon to Ian Smith: Rhodesian Alternate Histories of Zimbabwe." PhD diss., University of Connecticut, 2019.

Edward Guimont, "Hunting Dinosaurs in Central Africa," *Contingent Magazine*, March 18, 2019, accessed August 3, 2022, https://contingentmagazine.org/2019/03/18/hunting-dinosaurs-africa/.

Edward Guimont, "The Megalodon: A Monster of the New Mythology," *M/C Journal*, Vol. 24, No. 5, October 6, 2021, accessed January 12, 2022, https://journal.media-culture.org.au/index.php/mcjournal/article/view/2793.

Edward Guimont, "Squaring the Spheroid," *Lapsus Lima*, June 26, 2019, accessed June 26, 2019, http://www.lapsuslima.com/squaring-the-spheroid/.

Edward Guimont and Horace A. Smith, *When the Stars Are Right: H. P. Lovecraft and Astronomy* (New York: Hippocampus Press, 2023).

Gad Guterman, "Field Tripping: The Power of *Inherit the Wind*," *Theatre Journal*, Vol. 60, No. 4 (Dec., 2008), 563–83.

Jeff Guy, "Class, Imperialism and Literary Criticism: William Ngidi, John Colenso and Matthew Arnold," *Journal of Southern African Studies*, Vol. 23, No. 2 (Jun., 1997), 219–41.

Jeff Guy, "The Religious Thinking of J. W. Colenso: The Theology of a Heretic," *Religion in Southern Africa*, Vol. 4, No. 2 (Jul., 1983), 3–20.

Evelina Gužauskytė, *Christopher Columbus's Naming in the Diarios of the Four Voyages (1492–1504): A Discourse of Negotiation* (Toronto: University of Toronto Press, 2014).

Steven Hahn, *A Nation under Our Feet: Black Political Struggles in the Rural South from Slavery to the Great Migration* (Cambridge, MA: Harvard University Press, 2003).

Fayza M. Haikal, "Private Collections and Temple Libraries in Ancient Egypt," in Mostafa El-Abbadi and Omnia Fathallah, eds., *What Happened to the Ancient Library of Alexandria?* (Leiden: Brill, 2008), 39–54.

P. E. H. Hair, "Columbus from Guinea to America," *History in Africa*, Vol. 17 (1990), 113–29.

Joy Hakim, "A Founding Father of science," *Science Scope*, Vol. 37, No. 6 (Feb., 2014), 6–8.

Alex Haley, "Foreward," in Marcus H. Boulware, *The Oratory of Negro Leaders: 1900–1968* (Westport, CT: Negro Universities Press, 1969), xi-xv.

Alex Haley, "Manuscript by Alex Haley, John Jasper Annotated Typescript and Holograph Notes," Broward County Library Digital Archives, 1960; accessed June 26, 2020, https://digitalarchives.broward.org/digital/collection/alexhaley/id/1984.

Stephanie Halmhofer, "Manufacturing History: Atlantis, Aryans, and the Use of Pseudoarchaeology by the Far-Right ," in Christopher T. Conner; Matthew N. Hannah and Nicholas J. MacMurray, eds., *Conspiracy Theories and Extremism in New Times* (Lanham, MD: Rowman & Littlefield, 2024), 53–81.

Abbas Hamdani, "An Islamic Background to the Voyages of Discovery," in Salma Khadra Jayyusi, ed., *The Legacy of Muslim Spain* (Leidein: E. J. Brill, 1992), 273–304.

Abbas Hamdani, "Ottoman Response to the Discovery of America and the New Route to India," *Journal of the American Oriental Society*, Vol. 101, No. 3 (Jul.–Sep., 1981), 323–30.

Lawrence C. Hamilton, "Conspiracy vs. Science: A Survey of U.S. Public Beliefs," *Casey School of Public Policy National Issue Brief*, No. 162 (Spring, 2022), 1–6.

Lawrence C. Hamilton, "Trumpism, Climate and COVID: Social Bases of the New Science Rejection," University of New Hampshire Scholars Repository, August 6, 2022 (accessed May 19, 2025), https://scholars.unh.edu/faculty_pubs/1520/.

Barry Hankins, *Jesus and Gin: Evangelicalism, the Roaring Twenties and Today's Culture Wars* (New York: Palgrave Macmillan, 2010).

Jean and David Hankins, "Otisfield, Maine: A Short History," Otisfield Historical Society, June 17, 1999, accessed January 1, 2023, https://sites.rootsweb.com/~mecotisf/bicenhs.htm.

James Hannam, *God's Philosophers: How the Medieval World Laid the Foundations of Modern Science* (London: Icon Books, 2009).

Howard H. Harlan, *John Jasper: A Case History in Leadership* (Charlottesville: University of Virginia, 1936).

Howard H. Harlan, *Zion Town: A Study in Human Ecology* (Charlottesville: University of Virginia, 1935).

Keith Raymond Harris, "The Epistemology of QAnon," in Luke Ritter, ed., *American Conspiracism: An Interdisciplinary Exploration* (London: Routledge, 2024), 19–33.

Peter Harrison and David C. Lindberg, "Early Christianity," in John Hedley Brooke and Ronald L. Numbers, eds., *Science and Religion Around the World* (Oxford: Oxford University Press, 2011), 67–91.

Benjamin Hart, "Who Are You Going to Believe, Lin Wood or NASA?," *Intelligencer*, June 13, 2022, accessed March 1, 2023, https://nymag.com/intelligencer/2022/06/lin-wood-pro-trump-attorney-is-a-flat-earther.html.

Syed Nomanul Haq, "That Medieval Islamic Culture Was Inhospitable to Science," in Ronald L. Numbers, ed., *Galileo Goes to Jail and Other Myths about Science and Religion* (Cambridge, MA: Harvard University Press, 2010), 35–42.

Ikram Hawramani, "The Quran and the Shape of the Earth: Is It Round or Flat?," *Ikram Hawramani's Website*, November 30, 2018, accessed August 14, 2020, https://hawramani.com/the-quran-and-the-shape-of-the-earth-is-it-round-or-flat/.

William E. Hatcher, *John Jasper: The Unmatched Negro Philosopher and Preacher* (New York: Fleming H. Revell Company, 1908).

Peter Heather, *Christendom: The Triumph of a Religion, AD 300–1300* (New York: Alfred A. Knopf, 2023).

Robert A. Heinlein, *To Sail Beyond the Sunset: The Life and Loves of Maureen Johnson (Being the Memoirs of a Somewhat Irregular Lady)* (New York: Putnam Books, 1987).

Edward Hendrie, *The Greatest Lie on Earth: Proof That Our World Is Not a Moving Globe* (Great Mountain Publishing, 2016).

Irving Hexham, *The Irony of Apartheid: The Struggle for National Independence of Afrikaner Calvinism Against British Imperialism* (Lewiston, NY: Edwin Mellen Press, 1981).

John D. Hicks, "The Political Career of Ignatius Donnelly," *The Mississippi Valley Historical Review*, Vol. 8, No. 1/2 (Jun.–Sep., 1921), 80–132.

Mary E. Hicks, *Captive Cosmopolitans: Black Mariners and the World of South Atlantic Slavery* (Chapel Hill: University of North Carolina Press, 2024).

"Higher Degrees in Sociology Conferred in 1941," *American Journal of Sociology*, Vol. 48, No. 1 (Jun., 1942): 87–96.

Sharon A. Hill, "Flat-earthers as scientifical Americans: One message from 'Behind the Curve'," *I Doubt It*, March 21, 2019, accessed March 15, 2022, https://idoubtit.wordpress.com/2019/03/21/flat-earthers-as-scientifical-americans-one-message-from-behind-the-curve/.

Sharon A. Hill, *Scientifical Americans: The Culture of Amateur Paranormal Researchers* (Jefferson, NC: McFarland & Company, 2017).

Christopher Z. Hobson, "The Lord Is a Man of War: John Jasper, Covenant, and Apocalypse," *African American Review*, Vol. 44, No. 4 (Winter, 2011), 619–31.

Holger Hoock, *Empires of the Imagination: Politics, War, and the Arts in the British World, 1750–1850* (London: Profile Books, 2010).

J. C. Hollmann, "'The Sky's Things': |xam Bushman 'Astrological Mythology' as recorded in the Bleek and Lloyd Manuscripts," *African Skies/Cieux Africains*, No. 11 (Jul., 2007), 8–12.

Matthew Hongoltz-Hetling, *If It Sounds Like a Quack…: A Journey to the Fringes of American Medicine* (New York: PublicAffairs, 2023).

A. G. Hopkins, *An Economic History of West Africa, Second Edition* (Abingdon: Routledge, 2019).

H. E. Houghton, "Some South African Amateur Astronomers," *Monthly Notes of the Astronomical Society of South Africa*, Vol. 6, No. 5 (May, 1947), 39–42.

A. Hottentot, "Audi Alteram Partem," *The Earth not a Globe Review*, No. 6 (Mar., 1894), 135.

A. Hottentot, "'Parallax' Triumphant," *The Earth not a Globe Review*, No. 3 (Jul., 1893), 17–20.

Terrence Howard, *One Times One Equals Two* (2019), 11, 21. Accessed 3 June 2024, https://tcotlc.com/wp-content/uploads/2021/10/OTOET_PREVIEW_062_October_03_2021.pdf.

Cynthia Ann Humes, "Hindutva, Mythistory, and Pseudoarchaeology," *Numen*, Vol. 59, No. 2/3 (2012), 178–201.

David Hutchings and James C. Ungureanu, *Of Popes and Unicorns: Science, Christianity, and How the Conflict Thesis Fooled the World* (Oxford: Oxford University Press, 2022).

Richard Huzzey, "Minding Civilisation and Humanity in 1867: A Case Study in British Imperial Culture and Victorian Anti-Slavery," *The Journal of Imperial and Commonwealth History*, Vol. 40, No. 5 (Nov., 2012), 807–25.

Roland Imhoff et al, "Conspiracy mentality and political orientation across 26 countries," *Nature Human Behaviour*, Vol. 6 (2022), 392–403.

Roland Imhoff and Pia Lamberty, "Conspiracy Theories as Psycho-Political Reactions to Perceived Power," in Michael Butter and Peter Knight, eds.,

Routledge Handbook of Conspiracy Theories (London: Routledge, 2020), 192–205.

Christopher Ingraham, "Minnesota Republicans elect a flat-earther to a party leadership post," *Minnesota Reformer*, April 16, 2025, accessed April 17, 2025, https://minnesotareformer.com/2025/04/16/minnesota-republicans-elect-a-flat-earther-to-a-party-leadership-post/.

Washington Irving, *A History of the Life and Voyages of Christopher Columbus, Vol. I* (New-York, G. & C. Carvill, 1828).

Washington Irving, *A History of New-York from the Beginning of the World to the End of the Dutch Dynasty, by Diedrich Knickerbocker* (Philadelphia: J. B. Lippincott & Co., 1871 [1809]).

Washington Irving, *Voyages and Discoveries of the Companions of Columbus* (Philadelphia: Carey and Lea, 1831).

"Is the Earth round or flat? Culture Ministry chief official says proof is still needed," *The Slovak Spectator*, May 3, 2024, accessed May 15, 2024, https://spectator.sme.sk/c/23325477/is-the-earth-round-or-flat-culture-ministry-chief-official-says-proof-is-still-needed.html?ref=njctse.

Frank Jacobs, "Will COVID-19 kill off flat Earthism once and for all?," *Big Think*, March 20, 2020, accessed April 1, 2020, https://bigthink.com/strange-maps/end-of-flat-earth/.

Mark Jacobson, *Pale Horse Rider: William Cooper, the Rise of Conspiracy, and the Fall of Trust in America* (New York: Blue Rider Press, 2018).

Susan Jacoby, *The Age of American Unreason* (New York: Pantheon Books, 2008).

Isaac James, *"The Sun Do Move:" The Story of the Life of John Jasper* (Richmond: Hunter's Letter Shop, 1954).

John Jasper, "De Sun Do Move," in Langston Hughes and Arna Bontemps, eds., *The Book of Negro Folklore* (New York: Dodd, Mead & Company, 1958), 225–33.

John Jasper, *"The Sun Do Move!" The Celebrated Theory of the Sun's Rotation Around the Earth* (New York: Brentano's Literary Emporium, 1882).

Andrew Jewett, *Science under Fire: Challenges to Scientific Authority in Modern America* (Cambridge, MA: Harvard University Press, 2020).

Daniel Jolley, Silvia Mari and Karen Douglas, "Consequences of conspiracy theories," in Michael Butter and Peter Knight, eds., *Routledge Handbook of Conspiracy Theories* (London: Routledge, 2020), 231–41.

Charles K. Johnson, "Anti-Apartheid Act lifted in South Africa," *Flat Earth News*, No. 79 (Sep., 1990), 1.

Charles K. Johnson, "Australia Not Down Under," *Flat Earth News* (May, 1978), 1.

Charles K. Johnson, "Babylon-Babel Revisited," *Flat Earth News*, No. 87 (Dec., 1993), 1.

Charles K. Johnson, “Backward is forward or this Flat Earth is a spinning ball,” *Flat Earth News* (Mar., 1978), 4.
Charles K. Johnson, “Columbus Proved Earth Flat,” *Flat Earth News*, No. 83 (Sep., 1992), 1.
Charles K. Johnson, “Columbus Proved World Flat,” *Flat Earth News* (Dec., 1978), 3.
Charles K. Johnson, “Columbus Proved World Flat,” *Flat Earth News* (Sep., 1979), 2.
Charles K. Johnson, “Columbus Proved World Flat Didn’t Go Over Edge of Ball,” *Flat Earth News*, No. 79 (Sep., 1990), 1.
Charles K. Johnson, “The fall of White South Africa,” *Flat Earth News*, No. 75 (Mar., 1988), 1.
Charles K. Johnson, “Outgoing Correspondence 2348,” The Flat Earth Society Library, March 12, 2014, accessed November 30, 2019, https://www.theflatearthsociety.org/library/correspondence/Charles%20K.%20Johnson%20-%20Outgoing%20Correspondence%202348.pdf
Charles K. Johnson, “Police Chief Gates saved the United States,” *Flat Earth News*, No. 83 (Sep., 1992), 2.
Charles K. Johnson, “Proof in Egypt,” *Flat Earth News*, No. 83 (Sep., 1992), 1–2.
Charles K. Johnson, “Rivers,” *Flat Earth News* (Sep., 1979), 2.
Charles K. Johnson, “Scientists claim their word is “Truth”,” *Flat Earth News*, No. 84 (Dec., 1992), 1.
Charles K. Johnson, “The Wings of the Earth,” *Flat Earth News*, No. 82 (Dec., 1992), 2.
Charles W. Jones, “The Flat Earth,” Thought: Fordham University Quarterly, Vol. 9, No. 2 (Sep., 1934), 296–307.
E. J. P. Jorissen, *Transvaalsche Herinneringen 1876–1896* (Pretoria: J. H. de Busy, 1897).
Aditi Joshi, “Islam’s Flat Earth Controversy: Unveiling The Debate Over Earth’s Shape And The Role Of Fatwas In Islamic Thought,” *The Jaipur Dialogues*, August 15, 2023, accessed August 15, 2023, https://www.thejaipurdialogues.com/islam/islams-flat-earth-controversy-unveiling-the-debate-over-earths-shape-and-the-role-of-fatwas-in-islamic-thought/.
Journal of the Seventy-Fourth National Encampment of the Grand Army of the Republic (Washington, DC: United States Government Printing Office, 1941).
Fernanda Pérez-Gay Juárez, Lina Khayyat, Michael Ronca, and Ian Gold, “Viral Belief: The Psychology of COVID Conspiracy Theories,” in Luke Ritter, ed., *American Conspiracism: An Interdisciplinary Exploration* (London: Routledge, 2024), 53–79.

Jacob W. Justice and Brett J. Bricker, "Hacked: Defining the 2016 Presidential Election in the Liberal Media," *Rhetoric & Public Affairs*, Vol. 22, No. 3 (Fall, 2019), 389–420.

Walima T. Kalusa, "Death, Christianity, and African Miners: Contesting Indirect Rule in the Zambian Copperbelt, 1935—1962," *The International Journal of African Historical Studies*, Vol. 44, No. 1 (2011), 89–112.

Ousmane Oumar Kane, *Beyond Timbuktu: An Intellectual History of Muslim West Africa* (Cambridge, MA: Harvard University Press, 2016).

Alex Kaplan, "Rumble promotes the flat Earth conspiracy theory among its "editor picks"," Media Matters for America, February 28, 2024, accessed March 16, 2025, https://www.mediamatters.org/rumble/rumble-promotes-flat-earth-conspiracy-theory-among-its-editor-picks.

Alex Kaplan, Twitter, February 2, 2021, accessed February 18, 2024, https://twitter.com/alkapdc/status/1356612532404969476.

Roshni Kapur, "The Afghan Taliban and Covid-19: Leveraging the Crisis or a Change of Heart?," Middle East Institute, April 13, 2021, accessed January 31, 2024, https://www.mei.edu/publications/afghan-taliban-and-covid-19-leveraging-crisis-or-change-heart.

Rita Katz, *Saints and Soldiers: Inside Internet-Age Terrorism, From Syria to the Capitol Siege* (New York: Columbia University Press, 2022).

Emma Keates, "Google becomes latest tech giant to send Trump money," The A. V. Club, January 9, 2025, accessed January 12, 2025, https://www.avclub.com/google-trump-donation.

James B. Kelley, "Song, Story, or History: Resisting Claims of a Coded Message in the African American Spiritual "Follow the Drinking Gourd"," *The Journal of Popular Culture*, Vol. 41, No. 2 (Apr., 2008), 262–80.

Robin D. G. Kelley, *Freedom Dreams: The Black Radical Imagination* (Boston: Beacon Press, 2002).

Ibram X. Kendi, "The Hopefulness and Hopelessness of 1619," *The Atlantic*, August 20, 2019, accessed August 21, 2019, https://www.theatlantic.com/ideas/archive/2019/08/historical-significance-1619/596365/.

Dane Kennedy, *The Last Blank Spaces: Exploring Africa and Australia* (Cambridge: Harvard University Press, 2013).

Daniel Kennefick, *No Shadow of a Doubt: The 1919 Eclipse That Confirmed Einstein's Theory of Relativity* (Princeton, NJ: Princeton University Press, 2019).

Langston Kerman and Troy Walker, "Talk That Ship To Me," *My Momma Told Me*, March 22, 2022, accessed March 22, 2022, https://www.iheart.com/podcast/1119-my-momma-told-me-with-lan-69919303/episode/talk-that-ship-to-me-with-94482894/.

Joan Kerr, "A. Constance Roth," *Design & Art Australia Online*, September 30, 2014, accessed August 27, 2022, https://www.daao.org.au/bio/a-constance-roth/biography/.

Amira Kharroubi and Jamel Touir, "The Geocentric Model of the Earth: Physics and Astronomy Arguments," *The International Journal of Science & Technoledge*, Vol. 4, No. 8 (Aug., 2016), 57–62.

David Killingray, "Black Baptists and Pan-Africanism in Britain, 1890–1913," *Immigrants & Minorities*, Vol. 35, No. 1–2 (Sep., 2020), 105–30.

David Killingray, "Rights, Land, and Labour: Black British Critics of South African Policies before 1948," *Journal of Southern African Studies*, Vol. 35, No. 2 (Jun., 2009), 375–98.

David Killingray, "The Telephone," in David Dabydeen, John Gilmore, and Cecily Jones, eds., *The Oxford Companion to Black British History* (Oxford: Oxford University Press, 2007), 475–76.

Ben Kissel, Marcus Parks, and Henry Zebrowski, "Episode 334: Flat Earth," *The Last Podcast on the Left*, September 29, 2018, accessed July 23, 2024, https://www.audible.com/podcast/Episode-334-Flat-Earth/B09RKFCLTB.

Dennis King, *Lyndon LaRouche and the New American Fascism* (New York: Doubleday, 1989).

Asma Shakir Khawaja and Asma Hussain Khan, "Media Strategy of ISIS," *Strategic Studies*, Vol. 36, No. 2 (Summer, 2016), 104–21.

Jim al-Khalili, *The House of Wisdom: How Arabic Science Saved Ancient Knowledge and Gave Us the Renaissance* (New York: Penguin Books, 2011).

Jagdish Khubchandani, Elizabeth Bustos, Sabrina Chowdhury, Nirbachita Biswas and Teresa Keller, "COVID-19 Vaccine Refusal among Nurses Worldwide: Review of Trends and Predictors," *Vaccines*, Vol. 10, No. 2 (Feb., 2022), 1–13.

Rudyard Kipling, "The Village that Voted the Earth was Flat," in Stephen Jones, ed., *The Mark of the Beast and other Fantastical Tales* (London: Gollancz, 2007), 637–74.

Damon Knight, *Charles Fort: Prophet of the Unexplained* (New York: Doubleday & Company, 1970).

Toke Lindegaard Knudsen, "Three Purāṇic Statements on the Shape of the Earth," *History of Science in South Asia*, Vol. 9 (2021), 126–66.

Annette Kolodny, *In Search of First Contact: The Vikings of Vinland, the Peoples of the Dawnland, and the Anglo-American Anxiety of Discovery* (Durham: Duke University Press, 2012).

W. P. Koorts and A. Slotegraaf, "/Xam astronomical references in G R von Wielligh's "*Boesman-Stories*"," *African Skies/Cieux Africains*, No. 11 (Jul., 2007), 13–15.

S. L. Kotar and J. E. Gessler, *Ballooning: A History, 1782–1900* (Jefferson, NC: McFarland & Company, 2010).

Sarah Krasnostein, *The Believer: Encounters with the Beginning, the End, and our Place in the Middle* (Portland, OR: Tin House, 2022).
Lawrence M. Krauss, "I Have No Idea How I Ended Up in That Stupid Geocentrism Documentary," *Slate*, April 8, 2014. Accessed November 23, 2020. https://slate.com/technology/2014/04/lawrence-krauss-on-ending-up-in-the-geocentrism-documentary-the-principle.html.
Daniel Kreps, "Tucker Carlson Says He's 'Open' to the Flat Earth Theory," *Rolling Stone*, December 16, 2023, accessed December 16, 2023, https://www.rollingstone.com/politics/politics-news/tucker-carlson-flat-earth-theory-1234931330/.
J. D. Krige and E. J. Krige, "The Lovedu of the Transvaal," in Daryll Forde, ed., *African Worlds: Studies in the Cosmological Ideas and Social Values of African Peoples* (Oxford: James Currey Publishers, 1999 [1954]), 55–82.
Paul Kruger, ed. August Schowalter, trans. Alexander Teixeira de Mattos, *The Memoirs of Paul Kruger, Four Times President of the South African Republic, Told by Himself* (New York: The Century Co., 1902).
Frank T. Kryza, *The Race for Timbuktu: In Search of Africa's City of Gold* (New York: HarperCollins, 2006).
"Krzysztof Bosak i jego teoria heliocentryzmu. "Nie ma fizycznych dowodów, na to które krąży wokół którego"," *Dziennik Gazeta Prawna*, March 26, 2018, accessed March 16, 2025, https://wiadomosci.dziennik.pl/wydarzenia/artykuly/571442,krzysztof-bosak-o-ziemi-i-sloncu.html.
Thomas S. Kuhn, *The Copernican Revolution: Planetary Astronomy in the Development of Western Thought* (Cambridge, MA: Harvard University Press, 1957).
Mustafa Küçük, "Erdoğan says America was discovered by Muslims, not Columbus," *Hürriyet Daily News*, November 17, 2014, accessed December 10, 2023, https://www.hurriyetdailynews.com/erdogan-says-america-was-discovered-by-muslims-not-columbus-74371.
Gerard P. Kuiper, "German Astronomy During the War," *Popular Astronomy*, Vol. 54, No. 6 (Jun., 1946), 263–86.
P. Pratap Kumar, "Hinduism in South Africa," in *The Wiley-Blackwell Companion to African Religions*, ed. Elias Kifon Bongmba (Hoboken, NJ: Wiley-Blackwell, 2012), 389–400.
Eric Kurlander, *Hitler's Monsters: A Supernatural History of the Third Reich* (New Haven: Yale University Press, 2017).
Vera M. Kutzinski, "The Descent of Nommo: Literacy as Method in Jay Wright's "Benjamin Banneker Helps to Build a City""," *Callaloo*, No. 19 (Autumn, 1983), 103–19.
Olena Kuzii and Andrij Rovenchak, "What the gravitation of a flat Earth would look like and why thus the Earth is not actually flat," *European Journal of Physics*, Vol. 40, No. 3 (Apr., 2019), 1–11.

M. L. W. Laistner, "The Library of the Venerable Bede," in Alexander Hamilton Thompson, ed., *Bede, His Life, Times, and Writings: Essays in Commemoration of the Twelfth Centenary of His Death* (Oxford: The Clarendon Press, 1935), 237–66.

Edward J. Larson, *Summer for the Gods: The Scopes Trial and America's Continuing Debate Over Science and Religion* (New York: Basic Books, 1997).

Cleophus J. LaRue, *The Heart of Black Preaching* (Louisville, KY: Westminster John Knox Press, 2000).

Jerome Lawrence and Robert E. Lee, ed. Alan Woods, *The Selected Plays of Jerome Lawrence and Robert E. Lee* (Columbus, OH: Ohio State University Press, 1995).

Gary Leiser, "Birdmen of the Middle East: Early Attempts at Human Flight," *Aerospace Historian*, Vol. 35, No. 3 (Fall, 1988), 173–80.

Toby Lester, *The Fourth Part of the World: The Race to the Ends of the Earth, and the Epic Story of the Map That Gave America Its Name* (New York: Free Press, 2009).

Philippa Levine, *The Amateur and the Professional: Historians, Antiquarians, and Archaeologists in Victorian England, 1838–1886* (Cambridge: Cambridge University Press, 1986).

Rebecca Lewis, "Alternative Influence: Broadcasting the Reactionary Right on YouTube," *Data & Society*, September 18, 2018, accessed February 19, 2021, https://datasociety.net/wp-content/uploads/2018/09/DS_Alternative_Influence.pdf.

Robert Lacey, *Inside the Kingdom: Kings, Clerics, Modernists, Terrorists, and the Struggle for Saudi Arabia* (London: Penguin Books, 2009).

Lucius Lactantius, "Divine Institutes, Book III (Of the False Wisdom of Philosophers)," *New Advent*, 2021, accessed June 16, 2023, https://www.newadvent.org/fathers/07013.htm.

Renée Landell, "Naming 'Paradise': The Adamic Imagination, Colonial Toponyms, and Remembering the Indigenous Caribbean," *Environmental History Now*, April 15, 2021, accessed January 10, 2022, https://envhistnow.com/2021/04/15/naming-paradise-the-adamic-imagination-colonial-toponyms-and-remembering-the-indigenous-caribbean/.

James E. Landing, *Black Judaism: Story of an American Movement* (Durham, NC: Carolina Academic Press, 2002).

Philippa Lang, *Science: Antiquity and its Legacy* (London: I. B. Tauris, 2015).

John Langdon, ""A Monument of Union": Social Change and Personal Experience at the Manea Fen Community, 1839–1841," *Utopian Studies*, Vol. 23, No. 2 (2012), 504–31.

Rodolfo Langhi, "Some Brief Reflexions About the Astronomy Education in Brazil," *Academia Letters*, No. 1088 (Jun., 2021), 1–5.

Talia Lavin, *Culture Warlords: My Journey Into the Dark Web of White Supremacy* (New York: Hachette Books, 2020).

Donald A. Laskey, "A Passage of Time: Joshua Slocum's Circumnavigation and the World Behind It" (MA diss., Central Michigan University, 2015).

Robert Gordon Latham, *Man and His Migrations* (London: John Van Voorst, 1851).

Bruno Latour, trans. Catherine Porter, *An Inquiry into Modes of Existence: An Anthropology of the Moderns* (Cambridge, MA: Harvard University Press, 2018).

Anne Lawrence-Mathers, *Medieval Meteorology: Forecasting the Weather from Aristotle to the Almanac* (Cambridge: Cambridge University Press, 2019).

Lauranett L. Lee and Shelby M. Driskill, *"Knowledge of This Cannot Be Hidden": A Report on the Westham Burying Ground* (Richmond: University of Richmond, 2019). Available online: https://equity.richmond.edu/about/foundational-work/inclusive-excellence/report-pdfs/burying-ground-report.pdf

David Lehmann, "Ritual, text and politics: the evangelical mindset and political polarisation," in Katerina Hatzikidi and Eduardo Dullo, eds., *A Horizon of (Im)possibilities: A Chronicle of Brazil's Conservative Turn* (London: University of London Press, 2021), 103–19.

Thomas Levenson, *The Hunt for Vulcan: …And How Albert Einstein Destroyed a Planet, Discovered Relativity, and Deciphered the Universe* (New York: Random House, 2015).

Bernard Lewis, "The Arab Destruction of the Library of Alexandria: Anatomy of a Myth," in Mostafa El-Abbadi and Omnia Fathallah, eds., *What Happened to the Ancient Library of Alexandria?* (Leiden: Brill, 2008), 213–17.

Godfrey Lien-Hardt, "The Shilluk of the Upper Nile," in Daryll Forde, ed., *African Worlds: Studies in the Cosmological Ideas and Social Values of African Peoples* (Oxford: James Currey Publishers, 1999 [1954]), 138–63.

"Lilia Lemoine asumió como secretaria de la Comisión de Ciencias: el video que se viralizó para criticar su nombramiento," *La Nacion*, May 8, 2024, accessed May 19, 2024, https://www.lanacion.com.ar/politica/lilia-lemoine-asumio-como-secretaria-de-la-comision-de-ciencias-el-video-que-se-viralizo-para-nid08052024/.

Kristen Lippincott, "Power and Politics: The Use of the Globe in Renaissance Portraiture," *Globe Studies*, No. 49/50 (2002), 121–38.

Maija Liuhto, "The Schools of the Taliban," *The Diplomat*, November 3, 2016, accessed February 10, 2021, https://thediplomat.com/2016/11/the-schools-of-the-taliban/.

Lloyd's Weekly News, 5 Mar,. 1911, 6.

Richard Adams Locke, "The Moon Hoax," *Amazing Stories*, Vol. 1, No. 6 (Sep., 1926 [1835]), 556–75.

Thomas J. Logan and Bruce Williams, "On the Meroe Observatory," *Beitraege zur Sudanforschung*, No. 7 (2000), 59–84.
Rayford Logan, *The Negro in American Life and Thought: The Nadir, 1877–1901* (New York: Dial Press, 1954).
Myra A. Lopes, *Captain Joshua Slocum: A Centennial Tribute* (New Bedford, MA: RPI Graphics, 1994).
Taylor Lorenz, *Extremely Online: The Untold Story of Fame, Influence, and Power on the Internet* (New York: Simon & Schuster, 2023).
Velda Love, "400 Years Commemoration 1619 – 2019 Africans in America," United Church of Christ, 2019, accessed August 13, 2020, https://www.ucc.org/400_years_commemoration_1619_2019_africans_in_america.
Paul E. Lovejoy, *Jihād in West Africa during the Age of Revolutions* (Athens, OH: Ohio University Press, 2016).
Jonathan Lyons, *The House of Wisdom: How the Arabs Transformed Western Civilization* (New York: Bloomsbury Press, 2009).
Tofigh Maboudi, "Reconstituting Tunisia: Participation, Deliberation, and the Content of Constitution," *Political Research Quarterly*, Vol. 73, No. 4 (Dec., 2020), 774–89.
Wyatt MacGaffey, "Who Owns Ancient Egypt?," *The Journal of African History*, Vol. 32, No. 3 (1991), 515–19.
Geo Maher, *Anticolonial Eruptions: Racial Hubris and the Cunning of Resistance* (Oakland, CA: University of California Press, 2022).
Shiraz Maher, *Salafi-Jihadism: The History of an Idea* (Oxford: Oxford University Press, 2016).
Rachel Mairs and Maya Muratov, *Archaeologists, Tourists, Interpreters: Exploring Egypt and the Near East in the Late 19th–Early 20th Centuries* (London: Bloomsbury Academic, 2015).
Sebastian Major, "Episode #65 – Was There a Real Hunt for the Fountain of Youth?," *Our Fake History*, May 2, 2018, accessed September 16, 2022, https://ourfakehistory.com/index.php/season-3/episode-65-was-there-a-real-hunt-for-the-fountain-of-youth/.
Sebastian Major, "Episode #114 – Who Was Mali's Richest King? (Part I)," *Our Fake History*, July 8, 2020, accessed July 8, 2020, https://ourfakehistory.com/index.php/season-5/episode-114-who-was-malis-richest-king-part-i/.
Ussama Makdisi, *Age of Coexistence: The Ecumenical Frame and the Making of the Modern Arab World* (Oakland, CA: University of California Press, 2019).
Ussama Makdisi, *Faith Misplaced: The Broken Promise of U.S.-Arab Relations: 1820–2001* (New York: PublicAffairs, 2010).
Tetu Makino, "On Adiabatic Oscillations of a Stratified Atmosphere on the Flat Earth," *arXiv* (Sep., 2020), 1–35.
Tetu Makino, "On Incompressible Vibrations of the Stratified Atmosphere on the Flat Earth," *arXiv* (Feb., 2020), 1–17.

J. McK Malville, R. Schild, F. Wendorf, and R. Brenmer, "Astronomy of Nabta Playa," *African Skies/Cieux Africains*, No. 11 (Jul., 2007), 2–7.

John Mandeville, *The Travels of Sir John Mandeville* (London: Macmillan and Co., 1900).

Jerre Mangione, *The Dream and the Deal: The Federal Writers' Project, 1935–1943* (Boston: Little, Brown and Company, 1972).

Gabriela Manley, "A Scottish Kind of Conspiracy," *Anthropology News*, July 12, 2019, accessed January 24, 2024, https://www.anthropology-news.org/articles/a-scottish-kind-of-conspiracy/.

Bertram Mapunda, "Debate: Why Study Precolonial African Technology and Material Culture?," *Technology and Culture*, Vol. 64, No. 3 (Jul., 2023), 665–76.

J. J. Maquet, "The Kingdom of Ruanda," in Daryll Forde, ed., *African Worlds: Studies in the Cosmological Ideas and Social Values of African Peoples* (Oxford: James Currey Publishers, 1999 [1954]), 164–89.

Robert A. Margo, *Race and Schooling in the South, 1880–1950: An Economic History* (Chicago: The University of Chicago Press, 1990).

Ben Markham, "The Challenge to 'Informal' Empire: Argentina, Chile and British Policy-Makers in the Immediate Aftermath of the First World War," *The Journal of Imperial and Commonwealth History*, Vol. 45, No. 3 (2017), 449–74.

Peter Martinez, "Nigcomsat-1 Launch Approaches," *African Skies*, No. 10 (Oct., 2006), 3.

Anthony W. Marx, *Making Race and Nation: A Comparison of South Africa, the United States, and Brazil* (Cambridge: Cambridge University Press, 1998).

Douglas S. Massey and Jacob S. Rugh, "The Great Recession and the Destruction of Minority Wealth," *Current History*, Vol. 117, No. 802 (Nov., 2018), 298–303.

J. R. Masson, "Geographical Knowledge and Maps of Southern Africa Before 1500 A.D.," *Terrae Incognitae: The Journal of the Society for the History of Discoveries*, Vol. 18, No. 1 (1986), 1–20.

Bruce Masters, *Christians and Jews in the Ottoman Arab World: The Roots of Sectarianism* (Cambridge: Cambridge University Press, 2001).

Meredith Mattlin, "Examining the Social Construction of Health, Illness, and Wellness in Anti-Science Communities," MA diss., Vanderbilt University, 2018.

Cristiano Mattos, Felipe Sanches Lopez, José Luis Ortega, and André Rodrigues, "The Public Discussion on Flat Earth Movement: An Analysis Based on the Esperantist-Epideictic Discourse," *Science & Education*, Vol. 31 (Jan., 2022), 1339–61.

Layla S. Mayboudi, *Geometry Creation and Import With COMSOL Multiphysics* (Dulles, VA: Mercury Learning and Information, 2019).

Adrienne Mayor, *The First Fossil Hunters: Paleontology in Greek and Roman Times* (Princeton: Princeton University Press, 2000).

Adrienne Mayor, *Fossil Legends of the First Americans* (Princeton: Princeton University Press, 2005).

John McAleer, "'Stargazers at the world's end': telescopes, observatories and 'views' of empire in the nineteenth-century British Empire," The British Journal for the History of Science, Vol. 46, No. 3 (Sep., 2013), 389–413.

Kelly McClure, "Trump presents interesting theory on how magnets work during rally in Iowa," *Salon*, January 6, 2024, accessed March 11, 2024, https://www.salon.com/2024/01/06/presents-interesting-theory-on-how-magnets-work-during-rally-in-iowa/.

Samuel McIlhagga, "The California militia ready for civil war," *UnHerd*, June 27, 2023, accessed February 18, 2024, https://unherd.com/2023/06/the-california-militia-ready-for-civil-war/.

Peter McCandless, "Mesmerism and Phrenology in Antebellum Charleston: "Enough of the Marvellous"," *The Journal of Southern History*, Vol. 58, No. 2 (May, 1992), 199–230.

T. C. McCaskie, "Cultural Encounters: Britain and Africa in the Nineteenth Century," in Philip D. Morgan and Sean Hawkins, eds., *Black Experience and the Empire* (Oxford: Oxford University Press, 2004), 166–93.

John McGrath, "Sixteenth-Century Florida in the European Imagination," *The Florida Historical Quarterly*, Vol. 91, No. 3 (Winter, 2013), 401–18.

Justin McHenry, *Lemuria: A True Story of a Fake Place* (Port Townsend, WA: Feral House, 2024).

Robbie McLaughlan, *Re-imagining the 'Dark Continent' in fin de siècle Literature* (Edinburgh: Edinburgh University Press, 2012).

Johannes Meintjes, *President Paul Kruger: A Biography* (London: Cassell, 1974).

Martin Meredith, *Diamonds, Gold, and War: The British, the Boers, and the Making of South Africa* (New York: PublicAffairs, 2007).

Anna Merlan, "The Brain Rot Cabinet," *Mother Jones*, May 8, 2025, accessed May 9, 2025, https://www.motherjones.com/politics/2025/05/trump-cabinet-conspiracies-missing-children-epstein/.

Anna Merlan, *Republic of Lies: American Conspiracy Theorists and Their Surprising Rise to Power* (New York: Metropolitan Books, 2019).

Thomas R. Metcalf, *Imperial Connections: India in the Indian Ocean Arena, 1860–1920* (Berkeley, CA: University of California Press, 2007).

John Michell, *Eccentric Lives and Peculiar Notions* (London: Thames and Hudson, 1984).

Nancy Smith Midgette, "In Search of Professional Identity: Southern Scientists, 1883–1940," *The Journal of Southern History*, Vol. 54, No. 4 (Nov., 1988), 597–622.

Walter D. Mignolo, "DELINKING: The rhetoric of modernity, the logic of coloniality and the grammar of decoloniality," *Cultural Studies*, Vol. 21, Nos. 2–3 (2007), 449–514.

Natasha L. Mikles and Joseph P. Laycock, "Tracking the Tulpa: Exploring the "Tibetan" Origins of a Contemporary Paranormal Idea," *Nova Religio*, Vol. 19, No. 1 (Aug., 2015), 87–97.

Tiya Miles, *Tales from the Haunted South: Dark Tourism and Memories of Slavery from the Civil War Era* (Chapel Hill: University of North Carolina Press, 2015).

Floyd J. Miller, *The Search for a Black Nationality: Black Emigration and Colonization, 1787–1863* (Urbana, IL: University of Illinois Press, 1975).

Jay Miller, "Why the World is on the Back of a Turtle," *Man*, New Series, Vol. 9, No. 2 (Jun., 1974), 306–08.

Giles Milton, *The Riddle and the Knight: In Search of Sir John Mandeville* (New York: Farrar, Straus and Giroux, 1996).

Nancy B. Miner, "De Sun Do Move: The Story of John Jasper," *Fluvanna History: A Biannual Publication of the Fluvanna County Historical Society*, No. 73 (2003), i-iv, 1–48.

Miracle Workers: Dark Ages, ep. 1, "Graduation" (dirs. Jarrad Paul & Andrew Mogel, writer Simon Rich), TBS, 28 January, 2020.

Pankaj Mishra, *Age of Anger: A History of the Present* (New York: Farrar, Straus and Giroux, 2017).

Pankaj Mishra, *From the Ruins of Empire: The Intellectuals Who Remade Asia* (New York: Farrar, Straus and Giroux, 2012).

Shaheed N. Mohammed, "Conspiracy Theories and Flat Earth Videos on YouTube," *The Journal of Social Media in Society*, Vol. 8, No. 2 (Fall, 2019), 84–102.

Violet Moller, *The Map of Knowledge: How Classical Ideas Were Lost and Found: A History in Seven Cities* (New York: Doubleday, 2019).

Marko Monteiro, "Science is a war zone: some comments on Brazil," *Tapuya: Latin American Science, Technology and Society*, Vol. 3, No. 1 (Jan., 2020), 4–8.

Amanda Montell, *Cultish: The Language of Fanaticism* (New York: Harper-Collins, 2021).

Monty Python and the Holy Grail. Directed by Terry Gilliam and Terry Jones. London: Python (Monty) Pictures, 1975.

T. Dunbar Moodie, *The Rise of Afrikanerdom: Power, Apartheid, and the Afrikaner Civil Religion* (Berkeley: University of California Press, 1975).

Patrick Moore, *Can You Speak Venusian? A Guide to the Independent Thinkers* (New York: W. W. Norton & Company, 1972).

Randy Moore, "Creationism in the United States: VII. The Lingering Impact of *Inherit the Wind*," *The American Biology Teacher*, Vol. 61, No. 4 (Apr., 1999), 246–50.

George Monbiot, "Conspiracist Thinking Is the Result of 40 Years of Social Atomization," *Jacobin*, September 27, 2021, accessed September 27, 2021, https://jacobin.com/2021/09/conspiracy-theories-qanon-covid-left-anarchism-counterculture-extreme-right-restoration-stories-bridging-communities/.

Jeffrey P. Moran, *The Scopes Trial: A Brief History with Documents* (Boston: Bedford/St. Martin's, 2002).

Jeffrey P. Moran, "The Scopes Trial and Southern Fundamentalism in Black and White: Race, Region, and Religion," *The Journal of Southern History*, Vol. 70, No. 1 (Feb., 2004), 95–120.

Kenneth Morgan, *Navigating by the Southern Cross: A History of the European Discovery and Exploration of Australia* (London: Bloomsbury Academic, 2021).

Robert C. Morris, *Reading, 'Riting, and Reconstruction: The Education of Freedmen in the South, 1861–1870* (Chicago: The University of Chicago Press, 1976).

Charles W. Morse, "Letters," *The Earth*, Vol. V, Nos. 52 & 53 (Nov./Dec., 1904), 55–56.

Terry Mortenson, "The Firmament: What Did God Create on Day 2?," *Answers Research Journal*, Vol. 13 (2020), 113–33.

Jean Dietz Moss, "Galileo's Letter to Christina: Some Rhetorical Considerations," *Renaissance Quarterly*, Vol. 36, No. 4 (Winter, 1983), 547–76.

Youssef Mroueh, "Precolumbian Muslims in the Americas," *Beautiful Islam*, 1996, accessed December 10, 2023, http://www.beautifulislam.net/history/muslims_americas.htm.

E. V. Mulgrave, "Letters," *The Earth*, Vol. III, Nos. 39 & 40 (Oct. and Nov., 1903), 291–92.

Bill V. Mullen, *Afro-Orientalism* (Minneapolis: University of Minneapolis Press, 2004).

Kathleen S. Murphy, *Captivity's Collections: Science, Natural History, and the British Transatlantic Slave Trade* (Chapel Hill, NC: University of North Carolina Press, 2023).

Angela Nagle, *Kill All Normies: Online Culture Wars from 4chan and Tumblr to Trump and the Alt-Right* (Winchester: Zero Books, 2017).

Joshua Nall, "Seafaring, enslavement, and the longitude problem," in Victoria Avery and Jake Subryan Richards, eds., *Black Atlantic: Power, People, Resistance* (London: Philip Wilson Publishers, 2023), 91–93.

Jayant V. Narlikar, *The Scientific Edge: The Indian Scientist from Vedic to Modern Times* (London: Penguin Books, 2003).

John M. Navroth, "Lovecraft and the Polar Myth," *Lovecraft Annual*, No. 3 (2009), 190–98.

Yaël Nazé, "A Doctoral Dissertation on a Geocentric Flat Earth: 'Zetetic' Astronomy at the University Level," *Skeptical Inquirer*, Vol. 42, No. 3 (May/Jun., 2018), 12–14.

Ryan M. Nefdt, "Decolonizing philosophy," *Africa is a Country*, June 30, 2017, accessed March 19, 2019, https://africasacountry.com/2017/06/decolonizing-philosophy.

William Cooper Nell, *The Colored Patriots of the American Revolution, With Sketches of Several Distinguished Colored Persons: To Which Is Added a Brief Survey of the Condition And Prospects of Colored Americans* (Boston: Robert F. Wallcut, 1855).

Ian Richard Netton, *Islam, Christianity and the Realms of the Miraculous: A Comparative Exploration* (Edinburgh: Edinburgh University Press, 2020).

Nicholas Nicastro, *Circumference: Eratosthenes and the Ancient Quest to Measure the Globe* (New York: St. Martin's Press, 2008).

Larry Niven, *Ringworld* (New York: Ballantine Books, 1970).

Sindiso M. Nleya and Siqabukile Ndlovu, "Audit of Mathematical Concepts in Pre-colonial Africa," in Artwell Nhemachena, Nokuthula Hlabangane, and Joseph Z. Z. Matowanyika, eds., *Decolonising Science, Technology, Engineering and Mathematics (STEM) in an Age of Technocolonialism: Recentring African Indigenous Knowledge and Belief Systems* (Bamenda: Langaa RPCIG, 2020), 125–50.

Claudio Nocelli, *The Navigator Who Crossed the Ice Walls: Worlds Beyond the Antarctica* (Buenos Aires: Claudio Nocelli, 2022).

John North, *Cosmos: An Illustrated History of Astronomy and Cosmology, Revised Edition* (Chicago: The University of Chicago Press, 2008).

David Northrup, "West Africans' Transatlantic Voyages and Trade, 1000–1800." Lecture, Ships and Seafarers in World History Symposium, Fairhaven, MA, 28 October, 2023.

C. Philipp E. Nothaft, "Zaccaria Lilio and the shape of the earth: A brief response to Allegro's "Flat earth science"," *History of Science*, Vol. 55, No. 4 (2017), 490–98.

Ronald L. Numbers, *The Creationists: The Evolution of Scientific Creationism* (New York: Alfred A. Knopf, 1992).

Ronald L. Numbers and Janet S. Numbers, "Science in the Old South: A Reappraisal," *The Journal of Southern History*, Vol. 48, No. 2 (May, 1982), 163–84.

Ronald L. Numbers and Lester D. Stephens, "Darwinism in the American South," in Ronald L. Numbers and John Stenhouse, eds., *Disseminating Darwinism: The Role of Place, Race, Religion, and Gender* (Cambridge: Cambridge University Press, 1999), 123–44.

"OANN's Posobiec Met With Polish Neo-Fascists and Amplified Their Messages Online," Southern Poverty Law Center, July 22, 2020, accessed March 16, 2025, https://www.splcenter.org/resources/hate-watch/oanns-posobiec-met-polish-neo-fascists-and-amplified-their-messages-online/.

Michael Oard, *The Weather Book* (Green Forest, AR: Master Books, 1997).

Kevin Ochieng Okoth, *Red Africa: Reclaiming Revolutionary Black Politics* (London: Verso, 2023).

John O'Connor, *The Secret History of Bigfoot: Field Notes on a North American Monster* (Naperville, IL: Sourcebooks, 2024).

Ali A. Olomi, "Baghdad: The Once and Future City of Stars," Public Books, February 8, 2022, accessed February 8, 2022, https://www.publicbooks.org/baghdad-medieval-muslim-astrology/.

Alex Olshansky, Robert M. Peaslee, and Asheley R. Landrum, "Flat-Smacked! Converting to Flat Eartherism," *Journal of Media and Religion*, Vol. 19, No. 2 (Jul., 2020), 46–59.

Benjamin B. Olshin, *Lost Knowledge: The Concept of Vanished Technologies and Other Human Histories* (Leiden: Brill, 2019).

H. M. Oluseyi, "Contributions from the First Half-Century of African American Astronomers with a Focus on Solar Physics," *African Skies/Cieux Africains*, No. 11 (Jul., 2007), 42–44.

Ibn Fadl Allah al-Omari, trans. Gaudefroy Demombynes, *Masalik al Absar fi Mamalik el-Amsar* (Paris: Lib. Orientaliste Paul Geuthner, 1927).

Naomi Oreskes, "Science in the Origins of the Cold War," in Naomi Oreskes and John Krige, eds., *Science and Technology in the Global Cold War* (Cambridge, MA: The MIT Press, 2014), 11–30.

Pete Orford, "Picturing John Jasper," *Cloisterham Tales*, July 31, 2013; accessed July 30, 2020, https://cloisterhamtales.wordpress.com/2013/07/31/picturing-john-jasper/.

George Orwell, *1984* (New York: Plume, 1983 [1949]).

George Orwell, "As I Please, *Tribune*, 21 December 1946," in Sonia Orwell and Ian Angus, eds., *The Collected Essays, Journalism and Letters of George Orwell, Volume IV: In Front of Your Nose, 1945–1950* (London: Secker & Warburg, 1968), 259–62.

Clare Oxby, "A Review of African Ethno-Astronomy: With Particular Reference to Saharan Livestock-Keepers," *La Ricerca Folklorica*, No. 40 (Oct., 1999), 55–64.

Oloyede S. Oyelekan and Julius B. Omiwale, "Trends in Nigeria's Science Education within the Global Context," in Femi S. Otulaja and Meshach B. Ogunniyi, eds., *The World of Science Education: Handbook of Research in Science Education in Sub-Saharan Africa* (Rotterdam: Sense Publishers, 2017), 43–64.

Kemal Özdemir, trans. Priscilla Mary Işın, *Ottoman Nautical Charts and the Atlas of Ali Macar Reis* (Istanbul: Creative Yayıncılık ve Tanitim, 1992).

Ricardo Padrón, *The Spacious Word: Cartography, Literature, and Empire in Early Modern Spain* (Chicago: The University of Chicago Press, 2004).

Nell Irvin Painter, *Exodusters: Black Migration to Kansas After Reconstruction* (New York: W. W. Norton & Company, 1976).

Ewan Palmer, "QAnon Telegram Account Is World's 'Largest Antisemitic Online Channel'," *Newsweek*, May 26, 2021, accessed February 18, 2024, https://www.newsweek.com/qanon-ghostezra-holocaust-denial-telegram-1594907.

Robert Palter, "Black Athena, Afrocentrism, and the History of Science," in Mary R. Lefkowitz and Guy MacLean Rogers, eds., *Black Athena Revisited* (Chapel Hill: University of North Carolina Press, 1996), 209–66.

John Pappademos, "An Outline of Africa's Role in the History of Physics," in Ivan Van Sertima, ed., *Blacks in Science: Ancient and Modern* (New Brunswick, NJ: Transaction Books, 1983), 177–96.

Tudor Parfitt, *Journey to the Vanished City: The Search for a Lost Tribe of Israel* (New York: St. Martin's Press, 1992).

Alex Park, "The blankest spot on Trump's world map," *Africa is a Country*, June 26, 2020, accessed July 1, 2020, https://africasacountry.com/2020/06/the-blankest-spot-on-trumps-world-map.

H. B. Parks, "Follow the Drinking Gourd," *Publications of the Texas Folk-Lore Society*, No. 7 (1928), 81–84.

Alex Pasternack, "The Moon-Walking, Alien-Hunting, Psychic Astronaut Who Got Sued By NASA," *Vice*, May 14, 2016, accessed January 24, 2024, https://www.vice.com/en/article/aek7ez/astronaut-edgar-mitchell-outer-space-inner-space-and-aliens.

Holly Patrick, "Tories are 'political wing of Flat Earth Society', taunts Starmer," *The Independent*, February 28, 2024, accessed March 27, 2024, https://www.independent.co.uk/tv/news/sunak-starmer-liz-truzz-deep-state-b2503941.html.

Roger Pearse, "Cosmas Indicopleustes, Christian Topography. Preface to the online edition.," *The Tertullian Project*, July 5, 2003, accessed June 22, 2023, https://www.tertullian.org/fathers/cosmas_00_0_eintro.htm.

Douglas T. Peck, "Anatomy of an Historical Fantasy: The Ponce de León-Fountain of Youth Legend," *Revista de Historia de América*, No. 123 (Jan.–Dec., 1998), 63–87.

Douglas T. Peck, "Re-Examination of Spanish Colonial Period Documents Related to Prehistoric Maya History and Mythology," *Revista de Historia de América*, No. 136 (Jan.–Dec., 2005), 21–35.

Janice Peck, *The Age of Oprah: Cultural Icon for the Neoliberal Era* (Boulder, CO: Paradigm Publishers, 2008).

Olaf Pedersen, trans. Richard North, *The First Universities:* Studium Generale *and the Origins of University Education in Europe* (Cambridge: Cambridge University Press, 1997).

Gordon Pentland, "The freethinkers' zetetic society: an Edinburgh radical underworld in the eighteen-twenties," *Historical Research*, Vol. 91, No. 252 (May, 2018), 314–32.

Joan Ruman Perkal, "Witchcraft Involves Millions in Africa," *Western Folklore*, Vol. 29, No. 4 (Oct., 1970), 284.

Sandra Perot. "Reconstructing Molly Welsh: Race, Memory, and the Story of Benjamin Banneker's Grandmother." MA thesis, University of Massachusetts Amherst, 2008.

Edward Pessen, *Jacksonian America: Society, Personality, and Politics* (Homewood, IL: Dorsey Press, 1969).

Tom Phillips, "Covid denialist and Bolsonaro ally Olavo de Carvalho died of virus, says daughter," *The Guardian*, January 25, 2022, accessed June 7, 2023, https://riotimesonline.com/brazil-news/opinion-editorial/brazils-bolsonaro-hows-he-ng-after-100-days-part-i-foreign-affairs/.

Tom Phillips and Dom Phillips, "'Unqualified, dangerous': the oddball officials running Bolsonaro's Brazil," *The Guardian*, January 2, 2020, accessed February 21, 2024, https://www.theguardian.com/world/2020/jan/02/bolsonaro-brazil-government-oddball-officials.

Rob Picheta, "The flat-Earth conspiracy is spreading around the globe. Does it hide a darker core?," CNN, November 18, 2019, accessed August 3, 2020, https://edition.cnn.com/2019/11/16/us/flat-earth-conference-conspiracy-theories-scli-intl/index.html.

Ruth Pike, "Sevillian Society in the Sixteenth Century: Slaves and Freedmen," *The Hispanic American Historical Review*, Vol. 47, No. 3 (Aug., 1967), 344–59.

David Pingree, "Astronomy and Astrology in India and Iran," *Isis*, Vol. 54, No. 2 (Jun., 1963), 229–46.

David Pingree, "Brahmagupta, Balabhadra, Pṛthūdaka and Al-Bīrūnī," *Journal of the American Oriental Society*, Vol. 103, No. 2 (Apr.–Jun., 1983), 353–60.

Steven Pinker, "Guess what? More people are living in peace now. Just look at the numbers," *The Guardian*, March 20, 2015, accessed July 10, 2020, https://www.theguardian.com/commentisfree/2015/mar/20/wars-john-gray-conflict-peace.

Karen C. Pinto, "Ecce! A Ninth-Century Isidorean T-O Map Labeled in Arabic," in Albrecht Classen, ed., *Globalism in the Middle Ages and the Early Modern Age: Innovative Approaches and Perspectives* (Berlin: Walter de Gruyter GmbH, 2023), 189–220.

Karen C. Pinto, "Fit for an Umayyad Prince: An Eighth- Century Map or the Earliest Mimetic Painting of the Moon?," *The Medieval Globe*, Vol. 4.2 (2018), 29–68.

Karen C. Pinto, *Medieval Islamic Maps: An Exploration* (Chicago: University of Chicago Press, 2016).

Points of History: The Inquisition; The Albigenses and the Waldenses; The Massacre of St Bartholomew; The Fifth of November or the Gunpowder Plot; Galileo and the Inquisition; Religious Toleration a Question of First Principles (Boston: Marlier & Co., 1901).

Tom Porter, "A fringe movement that claims bleach is a miracle cure is jubilant after Trump floated injecting disinfectant as a treatment for COVID-19," *Business Insider*, April 24, 2020, accessed February 23, 2024, https://www.businessinsider.com/mms-fans-celebrate-trump-claim-disinfectant-might-cure-coronavirus-2020-4.

Patrice Potvin, Marianne Bissonnette, Chirine Chamsine, Marie-Hélène Bruyère, Mohamed Amine Mahhou, Olivier Arvisais, Patrick Charland, Stéphane Cyr, and Vivek Venkatesh, "Science Education under a Totalitarian Theocracy: Analyzing the ISIS Primary Curriculum," *Journal of Research in Science, Mathematics and Technology Education*, Vol. 2, No. 3 (Sep., 2019), 179–200.

Theodore Powers, "Institutions, power and para-state alliances: a critical reassessment of HIV/AIDS politics in South Africa, 1999–2008," *The Journal of Modern African Studies*, Vol. 51, No. 4 (Dec., 2013), 605–26.

Cathryn J. Prince, *A Professor, A President, and A Meteor: The Birth of American Science* (New York: Prometheus Books, 2011).

Cyril Prinsloo, "Bolsonaro and the BRICS: Bull in a china shop?," *South African Institute of International Affairs*, Policy Briefing 180 (Mar., 2019), 1–15.

Jacopo Prisco, "The 'Mandela Effect' describes the false memories many of us share. But why can't scientists explain it?," CNN, September 18, 2023, accessed February 7, 2024, https://www.cnn.com/2023/09/18/world/mandela-effect-collective-false-memory-scn/index.html.

Donald R. Prothero, *Weird Earth: Debunking Strange Ideas about Our Planet* (Bloomington: Red Lightning Books, 2020).

Alicia Puglionesi, *Common Phantoms: An American History of Psychic Science* (Palo Alto, CA: Stanford University Press, 2020).

Lydia Pyne, *Endlings: Fables for the Anthropocene* (Minneapolis: University of Minnesota Press, 2022).

Q, "Dec 12 2018," Qmap, December 12, 2018, accessed August 24, 2021, https://qanon.pub/#2622.

Kinana Qaddour, "Inside ISIS' Dysfunctional Schools: The Failure of the Group's Educational System," *Foreign Affairs*, October 13, 2017, accessed February 21, 2021, https://www.foreignaffairs.com/articles/syria/2017-10-13/inside-isis-dysfunctional-schools.

Sarah A. Qidwai, "Reexamining Complexity: Sayyid Ahmad Khan's Interpretation of "Science" in Islam," in Bernard Lightman, ed., *Rethinking History, Science, and Religion: An Exploration of Conflict and the Complexity Principle* (Pittsburgh: University of Pittsburgh Press, 2019), 50–64.

Sarah Ahmed Qidwai. "Sir Syed (1817–1898) and Science: Popularization in Nineteenth Century India." PhD diss., University of Toronto, 2021.

John Edward Quinlan, *The Earth a Plane* (London: John Edward Quinlan, 1906).

Sadiah Qureshi, "Science, Empire and Globalization in the Nineteenth Century," in John Holmes and Sharon Ruston, eds., *The Routledge Research Companion to Nineteenth-Century British Literature and Science* (London: Routledge, 2017), 19–29.

Albert J. Raboteau, *Slave Religion: The "Invisible Institution" in the Antebellum South* (New York: Oxford University Press, 1978).

"Races disagree on impact of Simpson trial," CNN, October 6, 1995, accessed January 24, 2024, http://www.cnn.com/US/OJ/daily/9510/10-06/poll_race/oj_poll_txt.html.

Gideon Rachman, *The Age of the Strongman: How the Cult of the Leader Threatens Democracy Around the World* (New York: Other Press, 2022).

Gloria D. Rall, "Follow the Drinking Gourd," *The Planetarian*, Vol. 23, No. 3 (Sep., 1994), 8–12.

Nikki McCann Ramirez, "Georgia GOP Chair Goes Full Flat-Earth, Says Globes Are Part of a Conspiracy," *Rolling Stone*, May 24, 2023, accessed June 4, 2023, https://www.rollingstone.com/politics/politics-news/kandiss-taylor-globes-anti-flat-earth-brainwashing-1234741082/.

Edwin Archer Randolph, *The Life of Rev. John Jasper, Pastor of Sixth Mt. Zion Baptist Church, Richmond, Va., from His Birth to the Present Time, with His Theory on the Rotation of the Sun* (Richmond: R. T. Hill & Co., 1884).

Frederick N. Rasmussen, "Judge campaigns to restore reputation of city's first schools superintendent," *The Baltimore Sun*, August 28, 2010. Accessed January 29, 2021. https://www.baltimoresun.com/maryland/bs-xpm-2010-08-28-bs-md-backstory-john-mcjilton-20100828-story.html.

Peter Razzell, "The Protestant Ethic and the Spirit of Capitalism: A Natural Scientific Critique," *The British Journal of Sociology*, Vol. 28, No. 1 (Mar., 1977), 17–37.

Brian Regal, *The Battle over America's Origin Story: Legends, Amateurs, and Professional Historiographers* (London: Palgrave Macmillan, 2022).

Darren Reid, "British Humanitarianism, Indigenous Rights, and Imperial Crises: Assessing the Membership Base of the Aborigines' Protection Society, 1840–73," *Journal of British Studies*, Vol. 63, No. 3 (Jul., 2024), 625–42.

Report of the "Conditional Immortality" Conference, held at the Cannon Street Hotel, London, Monday, May 15th, 1876 (London: Kellaway & Co., 1876).

"Republican nominee for US Senate in Delaware has repeatedly promoted QAnon conspiracy theory, once called self 'flat earther'," *Chicago Tribune*, September 17, 2020, accessed August 24, 2021, https://www.chicagotribune.com/2020/09/17/republican-nominee-for-us-senate-in-delaware-has-repeatedly-promoted-qanon-conspiracy-theory-once-called-self-flat-earther/.

"Rev. Dr. Hatcher's Surprising Assertions," *Richmond Planet* (Jun. 23, 1894), 2; https://encyclopediavirginia.org/entries/rev-dr-hatchers-surprising-assertions-richmond-planet-june-23-1894/ (accessed 24 December 2022).

Darcy Ribeiro, trans. Gregory Ribassa, *The Brazilian People: The Formation and Meaning of Brazil* (Gainesville, FL: University of Florida Press, 2000 [1995]).

Heather Cox Richardson, *The Death of Reconstruction: Race, Labor, and Politics in the Post-Civil War North, 1865–1901* (Cambridge, MA: Harvard University Press, 2001).

Selden Richardson, *Built by Blacks: African American Architecture and Neighborhoods in Richmond* (Charleston, SC: The History Press, 2008).

Martin Ridge, *Ignatius Donnelly: The Portrait of a Politician* (Chicago: The University of Chicago Press, 1962).

Luke Ritter, "Nativism and the Roman Catholic Plot to Destroy America," in Luke Ritter, ed., *American Conspiracism: An Interdisciplinary Exploration* (London: Routledge, 2024), 177–96.

Stuart J. Robbins, "Episode 149 - Modern Flat Earth Thought, Part 2 (U.N. Flag and Airplane Flights)," *Exposing PseudoAstronomy*, October 1, 2016, accessed November 21, 2019, https://podcast.sjrdesign.net/shownotes_149.php.

C. Robertson, *The Mutual Relations of the Sun and the Earth: A Study in Physical Geography* (Edinburgh: St. Giles Printing Company, n.d.).

Michael F. Robinson, *The Lost White Tribe: Explorers, Scientists, and the Theory that Changed a Continent* (Oxford: Oxford University Press, 2016).

Michael F. Robinson, "Scientific Travel," in Nandini Das and Tim Youngs, eds., *The Cambridge History of Travel Writing* (Cambridge: Cambridge University Press, 2019), 493–95.

Camila Rocha, "From Orkut to Brasília: the origins of the New Brazilian Right," in Katerina Hatzikidi and Eduardo Dullo, eds., *A Horizon of (Im)possibilities: A Chronicle of Brazil's Conservative Turn* (London: University of London Press, 2021), 81–101.

Rocketman: Mad Mike's Mission to Prove the Flat Earth, directed by Toby Brusseau and Michael Linn (New York: Flix Premiere, 2019).

Duane W. Roller, *Through the Pillars of Herakles: Greco-Roman Exploration of the Atlantic* (New York: Routledge, 2006).

Marissa Rollnick, "The Growth of Science Education Organisations in the Old and New South Africa," in Femi S. Otulaja and Meshach B. Ogunniyi, eds.,

The World of Science Education: Handbook of Research in Science Education in Sub-Saharan Africa (Rotterdam: Sense Publishers, 2017), 27–41.

James S. Romm, *The Edges of the Earth in Ancient Thought: Geography, Exploration, and Fiction* (Princeton: Princeton University Press, 1992).

Allegra Rosenberg, "The Truth About Antarctica," *The Long Now Foundation*, February 21, 2023, accessed February 21, 2023, https://longnow.org/ideas/the-truth-about-antarctica/.

Mike Rothschild, *The Storm Is Upon Us: How QAnon Became a Movement, Cult, and Conspiracy Theory of Everything* (Brooklyn: Melville House Publishing, 2021).

Samuel Rowbotham, *Zetetic Astronomy: Earth Not a Globe, Third Edition* (London: Simpkin, Marshall, and Co., 1881).

Michael Royster, "Brazil's Bolsonaro: How's He Doing After 100 Days? Part I: Foreign Affairs," *The Rio Times*, April 10, 2019, accessed April 19, 2019, https://riotimesonline.com/brazil-news/opinion-editorial/brazils-bolsonaro-hows-he-ng-after-100-days-part-i-foreign-affairs/.

Britt Rusert, *Fugitive Science: Empiricism and Freedom in Early African American Culture* (New York: New York University Press, 2017).

Anne Spry Rush, *Bonds of Empire: West Indians and Britishness from Victoria to Decolonization* (Oxford: Oxford University Press, 2011).

Jeffrey Burton Russell, *Inventing the Flat Earth: Columbus and Modern Historians* (New York: Praeger, 1991).

Tim Russell, "Ringworld Ocean," The Universe of Larry Niven, n.d., accessed March 13, 2023, https://larryniven.net/files/ringworld_ocean-r.jpg.

Elias N. Saad, *Social History of Timbuktu: The Role of Muslim Scholars and Notables 1400–1900* (Cambridge: Cambridge University Press, 1983).

Abdelhamid Sabra, "An Eleventh-century Refutation of Ptolemy's Planetary Theory," in Erna and Frank D. Grande, eds., *Science and History: Studies in Honor of Edward Rosen* (Wrocław: The Polish Academy, 1978), 117–31.

Carl Sagan, *Cosmos* (New York: Random House, 1980).

Esmeralda Sánchez Salazar, Brandon Vaidyanathan, Elaine Howard Ecklund, and Adriana Garcia, "Challenging Evolution in Public Schools: Race, Religion, and Attitudes toward Teaching Creationism," *Socius: Sociological Research for a Dynamic World*, Vol. 5 (Sep., 2019), 1–13.

Andrea Salcedo, "Wisconsin pharmacist who destroyed more than 500 vaccine doses believes Earth is flat, FBI says," *The Washington Post*, February 1, 2021, accessed February 1, 2024, https://www.washingtonpost.com/nation/2021/02/01/wisconsin-pharmacist-vaccine-flatearth/.

Isaac Samuel, "a brief note on Africa's Scientific Manuscripts," *African History Extra*, September 30, 2023, accessed September 30, 2023, https://www.africanhistoryextra.com/p/a-brief-note-on-africas-scientific.

Isaac Samuel, "A complete history of Dogon country: Bandiagara from 1900 BC to 1900 AD: demystifying an ancient African society," *African History Extra*, March 5, 2023, accessed March 13, 2023, https://isaacsamuel.substack.com/p/a-complete-history-of-dogon-country.

Isaac Samuel, "Star-gazing from the cliffs of Bandiagara; truths and mysteries of Dogon astronomy in the west African context," *African History Extra*, October 28, 2022, accessed March 13, 2023, https://www.patreon.com/posts/star-gazing-from-73897317.

Shinya Sato, "Effects of localized source currents on magnetotelluric responses of a flat earth and spherical earth," *arXiv* (Jul. 2020), 1–11.

"Saudi Preacher Bandar Al-Khaybari: The Earth Does not Revolve around Itself - Extended Version," Middle East Media Research Institute, January 15, 2015, accessed January 25, 2024, https://www.memri.org/tv/saudi-preacher-bandar-al-khaybari-earth-does-not-revolve-around-itself-extended-version.

Robert J. Schadewald, "The Flat-out Truth: Earth Orbits? Moon Landings? A Fraud! Says This Prophet," *Science Digest*, Vol. 88, No. 1 (Jul., 1980), 58–63.

Robert J. Schadewald, "Is the World in Curious Shape?," *Isaac Asimov's Science Fiction Magazine*, Vol. 4, No. 12 (Dec., 1980), 97–106.

Robert J. Schadewald, *The Plane Truth: A History of the Flat Earth Movement* (Wendy S. Schadewald, 2015).

Robert J. Schadewald, "Scientific Creationism, Geocentricity and the Flat Earth," *The Skeptical Inquirer*, Vol. 6, No. 2 (Winter, 1981–82), 41–48.

Larry Schaaf, "Charles Piazzi Smyth's 1865 Conquest of the Great Pyramid," *History of Photography*, Vol. 3, No. 4 (Oct., 1979), 331–54.

Simon Schaffer, "The nebular hypothesis and the science of progress," in *History, Humanity, and Evolution: Essays for John C. Greene*, ed. James R. Moore (Cambridge: Cambridge University Press, 1989), 142–53.

Tom Schaller and Paul Waldman, *White Rural Rage: The Threat to American Democracy* (New York: Random House, 2024).

Cyrus Schayegh, "Three Questions for Historians of Science in the Modern Middle East and North Africa," *International Journal of Middle East Studies*, Vol. 47, No. 3 (Aug., 2015), 588–91.

Gerrit Schutte, "Abraham Kuyper and his South African Brethren," *Rozenberg Quarterly*, June 22, 2011, accessed March 5, 2024, https://rozenbergquarterly.com/abraham-kuyper-and-his-south-african-brethren/.

SciManDan, "Flat Earther Thinks They are Taking Over the World," YouTube, December 1, 2023, accessed December 2, 2023, https://www.youtube.com/watch?v=SVFl9DKf18o.

Nina Gilden Seavey, "True Conspiracies: The Legacy of J. Edgar Hoover, Dr. Martin Luther King, Jr., and James Earl Ray," in Luke Ritter, ed., *American Conspiracism: An Interdisciplinary Exploration* (London: Routledge, 2024), 80–97.

Mausi Segun, *"They Set the Classrooms on Fire": Attacks on Education in Northeast Nigeria* (New York City: Human Rights Watch, 2016).

David Sehat, "The Civilizing Mission of Booker T. Washington," *The Journal of Southern History*, Vol. 73, No. 2 (May, 2007): 323–62.

Avi Selk, "Why Neil deGrasse Tyson failed to prove Earth isn't flat," *The Washington Post*, March 12, 2018, accessed March 18, 2018, https://www.washingtonpost.com/news/speaking-of-science/wp/2018/03/12/why-neil-degrasse-tyson-failed-to-prove-earth-isnt-flat/?utm_term=.21e68d784588.

Namwali Serpell, "The Afronaut Archives: Reports from a Future Zambia," Public Books, March 28, 2019, accessed January 14, 2020, https://www.publicbooks.org/the-afronaut-archives-reports-from-a-future-zambia/.

Ivan Van Sertima, *They Came Before Columbus: The African Presence in Ancient America* (New York: Random House, 1976).

Muhammad Mann Shaaba, "Absolutism in Religion and Global Peace: Boko-Haram Factor in Nigeria and Its Educational Imperatives," *World Journal of Education*, Vol. 7, No. 2 (Apr., 2017), 11–18.

Manimugdha S. Sharma, "Faking history starts online," *The Times of India*, November 19, 2017, accessed February 25, 2024, https://timesofindia.indiatimes.com/home/sunday-times/faking-history-starts-online/articleshow/61705453.cms.

Ronald LaMarr Sharps, *Black Folklorists in Pursuit of Equality: African American Identity and Cultural Politics, 1893–1943* (Lanham, MD: Lexington Books, 2022).

William Shatner and Chris Kreski, *Star Trek Movie Memories* (New York: HarperCollins, 1995).

Konstantin Sheiko. "Lomonosov's Bastards: Anatolii Fomenko, Pseudo-History, and Russia's Search for a Post-Communist Identity." PhD diss., University of Wollongong, 2004.

Abubakar Shekau, "Hādhihi 'Aqīdatunā (This is Our Creed)," in Abdulbasit Kassim and Michael Nwankpa, eds., *The Boko Haram Reader: From Nigerian Preachers to the Islamic State* (Oxford: Oxford University Press, 2018), 139–46.

Abubakar Shekau, "Western Civilization is Atheism and Anti-Islam," in Abdulbasit Kassim and Michael Nwankpa, eds., *The Boko Haram Reader: From Nigerian Preachers to the Islamic State* (Oxford: Oxford University Press, 2018), 117–30.

Samuel Shenton, "Mr Shenton's Map with Apollo Flight Path," Flat Earth Society, 1972, accessed February 13, 2024, https://www.theflatearthsociety.org/home/index.php/featured/maps.

Samuel Shenton, "Outgoing Correspondence 4240," The Flat Earth Society Library, March 12, 2014, accessed November 30, 2019, https://www.theflatearthsociety.org/library/correspondence/Samuel%20Shenton%20-%20Outgoing%20Corresponcence%204240.pdf.

Samuel Shenton, "Outgoing Correspondence 5846," The Flat Earth Society Library, March 12, 2014, accessed November 30, 2019, https://www.theflatearthsociety.org/library/correspondence/Samuel%20Shenton%20-%20Outgoing%20Correspondence%205846.pdf.

Marika Sherwood, *Origins of Pan-Africanism: Henry Sylvester Williams, Africa, and the African Diaspora* (London: Routledge, 2010).

William Shoki, "The 'Mandela Effect' describes the false memories many of us share. But why can't scientists explain Political Struggle Is the Answer — Not Conspiracy Theories," *Jacobin*, January 20, 2020, accessed February 3, 2020, https://jacobin.com/2020/01/conspiracy-theories-nelson-mandela-anc-south-africa.

H. Colleen Sinclair, Andrew R. Burns, and Brett S. Burton, "Taking Conspiracies to Extremes: How Extremist Groups Use Conspiracy Narratives to Stoke Violence," in Luke Ritter, ed., *American Conspiracism: An Interdisciplinary Exploration* (London: Routledge, 2024), 137–158.

Merrill Singer, "The Southern Origin of Black Judaism," in Hans A. Baer and Yvonne Jones, eds., *African Americans in the South: Issues of Race, Class, and Gender* (Athens, GA: University of Georgia Press, 1992), 123–38.

P. W. Singer and Emerson T. Brooking, *LikeWar: The Weaponization of Social Media* (Boston: Houghton Mifflin Harcourt, 2018).

Justin Sink, "Obama mocks skeptics of climate change as 'flat-Earth society'," *The Hill*, June 25, 2013, accessed February 27, 2024, https://thehill.com/blogs/blog-briefing-room/news/154505-obama-mocks-skeptics-of-climate-change-as-flat-earth-society/.

Rudolph Simek, trans. Angela Hall, *Heaven and Earth in the Middle Ages: The Physical World before Columbus* (Woodbridge: The Boydell Press, 1996).

Laura Simmons, "Flat-Earthers Travel To Antarctica To Test Theories, But Are Quickly Humbled," IFLScience, December 17, 2024, accessed December 19, 2024, https://www.iflscience.com/flat-earthers-travel-to-antarctica-to-test-theories-but-are-quickly-humbled-77254.

William J. Simmons, *Men of Mark: Eminent, Progressive and Rising* (Cleveland: George M. Rewell & Co., 1887).

Thomas Simpson, "Historicizing Humans in Colonial India," in Efram Sera-Shriar, ed., *Historicizing Humans: Deep Time, Evolution, and Race in Nineteenth-Century British Sciences* (Pittsburgh: University of Pittsburgh Press, 2018), 113–37.

Harvard Sitkoff, *A New Deal for Blacks: The Emergence of Civil Rights as a National Issue, Volume I: The Depression Decade* (Oxford: Oxford University Press, 1978).

Martín Sivak, "Axel Kicillof: "Milei quiere resolver los problemas de Argentina con ideas ridículas del siglo XIX"," *El País*, October 5, 2023, accessed May 19, 2024, https://elpais.com/argentina/2023-10-05/axel-kircillof-milei-quiere-resolver-los-problemas-de-argentina-con-ideas-ridiculas-del-siglo-xix.html.

Sujit Sivasundaram, "A global history of science and religion," in Thomas Dixon, Geoffrey Cantor, and Stephen Pumfrey, eds., *Science and Religion: New Historical Perspectives* (Cambridge: Cambridge University Press, 2010), 177–97.

Lauren Rebecca Skarloff, *Black Culture in the New Deal: The Quest for Civil Rights in the Roosevelt Era* (Chapel Hill: University of North Carolina Press, 2009).

A. E. Skellam, *The Shape of the World* (London: Harrison & Sons, 1893).

Joshua Slocum, *Sailing Alone Around the World* (New York: Dover Publications, 1956 [1900]).

Victor Slocum, *Capt. Joshua Slocum: The Life and Voyages of America's Best Known Sailor* (New York: Sheridan House, 1950).

Albert Smith, "Geographical Longitudes," *The Earth not a Globe Review*, No. 5 (Jan., 1894), 106.

Albert Smith, "Notes," *The Earth not a Globe Review*, No. 5 (Jan., 1894), 111.

Albert Smith, "Professor Totten and Joshua's Long Day," *The Earth not a Globe Review*, No. 7 (May, 1894), 145–49.

Albert Smith, *The Sea-Earth Globe and its Monstrous Hypothetical Motions; or Modern Theoretical Astronomy* (Abington: Albert Smith, 1918).

Albert Smith, "The Shape of the Earth," *The Earth not a Globe Review*, No. 1 (Jan., 1893), 18–20.

Karl A. Smith and Albert Smith, *Is the Earth a Globe Whirling in Space?, Second Edition* (Abington: Albert Smith, 1918).

Ryan Smith, "Candace Owens Describes Science as 'Pagan Faith'," *Newsweek*, July 3, 2024, accessed July 5, 2024, https://www.newsweek.com/candace-owens-science-pagan-faith-christianity-podcast-1920510.

Charles Piazzi Smyth, *Our Inheritance in the Great Pyramid* (London: Alexander Strahan, 1864).

Dava Sobel, *Longitude: The True Story of a Lone Genius Who Solved the Greatest Scientific Problem of His Time* (New York: MJF Books, 1995).

Amara L. Solari, "The Relación Geográfica Map of Tabasco: Hybrid Cartography and Integrative Knowledge Systems in Sixteenth-Century New Spain," *Terrae Incognitae: The Journal of the Society for the History of Discoveries*, Vol. 41, No. 1 (2009), 38–58.

Will Sommer, *Trust the Plan: The Rise of QAnon and the Conspiracy That Unhinged America* (New York: HarperCollins, 2023).

Jon Soske, *Internal Frontiers: African Nationalism and the Indian Diaspora in Twentieth-Century South Africa* (Athens, OH: Ohio University Press, 2017).

Amelia Carolina Sparavigna, "The Science of Al-Biruni," *International Journal of Sciences*, Vol. 2, No. 12 (Dec., 2013), 52–60.

Anne Speckhard, Ardian Shajkovci, Claire Wooster and Neima Izadi, "Mounting a Facebook Brand Awareness and Safety Ad Campaign to Break the ISIS Brand in Iraq," *Perspectives on Terrorism*, Vol. 12, No. 3 (Jun., 2018), 50–66.

Avan Judd Stallard, "Origins of the Idea of Antipodes: Errors, Assumptions, and a Bare Few Facts," *Terrae Incognitae: The Journal of the Society for the History of Discoveries*, Vol. 42, No. 1 (2010), 34–51.

David Standish, *Hollow Earth: The Long and Curious History of Imagining Strange Lands, Fantastical Creatures, Advanced Civilizations, and Marvelous Machines Below the Earth's Surface* (Boston: De Capo Press, 2006).

Simona Stano, "The Internet and the spread of conspiracy content," in Michael Butter and Peter Knight, eds., *Routledge Handbook of Conspiracy Theories* (London: Routledge, 2020), 483–96.

Star Trek V: The Final Frontier. Directed by William Shatner. Los Angeles, CA: Paramount Pictures, 1989.

S. Frederick Starr, *Lost Enlightenment: Central Asia's Golden Age from the Arab Conquest to Tamerlane* (Princeton: Princeton University Press, 2013).

Eleanor Stephenson, "Maps, instruments, and maritime power," in Victoria Avery and Jake Subryan Richards, eds., *Black Atlantic: Power, People, Resistance* (London: Philip Wilson Publishers, 2023), 88–90.

Thomas Strange, "A black and white sermon: John Jasper's capitalization on white exploitation in late nineteenth-century Virginia," *American Nineteenth Century History*, Vol. 18, No. 3 (2017), 273–93.

Charles Stross, "Missile Gap," in Gardner Dozois, ed., *One Million A.D.* (Garden City, NY: Science Fiction Book Club, 2005), 283–347.

Asawin Suebsaeng and Kelly Weill, "Phat Earth, Feat. Dylan Gelula," *Fever Dreams*, February 23, 2022, accessed February 23, 2022, https://podcasts.apple.com/us/podcast/phat-earth-feat-dylan-gelula/id1558716930?i=1000551957397.

Robert Sungenis, *Flat Earth Flat Wrong: An Historical, Biblical & Scientific Analysis* (State Line, PA: Catholic Apologetics International Publishing, 2018).

Jeremi Suri, *Civil War by Other Means: America's Long and Unfinished Fight for Democracy* (New York: PublicAffairs, 2022).

Ellen E. Swartz, "Removing the Master Script: Benjamin Banneker "Re-Membered"," *Journal of Black Studies*, Vol. 44, No. 1 (Jan., 2013), 31–49.

Egerton Sykes, "Noah's Ark: The Secret of Mount Ararat," *Atlantean Research: Official Organ of the Research Centre Group*, Vol. 1, No. 6 (Mar., 1949), 84–90.

T. J. Tallie, "The Impossible Handshake: The Fault Lines of Friendship in Colonial Natal, 1850–1910," in Shannon Walsh and Jon Soske, eds., *Ties that Bind: Race and the Politics of Friendship in South Africa* (Johannesburg: Wits University Press, 2016), 100–24.

Liane Tanguay, "We, the Sheeple: Making Sense of Conspiracy Theory in the Context of Neoliberalism," *symplokē*, Vol. 29, Nos. 1–2 (2021), 665–76.

Alrutheus Ambush Taylor, *The Negro in the Reconstruction of Virginia* (New York: Russell and Russell, 1926).

Benjamin R. Teitelbaum, *War for Eternity: Inside Bannon's Far-Right Circle of Global Power Brokers* (New York: Dey Street, 2020).

Nand Teltumbde, "Saffron Science," *Economic and Political Weekly*, Vol. 50, No. 6 (7 Feb., 2015), 10–11.

Walter Magnes Teller, *The Search for Captain Slocum: A Biography* (New York: Charles Scribner's Sons, 1956).

Walter Magnes Teller, *The Voyages of Joshua Slocum* (New Brunswick, NJ: Rutgers University Press, 1958).

Edward E. Telles, *Race in Another America: The Significance of Skin Color in Brazil* (Princeton: Princeton University Press, 2006).

Robert K. G. Temple, *The Sirius Mystery* (New York: St. Martin's Press, 1976).

Alfred Lord Tennyson, "Ulysses," in *Alfred Lord Tennyson: Selected Poems*, ed. Christopher Ricks (London: Penguin Books, 2007), 49–50.

W. Bird Terwilliger, "John Nelson McJilton. Humorist, Divine, Educator," *Maryland Historical Magazine*, No. 32 (Dec., 1937), 301–31.

Christopher B. Teuton, *Deep Waters: The Textual Continuum in American Indian Literature* (Lincoln, NE: University of Nebraska Press, 2010).

"The "Earth is a Realm" Quote Misattributed to Nikola Tesla," FlatEarth.ws, November 8, 2018, accessed April 9, 2024, https://flatearth.ws/tesla-realm#more-343.

TheMatrixQ, "F.S. Cherry church 1886 founding year doubtful," *?verything!*, March 27, 2013, accessed June 8, 2023, https://farpointe.wordpress.com/2013/03/27/f-s-cherry-church-1886-founding-year-doubtful/.

Courtney E. Thompson, *An Organ of Murder: Crime, Violence, and Phrenology in Nineteenth-Century America* (New Brunswick, NJ: Rutgers University Press, 2021).

E. P. Thompson, "Time, Work-Discipline, and Industrial Capitalism," *Past & Present*, No. 38 (Dec., 1967), 56–97.

Alexander Thurston, *Boko Haram: The History of an African Jihadist Movement* (Princeton: Princeton University Press, 2018).

John H. Tice, "The Supposed Planet Vulcan," *Scientific American*, Vol. 35, No. 25 (Dec. 16, 1876), 389.

The Times, 2 Mar., 1911, 3.

The Times, 3 Mar., 1911, 3.

The Times, 4 Mar., 1911, 3.

The Times, 8 Mar., 1911, 3.

The Times, 13 Apr., 1911, 3.

Sophie Thompson, "Millie Bobby Brown says she no longer believes the Earth is flat," *indy100*, November 12, 2022, accessed September 16, 2024, https://www.indy100.com/video/millie-bobby-brown-flat-earth.

Sigurður Þórarinsson, trans. Jóhann Hannesson, *Hekla: A Notorious Volcano* (Reykjavík: Almenna bókafélagið, 1970).

Hulda Thórisdóttir, Silvia Mari and André Krouwel, "Conspiracy Theories, Political Ideology and Political Behaviour," in Michael Butter and Peter Knight, eds., *Routledge Handbook of Conspiracy Theories* (London: Routledge, 2020), 304–16.

Helen Tilley, *Africa as a Living Laboratory: Empire, Development, and the Problem of Scientific Knowledge, 1870–1950* (Chicago: The University of Chicago Press, 2011).

Aric Toler, ""No Safe Spaces on the Flat Earth"—Emergent Alt-Right Inspired Flat Earth Online Communities," Bellingcat, June 7, 2017, accessed December 10, 2019, https://www.bellingcat.com/resources/articles/2017/06/07/flat-earth-online-communities/.

Carlos de la Torre and Steve Striffler, "A New Nation," in Carlos de la Torre and Steve Striffler, eds., *The Ecuador Reader: History, Culture, Politics* (Durham, NC: Duke University Press, 2008), 99–101.

Charles A. L. Totten, *Joshua's Long Day and the Dial of Ahaz: A Scientific Vindication and "A Midnight Cry"* (New Haven: The Our Race Publishing Company, 1890).

John Tresch, *The Reason for the Darkness of the Night: Edgar Allan Poe and the Forging of American Science* (New York: Farrar, Straus and Giroux, 2021).

Christian W. Troll, *Sayyid Ahmad Khan: A Reinterpretation of Muslim Theology* (New Delhi: Vikas Publishing House, 1978).

"Tunisia's Constitution of 2014," Constitute Project, June 17, 2023, accessed February 4, 2024, https://www.constituteproject.org/constitution/Tunisia_2014.pdf.

TV Eye, ep. "Benn's Bandwagon" (prod. Jack Saltman), ITV, 28 May, 1981.

James Tynion IV and Martin Simmonds, "The Department of Truth #1," Image Comics, September 30, 2020, accessed January 23, 2025, https://imagecomics.com/comics/releases/the-department-of-truth-1.

Martha Ellicott Tyson, *A Sketch of the Life of Benjamin Banneker; from notes taken in 1836* (Baltimore: Maryland Historical Society, 1854).

A. A. Ubachukwu and P. N. Okeke, "Astronomy in Nigeria: Recent Developments," *African Skies*, No. 6 (Dec., 2001), 17–18.

"Václav Uruba: Objevitelé vesmíru / Člověk a kosmos," Ostravské sochy, n. d., accessed February 18, 2025, https://ostravskesochy.cz/dilo/752-Objevitele-vesmiru-Clovek-a-kosmos.

Goolam Vahed, "Gokhale, Polak, and the end of Indian indenture in South Africa, 1860–1911," in *South Asian Migrations in Global History: Labor, Law, and Wayward Lives*, ed. Neilesh Bose (London: Bloomsbury Academic, 2022), 37–62.

Jacques Vallee and Chris Aubeck, *Wonders in the Sky: Unexplained Aerial Objects from Antiquity to Modern Times and Their Impact on Human Culture, History, and Beliefs* (New York: Jeremy P. Tarcher/Penguin, 2009).

Immanuel Velikovsky, *Worlds in Collision* (New York: Macmillan Company, 1950).

Hebe Vessuri, "Milei charges against Argentine science," *Tapuya: Latin American Science, Technology and Society*, Vol. 7, No. 1 (Apr., 2024), 1–7.

The Viking. Directed by Roy William Neill. Beverly Hills, CA: Metro-Goldwyn-Mayer, 1928.

"Voto contra Bolsonaro no TSE tem menção a 'Terra plana'," *CartaCapital*, June 29, 2023, accessed July 5, 2023, https://www.cartacapital.com.br/politica/voto-contra-bolsonaro-no-tse-tem-mencao-a-terra-plana/.

Richard C. Wade, *Slavery in the Cities: The South 1820–1860* (New York: Oxford University Press, 1964).

Rachel E. Walker, *Beauty and the Brain: The Science of Human Nature in Early America* (Chicago: University of Chicago Press, 2022).

Paul Wallace, *Stars Beneath Us: Finding God in the Evolving Cosmos* (Minneapolis: Fortress Press, 2015).

Irving Wallace, *The Square Pegs: Some Americans Who Dared to Be Different* (New York: Alfred A. Knopf, 1957).

William Thomas Walsh, *Our Lady of Fátima* (New York: The Macmillan Company, 1951).

Booker T. Washington, "The Colored Ministry: Its Defects and Needs," *Christian Union*, Vol. 42, No. 7 (Aug. 14, 1890): 190–200.

Booker T. Washington, *The Future of the American Negro* (Boston: Small, Maynard & Company, 1900).

John M. Watanabe, "In the World of the Sun: A Cognitive Model of Mayan Cosmology," *Man*, New Series, Vol. 18, No. 4 (Dec., 1983), 710–23.

Nigel Watson, *UFOs of the First World War: Phantom Airships, Balloons, Aircraft and Other Mysterious Aerial Phenomena* (Stroud: The History Press, 2015).

Paul Webb, "Science Education in South Africa: Issues of Language and Scientific Literacy," in Femi S. Otulaja and Meshach B. Ogunniyi, eds., *The World of*

Science Education: Handbook of Research in Science Education in Sub-Saharan Africa (Rotterdam: Sense Publishers, 2017), 119–31.

Kelly Weill, "Anti-Vax Flat Earth Preacher Dies of COVID-19," *The Daily Beast*, October 14, 2021, accessed March 16, 2024, https://www.thedailybeast.com/flat-earth-preacher-rob-skiba-dies-of-covid-19.

Kelly Weill, *Off the Edge: Flat Earthers, Conspiracy Culture, and Why People Will Believe Anything* (Chapel Hill: Algonquin Books, 2022).

Carl R. Weinberg, *Red Dynamite: Creationism, Culture Wars, and Anticommunism in America* (Ithaca: Cornell University Press, 2021).

Nancy Joan Weiss, *Farewell to the Party of Lincoln: Black Politics in the Age of FDR* (Princeton, NJ: Princeton University Press, 1983).

Anton Weiss-Wendt, *Putin's Russia and the Falsification of History: Reasserting Control over the Past* (London: Bloomsbury Academic, 2020).

Mike Wendling, *Alt-Right: From 4Chan to the White House* (London: Pluto Press, 2018).

Roland Wenzlhuemer, *Doing Global History: An Introduction in 6 Concepts* (London: Bloomsbury Academic, 2019).

Mick West, *Escaping the Rabbit Hole: How to Debunk Conspiracy Theories Using Facts, Logic, and Respect* (New York: Skyhorse, 2018).

Mick West, "Soundly Proving the Curvature of the Earth at Lake Pontchartrain," *Metabunk*, July 22, 2017, accessed June 9, 2023, https://www.metabunk.org/threads/soundly-proving-the-curvature-of-the-earth-at-lake-pontchartrain.8939/.

Andrew Dickson White, *A History of the Warfare of Science with Theology in Christendom* (New York: D. Appleton and Company, 1896).

Kase Wickman, "Lauren Boebert Told Congress She Is Worried About Space Aliens Operating From Underwater "Bases"," *Vanity Fair*, November 14, 2024, accessed November 23, 2024, https://www.vanityfair.com/style/story/rep-lauren-boebert-secret-underwater-alien-base-congress-ufo-hearing.

Agus Widayoko, Praptika Septi Femilia, Albertus Djoko Lesmono, Heri Sudjatmi, Vicki Dian Prastiwi, and Ninik Munfarikha, "Description of Students' Scientific Literacy Competencies on the Scientific Issue of Flat Earth Theory," *Anatolian Journal of Education*, Vol. 4, No. 2 (Oct., 2019), 31–38.

Leo Wiener, *Africa and the Discovery of America, Vols. I-III* (Philadelphia: Innes & Sons, 1920–1922).

Mark Wilding, "Circle Or Sphere? Inside The UK's First 'Flat Earth' Conference," *Esquire*, October 9, 2018, accessed July 27, 2020, https://www.esquire.com/uk/culture/a22725244/inside-the-uks-first-flat-earth-conference/.

Isabel Wilkerson, *The Warmth of Other Suns: The Epic Story of America's Great Migration* (New York: Random House, 2010).

John Williams (a), "Isidore, Orosius and the Beatus Map," *Imago Mundi*, Vol. 49 (1997), 7–32.

John Williams (b), "The West Indies," *The Earth*, Vol. III, Nos. 25/26 (Jun., 1902), 17–19.

Brad Wilson, "Excellence in Video: "Sixth Mount Zion: The Sun Do Move but this Church Does Not"," YouTube, February 5, 2018, accessed October 14, 2023, https://www.youtube.com/watch?v=Bw9O0EbZ-bE.

Derek Wilson, *The Circumnavigators: A History* (New York: Carroll & Graf, 2003).

J. Leighton Wilson, *Western Africa: Its History, Condition, and Prospects* (New York: Harper & Brothers, 1856).

Brenda Wineapple, *Ecstatic Nation: Confidence, Crisis, and Compromise, 1848–1877* (New York: HarperCollins, 2013).

Thomas Winship, "The Contents of our Letter Box," *The Earth not a Globe Review*, No. 2, New Series (Jan., 1895), 44–45.

Thomas Winship, "A Defence of Philosophic Doubt," *The Earth*, Vol. III, Nos. 25/26 (Jun., 1902), 30–32.

Thomas Winship, "A Defence of Philosophic Doubt," *The Earth*, Vol. III, Nos. 27/28 (Jul., 1902), 49–52.

Thomas Winship, "A Defence of Philosophic Doubt," *The Earth*, Vol. III, Nos. 31/32 (Feb./Mar., 1903), 140–43.

Thomas Winship, "A Defence of Philosophic Doubt," *The Earth*, Vol. III, Nos. 33/34 (Apr./May, 1903), 173–77.

Thomas Winship, "A Defence of Philosophic Doubt," *The Earth*, Vol. III, Nos. 35/36 (Jun./Jul., 1903), 214–16.

Thomas Winship, "A Defence of Philosophic Doubt," *The Earth*, Vol. III, Nos. 39/40 (Oct./Nov, 1903), 280–83.

Thomas Winship, *Zetetic Cosmogony; or Conclusive Evidence that the World is not a Rotating Revolving Globe but a Stationary Plane Circle* (Durban: T. L. Cullingworth, 1899).

Brian Winter, "Jair Bolsonaro's Guru," *America's Quarterly*, December 17, 2018, accessed February 21, 2024, https://www.americasquarterly.org/article/jair-bolsonaros-guru/.

Brian Winter, "Messiah Complex: How Brazil Made Bolsonaro," *Foreign Affairs*, Vol. 99, No. 5 (Sep./Oct., 2020), 119–31.

J. D. Wolf, "Exclusive: Trump Shared Racist, Flat-Earth Facebook Account With South African President," Meidas+, May 22, 2025, accessed May 22, 2025, https://www.meidasplus.com/p/exclusive-trump-shared-racist-flat.

Audra J. Wolfe, *Freedom's Laboratory: The Cold War Struggle for the Soul of Science* (Baltimore: Johns Hopkins University Press, 2018).

George H. C. Wong, "China's Opposition to Western Science during Late Ming and Early Ch'ing", *Isis*, Vol. 54, No. 1 (Mar., 1963), 29–49.

Carter Godwin Woodson, *The History of the Negro Church* (Washington, D.C.: The Associated Publishers, 1921).

Carter Godwin Woodson, "John Jasper," *Negro History Bulletin*, Vol. 5, No. 4 (Jan., 1942), 95.

The World's Most Famous Court Trial: Tennessee Evolution Case (Cincinnati: National Book Company, 1925).

Robert F. Worth, "Narendra Modi's Populist Facade Is Cracking," *The Atlantic*, January 2, 2025, accessed January 2, 2025, https://www.theatlantic.com/magazine/archive/2025/02/modi-nationalism-demagoguery-limitations/681094/.

Jay Wright, *Transfigurations: Collected Poems* (Baton Rouge: LSU Press, 2000).

M. R. Wright, *Cosmology in Antiquity* (New York: Routledge, 1995).

Elly M. Wynia, *The Church of God and Saints of Christ: The Rise of Black Jews* (New York: Garland Publishing, 1994), 19–28.

Xiaoqiao Xu, "Stabilizing the Empire: Western Calendric Astronomy as Statecraft during the Kangxi Reign, 1667–1720," *The Quiet Corner Interdisciplinary Journal*, Vol. 3, No. 1 (Aug., 2022), 22–41.

Yachtsman, "Navigation Proves the World a Plane!," *The Earth not a Globe Review*, No. 5, New Series (Oct., 1895), 115–16.

Oksana Yakushko, *Scientific Pollyannaism: From Inquisition to Positive Psychology* (London: Palgrave Macmillan, 2019).

Eric S. Yellin, "The (White) Search for (Black) Order: The Phelps-Stokes Fund's First Twenty Years, 1911–1931," *The Historian*, Vol. 65, No. 2 (Winter, 2002): 319–52.

Paul Younger, *New Homelands: Hindu Communities in Mauritius, Guyana, Trinidad, South Africa, Fiji, and East Africa* (Oxford: Oxford University Press, 2010).

Hayrettin Yücesoy, "Translation as Self-Consciousness: Ancient Sciences, Antediluvian Wisdom, and the 'Abbāsid Translation Movement," *Journal of World History*, Vol. 20, No. 4 (Dec., 2009), 523–57.

Muhammed Yusuf, "Translation of Selections from Muhammed Yusuf: Hādhihi 'Aqīdatunā WaMinhāj Da'watinā (This is our Creed and Method of Proclamation)," in Abdulbasit Kassim and Michael Nwankpa, eds., *The Boko Haram Reader: From Nigerian Preachers to the Islamic State* (Oxford: Oxford University Press, 2018), 27–34.

Muhammed Yusuf and Isa Pantami, "Debate on the Status of Western Education and Working for the Nigerian Government Between Mallam Isa Ali Ibrahim Pantami and Mallam Muhammad Yusuf Maiduguri," in Abdulbasit Kassim and Michael Nwankpa, eds., *The Boko Haram Reader: From Nigerian Preachers to the Islamic State* (Oxford: Oxford University Press, 2018), 11–26.

Carlo Zaccagnini, "Maps of the World," in *Leggo! Studies Presented to Frederick Mario Fales on the occasion of his 65th birthday*, eds. Daniele Morandi Bonacossi et al (Wiesbaden: Harrassowitz Verlag, 2012), 865–74.

Brandy Zadrozny and Ben Collins, "How three conspiracy theorists took 'Q' and sparked Qanon," NBC News, August 14, 2018, accessed April 29, 2019, https://www.nbcnews.com/tech/tech-news/how-three-conspiracy-theorists-took-q-sparked-qanon-n900531.

Erik A. Zavrel, "How the Discovery Channel Television Show "Mythbusters" Accurately Depicts Science and Engineering Culture," *Journal of Science Education and Technology*, Vol. 20, No. 2 (Apr., 2011), 201–07.

Zetetic Society, *Chart & Compass, Sextant & Sundial, Latitudes & Longitudes, Plumbline & Pendulum, Globe or Plane?* (London: G. W. Cook, n. d.).

Charles U. Zug and Steven F. Pittz, "Good Conspiracism," in Luke Ritter, ed., *American Conspiracism: An Interdisciplinary Exploration* (London: Routledge, 2024), 159–74.

Index

E. Guimont, *The Power of the Flat Earth Idea*,
Palgrave Studies in the History of Science and Technology,
https://doi.org/10.1007/978-3-032-23981-5

The manufacturer's authorised representative in the EU is Springer Nature Customer Service Centre GmbH, Europaplatz 3, 69115 Heidelberg, Germany. If you have any concerns regarding our products, please contact ProductSafety@springernature.com

Printed and bound by CPI Group (UK) Ltd, Croydon, CR0 4YY
17/09/2026
02218454-0014